AF790023

Energy Systems

Series Editor:

Panos M. Pardalos, University of Florida, USA

For further volumes:
http://www.springer.com/series/8368

Energy Systems

Series Editor

Panos M. Pardalos, University of Florida, USA

For further volumes:
http://www.springer.com/series/8368

Qipeng P. Zheng • Steffen Rebennack •
Panos M. Pardalos • Mario V. F. Pereira •
Niko A. Iliadis
Editors

Handbook of CO_2 in Power Systems

Springer

Editors
Qipeng P. Zheng
West Virginia University
Department of Industrial &
Management Systems Engineering
Morgantown, West Virginia
USA

Dr. Steffen Rebennack
Colorado School of Mines
Division of Economics and
Business Engineering Hall
Golden Colorado
USA

Panos M. Pardalos
University of Florida
Dept. Industrial & Systems
Engineering
Gainesville Florida
USA

Dr. Mario V. F. Pereira
Centro Empresarial
Rio de Janeiro
Brazil

Dr. Niko A. Iliadis
EnerCoRD - Energy Consulting,
Research & Development
Athens
Nea Smyrni
Greece

ISSN 1867-8998 ISSN 1867-9005 (electronic)
ISBN 978-3-642-27430-5 ISBN 978-3-642-27431-2 (eBook)
DOI 10.1007/978-3-642-27431-2
Springer Heidelberg New York Dordrecht London

Library of Congress Control Number: 2012936489

© Springer-Verlag Berlin Heidelberg 2012
This work is subject to copyright. All rights are reserved by the Publisher, whether the whole or part of the material is concerned, specifically the rights of translation, reprinting, reuse of illustrations, recitation, broadcasting, reproduction on microfilms or in any other physical way, and transmission or information storage and retrieval, electronic adaptation, computer software, or by similar or dissimilar methodology now known or hereafter developed. Exempted from this legal reservation are brief excerpts in connection with reviews or scholarly analysis or material supplied specifically for the purpose of being entered and executed on a computer system, for exclusive use by the purchaser of the work. Duplication of this publication or parts thereof is permitted only under the provisions of the Copyright Law of the Publisher's location, in its current version, and permission for use must always be obtained from Springer. Permissions for use may be obtained through RightsLink at the Copyright Clearance Center. Violations are liable to prosecution under the respective Copyright Law.
The use of general descriptive names, registered names, trademarks, service marks, etc. in this publication does not imply, even in the absence of a specific statement, that such names are exempt from the relevant protective laws and regulations and therefore free for general use.
While the advice and information in this book are believed to be true and accurate at the date of publication, neither the authors nor the editors nor the publisher can accept any legal responsibility for any errors or omissions that may be made. The publisher makes no warranty, express or implied, with respect to the material contained herein.

Printed on acid-free paper

Springer is part of Springer Science+Business Media (www.springer.com)

Handbook of CO_2 in Power Systems: Preface

After the sequels of "Handbook of Power Systems I & II" were published by Springer in 2010, we focused our efforts on specific aspects of power systems, which lead to a series of handbooks such as "Handbook of Networks in Power Systems" and "Handbook of CO_2 in Power Systems." This handbook focuses on the aspects of power systems related to CO_2.

Global warming has become a critical issue to our planet, Earth. The continuing rise in the average temperature of the atmosphere and ocean causes serious problems such as: glacier shrinkage, sea level elevation, species extinction, agricultural output decrease, and extreme weather events. It is an indisputable fact that greenhouse gas concentrations resulting from human activities continue to exacerbate global warming. Although water vapor contributes the most to the greenhouse effect, (i.e., the effect of solar heat remaining on our planet), due to industrialization the dramatic increase of carbon dioxide (CO_2) in the atmosphere accounts for the biggest percentage contribution in the temperature rise. According to historic studies regarding the contents of the atmosphere, the mole fraction of CO_2 has increased around 36% percent from its preindustrial level; the last 40 years account for more than half of the increase.

Many countries, territories, and international organizations have joined together to curb the global warming trend. Among the earliest efforts, the United Nations Framework Convention on Climate Change (UNFCCC), which began in 1992, encompasses the largest number of countries signed to fight climate change. This is despite the fact that the convention is a non-binding treaty with no mandatory limits on emissions. Similar to the UNFCCC protocol, the Kyoto protocol, initiated in 1997, consists of nations around the globe that have signed to reduce their emissions from 1990 benchmark emission levels. Many policies, economic stimuli and technologies have been developed to reduce carbon emissions, considered to be the major factor in global warming. Among the many policies, the three most often debated and tested are that of the carbon tax, the cap-and-trade system, and of renewable portfolio standards. Carbon capture and storage technologies are considered a panacea in reducing the emissions from extensive use of fossil fuels.

World energy consumption has been increasing steadily since industrialization, and especially within the last 30 years; this recent increase is also the major cause for the increase in CO_2 concentration in the atmosphere. Fossil fuels, i.e., oil, coal and natural gas, continue to play a central role in our energy consumption. In order to reduce global warming, it is important to focus on global energy production and consumption. Since the introduction of daily use of electrical bulb for lighting, our society has been increasingly relying on electricity as an energy carrier. Since this reliance, the percentage of electrical power as a total supply of and demand for energy has grown exponentially; this trend is unlikely to be reversed in the short term. It is no doubt that CO_2 policies, technologies, and operations in power systems must play a prominent role in reducing total CO_2 emissions.

This book presents the state of the art on how CO_2 issues are considered in power systems. The book encompasses three major areas: (1) CO_2 policies and markets; (2) the integration of CO_2 mitigation in power systems operations; and (3) CO_2 capture and storage technologies and their impacts on power systems. The main purpose of this book is to promote new studies related to the integration of CO_2 issues in power systems by presenting a collection of articles which provides mathematical modeling, optimization techniques, and economic analysis of these problems.

We would like to thank all the authors for their in-depth studies of CO_2 issues in power systems, all the referees for their careful and constructive reviews, and the publisher for creating this volume.

Qipeng P. Zheng
Steffen Rebennack
Panos M. Pardalos
Niko A. Iliadis
Mario V. Pereira

Contents

Variational Inequality Formulations for Oligopolistic Electricity Models with Marketable CO_2 Emission Permits

William Chung

Abstract In this paper, we first provide an introduction of the electricity market and its major players. Then, equilibrium models of the CO_2 emission permits market are discussed. We consider a variational inequality (VI) approach to model an oligopolistic competition in the market of CO_2 emission permits and electric power. A Cournot model of electricity market is described and a variational inequality framework is developed for oligopolistic electricity models with marketable CO_2 permits. Properties of the model are provided and some solution methods are discussed. A simple example is presented and implemented in a PC and solved by a built-in solver powered by Newton's method of VI. Discussions of possible extensions of the model are given, like using conjectured supply function approach and including non-electric power participants of CO_2 trading.

Keywords Variational inequality • Oligopolistic • Electricity • CO_2 • Emission permits

1 Introduction

1.1 *Marketable CO_2 Permits in Power System*

The marketable pollution permit system (called emissions trading system, or pollution cap-and-trade system) is one of the policy instruments offering incentives to reduce pollution. Compared with other taxable economic instruments, the emissions trading system allows government authorities to determine and set the

W. Chung (✉)
Department of Management Sciences, City University of Hong Kong, Kowloon Tong, Kowloon, Hong Kong
e-mail: william.chung@cityu.edu.hk

Q.P. Zheng et al. (eds.), *Handbook of CO_2 in Power Systems*, Energy Systems,
DOI 10.1007/978-3-642-27431-2_1, © Springer-Verlag Berlin Heidelberg 2012

total quality of emissions for a certain period before the introduction of the system. The initial concept of emissions trading can be found in Crocker [5] for air pollution permits. The main idea behind emissions trading is to initially allocate emission permits to firms either by auction or by a free-of-charge "grandfathering mechanism." Firms are then allowed to trade permits among themselves, with the intention that firms should hold a number of permits at least equal to their emissions. Hence, firms that have excess permits can sell their permit on the market, whereas firms who need additional permits can purchase permits available in the market. Obviously, firms need to make their abatement decisions by considering the difference between the marginal abatement costs and the market price of the emission permits.

The first large-scale emissions trading system was implemented in the United States (US) electric sector SO_2 emissions through the 1990 Clean Air Act Amendments. Emissions trading was later adopted in the US for NOx. In Europe, the Kyoto Protocol [19] allowed burden sharing between the 15 European Union (EU) countries with a total emissions reduction target of 8%, compared to the emission level in 1990. The EU's reduction target is 20% by 2020. To meet these goals, the EU has adopted the emission trading system for the greenhouse gas CO_2 and introduced the EU Emissions Trading Scheme (ETS). There are three periods of emissions trading: 2005–2007, 2008–2012, and the Post-Kyoto period. According to Geogopoulou et al. [10], at the start of each period, all ETS countries need to present their individual national allocation plan, which defines how much emission allowance a major emitter, such as an electric power plant, will be allotted in this period. It should be noted that the volume of other greenhouse gases are rated with the weight reflecting its potential global warming damage compared to CO_2.

There are other vital issues concerning the design of emission allocations systems, including transparency and transactions costs, international competitiveness of affected industries, the effect of mechanisms designed to stabilize prices such as price ceilings, the value of "banking" schemes to buffer inter-annual variations in emissions, and efficiency implications of different ways to dispose of auction revenues. Some of these issues are discussed in Sect. 6 from the modeling perspective.

1.2 The Electricity Market

For the past two decades, electricity systems were being restructured in many countries and regions, including the US, England, Wales, and the Nordic countries. The electricity systems were moving towards deregulation and competition to improve economic efficiency.

The supply chain of the electricity market consists of three functions: generation, transmission, and distribution. Before being restructured, full vertical integration in the control and ownership of these three functions had been the dominant structure. To introduce competition in the electricity systems, separation of the control and ownership became the common practice. Specifically, competition was introduced in generation, while transmission and distribution remained regulated monopolies. Generators still need to have access to transmission and distribution services in order to reach their customers, since they do not have the ownership to control the transmission and distribution functions. On the other hand, generators also need to deal with customers (and/or traders) in the energy markets. To enhance the market structure, brokers and aggregators would allow the buying or selling of power. In general, there are two basic models for the energy market structure that were stated by Hobbs [13] and Shaidehpour et al. [25]: the pooling company (PoolCo) market and bilateral contract.

A PoolCo is defined as a central auction marketplace that clears the energy market for buyers and sellers. In this model, buyers (brokers/aggregators) and sellers (brokers/aggregators/generators) submit bids to the pool on the amount of power that they are willing to trade in the market. If a seller bids too high, it may not be able to sell. On the other hand, buyers compete for buying power. Similarly, if a buyer's bid is too low, it may not be able to purchase. In addition, sellers compete for the transmission services provided by the grid owner to supply energy to the grid. Hence, the grid owner (or administrator) would price scare transmission capacity and should function independent of any market participants in order to efficiently ration the transmission capacity. Normally, an independent system operator (ISO) would be established as the grid owner. An ISO within a PoolCo would also implement the economic dispatch and produce a single spot price, which includes the cost of transmission services, in order for electricity to clear the market.

Bilateral contracts are negotiable agreements on delivery and receipt of power between two participants in the market. Although the details of each contract are set independent of the ISO, the ISO must verify if there is sufficient transmission capacity, and may provide a reference market price.

There are different kinds of energy market structures, depending on the functions provided by the ISO and designed by regulators. The Pennsylvania-New Jersey-Maryland Interconnection ISO (PJM) in the US and the National Grid Company (NGC) in the UK are some examples of how authority and control functions differ.

1.3 *Equilibrium Models*

There are not many references in the literature of power market models that adequately include the market mechanism of CO_2 emission permits trading. This may be because the initial model developments of the restructured electric power market and of the CO_2 emission permits market have followed different paths. In addition, the participants of the CO_2 emission permits market are not limited to generators. Other energy-intensive industries, including the aluminum, chemical, paper and steel industries, are the key players in the permits market. Due to the oligopolistic competition in both markets, the practice of using the oligopolistic market model to simulate the players' behavior of the electricity market and to model CO_2 emission permits market has become common.

In the oligopolistic model of power markets, the most popular type is based on Cournot competition, in which firms compete in quantity strategies [27]. However, Day et al. [7] stated that Cournot models do not give meaningful equilibria when price elasticity of demand is low or zero, as they often are for short-run power demands, ancillary services, and short-run supplies of transmission capacity. This is because, under the Cournot approach, equilibrium prices are only determined by the demand function, and are therefore highly sensitive to demand representation and are usually higher than those observed in reality. To overcome this drawback, Day et al. proposed to use the Conjectured Supply Function instead of demand function to determine equilibrium prices in Cournot-based models.

Regarding the CO_2 emission permits market, a very thorough review of large-scale energy-economy-environment models can be found in Huntington and Weyant [16], showing that little detail for the electricity sector is normally included in the models. For instance, Jensen and Skytte [18] and Unger and Ahlgren [26] examined the specific impact of CO_2 emission permits trading in the Nordic countries' electricity sector. It should be noted that such works inadequately represent electricity sectors under oligopolistic competition. On the other hand, Chen and Hobbs [1] provided an oligopolistic power market model with tradable NOx permits (not CO_2, however) formulated as complementarity problems.

In this paper, we present a variational inequality (VI) approach to model an oligopolistic competition in the market of CO_2 emission permits and electric power, since the VI approach is able to cope with large-scale problems. Similar to Hobbs [13], the model includes a congestion-pricing scheme for transmission, in which load flows are modeled considering both the first and the second Kirchhoff laws by means of a linearized formulation. This chapter is organized as follows: Sect. 2 describes a Cournot model of electricity market, Sect. 3 provides a VI framework and the development of an oligopolistic electricity models with marketable CO_2 emission permits, Sects. 4 and 5 present the algorithm and an example, respectively, Sect. 6 remarks on other possible extensions of the model, and Sect. 7 presents the conclusion.

2 A Cournot Model of Electricity Market with CO_2 Emission Permits

Since this chapter illustrates the market mechanism of CO_2 emission permits trading and the electric power system, we use a Cournot model of electricity market with simple linear demand functions, instead of the conjectured supply functions. Note that it is easy to follow the same modeling approach to adopt the conjectured supply functions. Based on the power market equilibrium problem of Hobbs [13], we can have the following descriptions of the electricity market modeling structure: consumer side (consumers), producer side (generators), and a transmission network.

For consumer side, we first assume that consumers at a node i consume q_i MW, which is price responsive with linear demand functions $p_i(q_i) = P_{i0} - \left(\frac{P_{i0}}{Q_{i0}}\right)q_i$ \$/MWh, where P_{i0}\$/MWh and Q_{i0}MW are the price and quantity intercepts, respectively.

For producer side, each firm f owns several generating units, $h = 1, ...H(f, i)$ located at node i. $C_{f,i,h}$ is the per MWh cost of power generation $x_{f,i,h}$MW. The capacity of a generator is $X_{f,i,h}$MW.

For transmission network, the generating units (h) and the nodes (i) are connected by transmission lines to form a network. We assume that transmission is priced using a congestion pricing scheme mentioned in Schweppe et al. [24]. The owner of the grid charges a congestion-based fee, w_i \$/MWh, for transmitting power from an arbitrary hub node to node i. The network is assumed to be represented by a linear DC network and, therefore, all generation and sales can be modeled as being routed through the hub node at which there is neither generation nor consumption. A firm pays $-w_i$ to get power to the hub from a generator at node i, and then pays $+w_j$ to convey power for sale from the hub to customers at node j (the indices i and j designate nodes). Hence, the corresponding cost of transmitting power from a generator at node i to the customers at node j is $-w_i + w_j$.

Let y_i MW be power transmitted from the hub to a node i at zero cost (note that y_i may be negative). Since the power flowing through the transmission lines is subject to thermal line limits, there is limited transmission capacity. The owner of the grid is assumed to ration this limited transmission capacity to maximize the value of the transmission services, which is equivalent to having the grid choose values of y_i to maximize its revenue $\sum_i w_i y_i$. Note that w_i is a decision variable, which is the dual price of the flow balancing constraint at node i in the equilibrium model.

The electricity market (no-arbitrage) is assumed to be a bilateral market in which $s_{f,i}$ MW is the quantity sold by firm f to consumers at node i. There is no arbitrage; therefore, market clearing can be represented by $\sum_f s_{f,i} = s_i = q_i$. Hence, the firms determine the level of sales to each node, and then request transmission service from the grid, subject to the constraints of the energy balance and the generators' capacities, in order to maximize their profits. The owner of the grid determines the congestion price (w_i) and the transmission flow (y_i), subject to the constraint of the transmission capacity, in order to maximize the transmission revenue.

Constraint and cost of the license for firm f assumes that the initial emission permits (or licenses), l_{f0}, are either by sold (i.e., by auction) or given free of charge (i.e., by a "grandfathering mechanism" for a certain period) to firms. Hence, the value of a firm's initial endowment of licenses is given by $w_l^* l_{f0}$, where w_l^* denotes the given market price of a license under the assumption of perfect competition in the license market. At the end of the period, the firm f has to hold at least the same quantity of emission license, l_f, as the amount of CO_2 released. That is, the firm f needs to face the additional emission constraints

$$\sum\nolimits_{i,h} E_{f,i,h} x_{f,i,h} \leq l_f \text{ and } l_f \geq 0,$$

where $E_{f,i,h}$ is the emission rate of the h generator of firm f, node i. With the market price of a license being w_l^*, the cost of purchasing licenses is given by $w_l^*(l_f - l_{f0})$.

3 The VI Model Formulations

We first present the producers' and the grid owner's optimization problems and their variational inequality formulations. With market clearing conditions, these optimization problems can be converted into variational inequality (VI) formulations. To provide some key concepts of variational inequality representations to readers who are familiar with constrained convex optimization models, a brief introduction to variational inequalities is included in the Appendix.

3.1 *Sub-Model of Firm* f

We assume that each firm in the oligopoly is profit-maximizing. The firm's profit function u_f can be expressed as the following:

$$\begin{aligned} u_f &= u_f(s_i, x_{f,i,h}, l_f) \\ &= \sum\nolimits_i \left[\left(P_{i0} - \left(\frac{P_{i0}}{Q_{i0}} \right) \left(\sum\nolimits_f s_{f,i} \right) \right) - w_i^* \right] s_{f,i} - \sum\nolimits_{i,h} (C_{f,i,h} - w_i^*) x_{f,i,h} \\ &\quad - w_l^*(l_f - l_{f0}) \end{aligned} \tag{F1}$$

An oligopolistic firm's optimization problem is then expressed as the following:

$$Max\ u_f(s_i, x_{f,i,h}, l_f)$$

subject to

$$x_{f,i,h} \le X_{f,i,h} \quad \forall i, h \quad \text{(Generator capacity constraints)} \tag{F2}$$

$$\sum\nolimits_i s_{f,i} = \sum\nolimits_{i,h} x_{f,i,h} \quad \text{(Energy balance)} \tag{F3}$$

$$\sum\nolimits_{i,h} E_{f,i,h} x_{f,i,h} \le l_f \quad \text{(Emissions constraints)} \tag{F4}$$

$$\forall s_{f,i}, x_{f,i,h}, l_f \ge 0 \tag{F5}$$

where w_i^* and w_l^* are the market prices of transmission services and licenses, respectively.

Under the Nash-Cournot equilibrium assumption, firm f considers that $s_{g \neq f,i}$, w_i^*, and w_l^* are fixed. That is, each firm f selects its sales quantities ($s_{f,i}$), generation level ($x_{f,i,h}$), and emissions level (l_f), in order to maximize profit, given the sales quantities vector decisions of the other firms and subject to the generator capacity constraints, energy balance requirement, and emissions constraints. Hence, we can have the following definitions:

3.1.1 Definition of Nash-Cournot Equilibrium

In the Nash-Cournot equilibrium, $(s_{f,i}^*, \hat{s}_{f,i}^*, x_{f,i,h}^*, l_f^*)$ such that $u_f(s_{f,i}^*, \hat{s}_{f,i}^*, x_{f,i,h}^*, l_f^*) \geq u_f(s_{f,i}, \hat{s}_{f,i}^*, x_{f,i,h}, l_f) \forall s_{f,i}, x_{f,i,h}, l_f$, satisfying constraints (F2), (F3), (F4), and (F5) for all firms f, where $\hat{s}_{f,i}^* = (s_{1,i}^*, \ldots, s_{f-1,i}^*, s_{f+1,i}^*, \ldots, s_{F,i}^*)$, and there are F firms in the market.

For the solution approach, Harker [11] discussed a variational inequality approach for the determination of Nash-Cournot equilibrium.

On the other hand, it should be noted that in the equilibrium model expressed in VI, $s_{g \neq f,i}$, w_i^*, and w_l^* are decision variables.

3.1.2 VI of a Firm f's Sub-Model

Given w_i^*, w_l^*, and $\hat{s}_{f,i}^*$ are fixed, we can have

$$\nabla_{s_{f,i}} u_f = \frac{\partial u_f}{\partial s_{f,i}} = \sum\nolimits_i \left[P_{i0} - \left(\frac{P_{i0}}{Q_{i0}} \right) \left(2s_{f,i} + \sum\nolimits_{g \neq f} s_{g,i}^* \right) - w_i^* \right]$$

$$\nabla_{x_{f,i,h}} u_f = \frac{\partial u_f}{\partial x_{f,i,h}} = -\sum\nolimits_{i,h} \left(C_{f,i,h} - w_i^* \right)$$

$$\nabla_{l_f} u_f = \frac{\partial u_f}{\partial l_f} = -w_l^*$$

Hence, we can have variational inequalities of a firm f as

[VI-Firm f]

Find $\left(s_{f,i}^*, x_{f,i,h}^*, l_f^* \right) \in K_f$ such that

$$-\sum\nolimits_i \left[P_{i0} - \left(\frac{P_{i0}}{Q_{i0}} \right) \left(2s_{f,i}^* + \sum\nolimits_{g \neq f} s_{g,i}^* \right) - w_i^* \right] \times \left[s_{f,i} - s_{f,i}^* \right]$$

$$+\sum_{i,h}\left[C_{f,i,h}-w_i^*\right]\times\left[x_{f,i,h}-x_{f,i,h}^*\right]+w_l^*\times\left[l_f-l_f^*\right]\geq 0$$

$$\forall\left(s_{f,i},x_{f,i,h},l_f\right)\in K_f=\{\left(s_{f,i},x_{f,i,h},l_f\right)|$$

$$x_{f,i,h}\leq X_{f,i,h}\quad\forall i,h$$

$$\sum_i s_{f,i}=\sum_{i,h}x_{f,i,h}$$

$$\sum_{i,h}E_{f,i,h}x_{f,i,h}\leq l_f$$

$$\forall s_{f,i},x_{f,i,h},l_f\geq 0\}$$

3.2 Sub-Model of the Owner of the Grid

After restructuring, the electricity market consists of other players, including the owner of the grid network. The above-mentioned Nash-Cournot equilibrium assumption is no longer valid due to the transmission capacity and the consequent congestion cost (see [22]). Again, given a fixed transmission fee (w_i^*), we assume that the grid owner is profit-maximizing and the corresponding optimization problem can be expressed as

$$Max\ u_G(y_i)=\sum_i w_i^* y_i \qquad \text{(G1)}$$

subject to

$$-\sum_i PTDF_{i,k}y_i\leq T_{k-}\quad(\lambda_{\mathrm{k}-})\quad\forall k\quad\text{(Interface constraints)}\qquad\text{(G2)}$$

$$\sum_i PTDF_{i,k}y_i\leq T_{k+}\quad(\lambda_{\mathrm{k}+})\quad\forall k\quad\text{(Interface constraints)}\qquad\text{(G3)}$$

where $PDTF_{i,k}$ are power transfer distribution factors at node i with interfaces k.

By using Power Transfer Distribution Factors (*PDTF*s), the DC load flow network provides an approximate solution for a network carrying alternating current (AC) power. That is, the direct current (DC) load flow network is used to approximate relationships between generation and demand levels at the node, and real power flows through the lines. Flows through interfaces k are modeled using *PDTF*s. Hence, the net MW flow through k is $\sum_i PTDF_{i,k}y_i$. The lower and upper bounds on real power flows through an interface k are $-T_{k-}$ and T_{k+}. We assume no losses and that congestion is only basis for pricing. The impact on the system transmission line flows for a bilateral transaction from a specified buyer/seller combination can be determined from the *PTDF* matrix ($PTDF_{i,k}$). Hence, all generation and sales can be modeled as being routed through the hub node [24].

3.2.1 VI of the Grid Owner's Sub-Model

Given w_i^*, we can have

$$\nabla_{y_i} u_G = \frac{\partial u_G}{\partial y_i} = \sum_i w_i^*.$$

Similarly, we can have variational inequalities of the owner of the grid as

[VI-Grid]

Find $y_i^* \in K_G$ such that

$$-\sum_i w_i^* \times \left[y_i - y_i^*\right] \geq 0 \forall y_i \in K_G = \{y_i|$$

$$-\sum_i PTDF_{i,k} y_i \leq T_{k-} \quad (\lambda_{k-})\ \forall k$$

$$\sum_i PTDF_{i,k} y_i \leq T_{k+} \quad (\lambda_{k+})\ \forall k\}$$

3.3 *The Equilibrium Market Model of CO_2 Trading and Power*

In case of a Nash-Cournot equilibrium, each of the firms and the grid owner optimizes unilaterally. However, with the following two market clearing conditions (M1 and M2), w_i, w_l, and $s_{g \neq f,i}$ become market equilibrium variables.

3.3.1 Market Clearing Condition for Power

The market clearing condition for power is simply a flow balancing equation stating that the transmitted power y_i at node i must equal the total transmission service requested by generators from the hub to node i.

$$\sum_f s_{f,i} - \sum_{f,h} x_{f,i,h} = y_i \quad (w_i) \quad \forall i \tag{M1}$$

3.3.2 Market Clearing Condition for License

Let l_f^* be an equilibrium value of the emissions licenses. We can use the system of equalities and inequalities to govern the quantities and prices of licenses at equilibrium.

$$\sum_f [l_{f0} - l_f^*] = \begin{cases} = 0, \text{ if } w_l^* > 0, \\ \geq 0, \text{ if } w_l^* = 0. \end{cases} \tag{M2}$$

The equation above states that if the price of a license is positive, the market for licenses must clear in equilibrium. On the other hand, if there is an excess supply of licenses, the price of a license must be zero.

3.3.3 VI of Power and CO_2 Market

By adding **[VI-Firm *f*]**, **[VI-Grid]**, and both market clearing conditions (M1) and (M2), we can derive the variational inequality formulation of an equilibrium problem that represents a market of electric power and CO_2 emission permits.

[VI-Market Equilibrium]

Find $\left(s_{f,i}^*, x_{f,i,h}^*, l_f^*, w_l^*, y_i^*, w_i^*\right) \in K_M$ such that

$$\begin{aligned}
&-\sum_{f,i}\left[P_{i0} - \left(\frac{P_{i0}}{Q_{i0}}\right)\left(2s_{f,i}^* + \sum_{g\neq f} s_{g,i}^*\right) - w_i^*\right] \times \left[s_{f,i} - s_{f,i}^*\right] \\
&+\sum_{f,i,h}\left[C_{f,i,h} - w_i^*\right] \times \left[x_{f,i,h} - x_{f,i,h}^*\right] + \sum_f w_l^* \times \left[l_f - l_f^*\right] \\
&-\sum_i w_i^* \times \left[y_i - y_i^*\right] - \sum_i \left[\sum_f s_{f,i}^* - \sum_{f,h} x_{f,i,h}^* - y_i^*\right] \times \left[w_i - w_i^*\right] \\
&+\left[\sum_f \left(l_{f0} - l_f^*\right)\right] \times \left[w_l - w_l^*\right] \geq 0,
\end{aligned}$$

$$\forall\left(s_{f,i}, x_{f,i,h}, l_f, w_l, y_i, w_i\right) \in K_M = \left\{\left(s_{f,i}, x_{f,i,h}, l_f, w_l, y_i, w_i\right)\right|$$

$$x_{f,i,h} \leq X_{f,i,h} \quad \forall f, i, h$$

$$\sum_i s_{f,i} = \sum_{i,h} x_{f,i,h} \quad \forall f$$

$$\sum_{i,h} E_{f,i,h} x_{f,i,h} \leq l_f \quad \forall f$$

$$-\sum_i PTDF_{i,k}y_i \leq T_{k-} \quad \forall k$$

$$\sum_i PTDF_{i,k}y_i \leq T_{k+} \quad \forall k$$

$$\forall s_{f,i}, x_{f,i,h}, l_f, w_l \geq 0\}$$

4 Properties of the Model

Some properties related to the problem [VI-Market Equilibrium] are excerpted from Harker and Pang [12] presented in this section. The solution methods are then discussed.

4.1 Existence and Uniqueness

Definition 4.1. Let K be a nonempty subset of $\mathbf{R}^n$ and let F be a mapping from $\mathbf{R}^n$ into itself. The variational inequality problem, denoted by VI(K,F), is to find a vector $z^* \in K$ such that $F(z^*)^T(z - z^*) \geq 0 \forall z \in K$.

Typically, we assume that the set K is closed and convex. Indeed, K is often polyhedral in application like K_M in [VI-Market Equilibrium].

Definition 4.2. The mapping $F : \mathbf{R}^n \to \mathbf{R}^n$ is said to be

(a) Monotone over a set K if

$$[F(z) - F(y)]^T(z - y) \geq 0 \quad \forall z, y \in K;$$

(b) Pseudo-monotone over K if

$$F(y)^T(z - y) \geq 0 \text{ implies } F(z)^T(z - y) \geq 0 \quad \forall z, y \in K;$$

(c) Strictly monotone over K if

$$[F(z) - F(y)]^T(z - y) > 0 \quad \forall z, y \in K, z \neq y;$$

(d) Strongly monotone over K if there exists an $\alpha>0$ such that

$$[F(z) - F(y)]^T(z - y) \geq \alpha||z - y||^2 \quad \forall z, y \in K;$$

(e) Coercive with respect to K if there exists a vector $z^0 \in K$ such that

$$\lim_{z \in K, ||z|| \to \infty} F(z)^T(z - z^0)/||z|| = \infty$$

where $||\cdot||$ denotes any vector norm in $\mathbf{R}^n$.

Among these properties, it is clear that pseudo-monotonicity is the weakest, followed in order by monotonicity, strict monotonicity and strong monotonicity. If F is strongly monotone over K, then it is coercive with respect to K.

Proposition 4.1. *Let K be a nonempty, closed and convex subset of* $\mathbf{R}^n$ *and let F be a continuous, pseudo-monotone mapping from K into* $\mathbf{R}^n$. *Then* z^* *solves the problem VI(K,F) if and only if* $z^* \in K$ *and*

$$F(z)^T(z - z^*) \geq 0 \quad \forall z \in K$$

In particular, the solution set of VI(K,F) is convex if it is nonempty.

In general, the VI(K,F) can have more than one solution. If F is strictly monotone, then the VI(K,F) has at most one solution:

Proposition 4.2. *If F is strictly monotone on K, then the problem VI(K,F) has at most one solution.*

Theorem 4.1. *Let K be a nonempty, closed and convex subset of* $\mathbf{R}^n$ *and let F be a continuous mapping from K into* $\mathbf{R}^n$. *If F is coercive with respect to K, then the problem VI(K,F) has a nonempty, compact solution set.*

Since strong monotonicity implies both coercivity and strict monotonicity, we can have the following corollary.

Corollary 4.1. *Let K be a nonempty, closed and convex subset of* $\mathbf{R}^n$ *and let F be a continuous mapping from K into* $\mathbf{R}^n$. *If F is strongly monotone with respect to K, then there exists a unique solution to the problem VI(K,F).*

4.2 Solution Methods

A general approach for solving the problem VI(K,F) consists of creating a sequence $\{z^k\} \subseteq K$ such that each z^{k+1} solves problem VI(K,F^k):

$$F^k(z^{k+1})^T(z - z^{k+1}) = F(z^k) + A(z^k)(z - z^k)$$

where $A(z^k)$ is an $n \times n$ matrix, several methods exist which differ in the choice of $A(z^k)$:

$$\begin{aligned}
A(z^k) &= \nabla F(z^k) && \text{(Newton's method)} \\
&\approx \nabla F(z^k) && \text{(Quasi - Newton)} \\
&= D(z^k) && \text{(Linearized Jacobi)} \\
&= \begin{cases} L(z^k) + D(z^k)/\omega* \\ U(z^k) + D(z^k)/\omega* \end{cases} && \text{(Successive Overrelaxation (SOR))} \\
&= \tfrac{1}{2}[\nabla F(z^k) + \nabla F(z^k)^T] && \text{(Symmetrized Newton)} \\
&= G && \text{(Projection)}
\end{aligned}$$

where

$D(z^k) \equiv$ the diagonal part of $\nabla F(z^k)$,
$L(z^k) \equiv$ the lower triangular part of $\nabla F(z^k)$,
$U(z^k) \equiv$ the upper triangular part of $\nabla F(z^k)$,
$\omega* \equiv$ a scalar parameter $\in (0, 2)$,
$G \equiv$ a fixed, symmetric, positive definite matrix.

Note that Newton's method is the most powerful of the abovementioned methods. According to Theorem 4.1 of Harker and Pang [12], if the Jacobian matrix $\nabla F(z^*)$ is positive definite, there exists a neighborhood N of z^* such that whenever the initial vector z^0 is chosen in N, the entire sequence $\{z^k\}$ generated by Newton's method is well-defined and converges to z^*. Obviously, in the case of solving a system of nonlinear equations, evaluating the $n \times n$ Jacobian matrix $\nabla F(z^k)$ at each iteration becomes a well-recognized drawback.

In order to overcome this drawback, a class of linear approximation methods is proposed to update the matrix $A(z^k)$ from one iteration to the next. They are quasi-Newton, linearized Jacobi, SOR, symmetrized Newton, and projection. However, the convergence of these methods typically requires much stronger properties on the mapping F. For instance, the projection method is only applicable to the problem VI(K,F) when F is strongly monotone. In particular for nonlinear approximation schemes, the most popular is the nonlinear Jacobi (also called relaxation or diagonalization) algorithm whose local convergence is analogous to the linearized Jacobi method in Theorem 4.2 of Harker and Pang [12]: the more "diagonally dominant" $\nabla F(z^*)$ is, the better the method can be expected to perform. The details of the *nonlinear Jacobi* method can be found in Pang and Chan [23] and Dafermos [6] which discussed the convergence of this method. Note that there are other algorithms for solving the **[VI-Market Equilibrium]**, like the modified projection method mentioned in Nagurney [21].

We now put [VI-Market Equilibrium] into standard form of VI(K,F). Define the column vector $z = (s_{f,i}, x_{f,i,h}, l_f, y_i, w_i, w_l)$ and $F(z) = (Fs_{f,i}(z), Fx_{f,i,h}(z), Fl_f(z), Fy_i(z), Fw_i(z), Fw_l(z))$, where $Fs_{f,i}(z) = P_{i0} - \left(\frac{P_{i0}}{Q_{i0}}\right)\left(2s_{f,i} + \sum_{g \neq f} s_{g,i}\right) - w_i \forall f, i$; $Fx_{f,i,h}(z) = C_{f,i,h} - w_i \forall f, i, h$; $Fl_f(z) = w_l \forall l$; $Fy_i(z) = w_i \forall i$; $Fw_i(z) = \sum_f s_{f,i} - \sum_{f,h} x_{f,i,h} - y_i \forall i$; and $Fw_l(z) = \sum_f (l_{f0} - l_f) \forall l$.

The model [VI-Market Equilibrium] can now be expressed as a single VI(K_M,F) problem that can be solved by some existing solvers such as PATH based on Newton's method.

Lemma 4.1. *F(z) in the model [VI-Market Equilibrium] is strictly monotone; that is,*

$$[F(z^1) - F(z^2)]^T (z^1 - z^2) > 0, \quad \forall z^1, z^2 \in K_M, z^1 \neq z^2.$$

Proof. For any $z^1, z^2 \in K_M$ and $z^1 \neq z^2$, we have

$$(F(z^1) - F(z^2))^T (z^1 - z^2)$$

$$= \sum_{f,i} -\Big[(P_{i0} - \left(\frac{P_{i0}}{Q_{i0}}\right)\left(2s^1_{f,i} + \sum_{g \neq f} s^1_{g,i}\right) - w^1_i)$$
$$-(P_{i0} - \left(\frac{P_{i0}}{Q_{i0}}\right)\left(2s^2_{f,i} + \sum_{g \neq f} s^2_{g,i}\right) - w^2_i)\Big] \times \left[s^1_{f,i} - s^2_{f,i}\right]$$
$$+ \sum_{f,i,h} [(C_{f,i,h} - w^1_i) - (C_{f,i,h} - w^2_i)] \times \left[x^1_{f,i,h} - x^2_{f,i,h}\right]$$

$$+ \sum_f [w^1_l - w^2_l] \times \left[l^1_f - l^2_f\right] - \sum_i [w^1_i - w^2_i] \times \left[y^1_i - y^2_i\right]$$

$$- \sum_i \left[\left(\sum_f s^1_{f,i} - \sum_{f,h} x^1_{f,i,h} - y^1_i\right) - \left(\sum_f s^2_{f,i} - \sum_{f,h} x^2_{f,i,h} - y^2_i\right)\right] \times \left[w^1_i - w^2_i\right]$$
$$+ \left[\sum_f (l_{f0} - l^1_f) - \sum_f (l_{f0} - l^2_f)\right] \times \left[w^1_l - w^2_l\right]$$

$$= \sum_{f,i} -\left[\left(P_{i0} - \left(\frac{P_{i0}}{Q_{i0}}\right)\left(2s^1_{f,i} + \sum_{g \neq f} s^1_{g,i}\right)\right)\right.$$
$$\left. -\left(P_{i0} - \left(\frac{P_{i0}}{Q_{i0}}\right)\left(2s^2_{f,i} + \sum_{g \neq f} s^2_{g,i}\right)\right)\right] \times \left[s^1_{f,i} - s^2_{f,i}\right]$$

$$= \sum_{f,i} \left(\frac{P_{i0}}{Q_{i0}}\right)\left(s^1_{f,i} - s^2_{f,i}\right)^2 + \sum_i \left(\frac{P_{i0}}{Q_{i0}}\right)\left(\sum_f s^1_{f,i} - \sum_f s^2_{f,i}\right)^2 > 0.$$

Lemma 4.2. *The Jacobian matrix $\nabla F(z)$ in the model [VI-Market Equilibrium] is strictly is positive definite.*

Proof. Take a nonzero z, one easily has $z^T \nabla F(z) z > 0$.

5 An Example

A simple example model adopted from Hobbs [13] is modified with a CO_2 emission permits market. In the model, there are three buses, $i = 1, 2, 3$, each of which has customers. Generation, however, occurs only at buses $i = 1, 2$. Each pair of buses is interconnected by a single transmission line (l). All three lines have equal impedances. The demand functions are $p_i(q_i) = 40 - 0.08q_i$ for $i = 1, 2$, and $p_3(q_3) = 32 - 0.0516q_3$ \$/MWh. Note that demand is more elastic at the demand-only node (Bus 3). There are two producers $f = 1, 2$, each with one generator. Both generators have limited capacity of 1,000 MW, and a constant marginal cost of \$15/MWh and \$20/MWh for Firm 1 and Firm 2, respectively. Infinite transmission capacity is assumed. The PTDF is given in Table 1.

Note that if transmission capacity is limited, the transmission cost arises from congestion. The emission rate, E, is 0.930 tCO_2/kWh for all generators.

To find the reference level of total CO_2 emissions, a reference model **[Reference]** without CO_2 emissions trading and constraints has been developed, which has determined the reference level to be 523 tCO_2. Based on this level, we can allocate the initial licenses, l_{f0}, for each firm accordingly. In our example, we simply allocate 40% of the reference level as the firms' initial licenses. The results of the model **[Reference]** and of the 20% reductions model **[CO2 trading]** are shown in Tables 2 and 3, respectively. Table 4 presents the results of CO_2 emissions levels. The market price of the emissions licenses is \$4.087/tCO_2.

Table 1 Values of PTDFs

	Bus $i = 1$	2	3
Transmission line $l = 1$	0.33	−0.33	0
2	0	0.33	−0.33
3	−0.33	0	0.33

Table 2 Results of the model **[Reference]**

		Node, i =1	2	3
Sales, $s_{f,i}$	Firm 1	125	125	142.119
	Firm 2	62.5	62.5	45.22
Generation, $x_{f,i,1}$	Firm 1	392.119	–	–
	Firm 2	–	170.22	–
Transmitted power, y_i	–	−204.619	17.28	187.339
Equilibrium price, p_i	–	25	25	22.333

Table 3 Results of the model **[CO2 trading]**

		Node, i =1	2	3
Sales, $s_{f,i}$	Firm 1	109.161	109.161	117.563
	Firm 2	46.661	46.661	20.664
Generation, $x_{f,i,1}$	Firm 1	335.885	–	–
	Firm 2	–	113.986	–
Transmitted power, y_i	–	−180.063	41.836	138.227
Equilibrium price, p_i	–	27.534	27.534	24.868

Table 4 Results of CO_2 emissions

	[Reference]	[CO_2 trading]
Firm 1	364.671	312.373
Firm 2	158.305	106.007
Total	522.976	418.38

In our implementation, the Reference model and the CO_2 trading model are coded into the GAMS program (executed on a PC with Intel Core 2 Duo Processor and 4 GB memory) and solved by the built-in solver PATH 4.2.

6 Remarks

6.1 For an Arbitraged Market Equilibrium Problem

If there is arbitrage, substitute $\sum_f s_{f,i} + a_i$ for $\sum_f s_{f,i}$ as the quantity consumed in the demand function and the market clearing condition (M1), where a_i MW is the net amount of power sold by arbitragers to node i, and $\sum_i a_i = 0$. Obviously, the arbitrage marketing clearing price is the difference between the electricity price and the transmission cost, $p_i - w_i$.

6.2 For Conjectured Supply Function Approach

When price elasticity of demand is low, Cournot models do not give meaningful equilibria for the power market. This is because equilibrium prices are determined only by the demand function, and therefore are usually higher than those observed in reality. To enhance Cournot models, Day et al. [7] proposed to use the Conjectured Supply Function instead of demand function to determine equilibrium prices in Cournot-based models. According to Hobbs and Pang [15], the model postulates that the rival firms' sales $s_{-f,i} \equiv \sum_{g \neq f} s_{g,i}$ are related to the price $p_{f,i}$ via the following linear expression:

$$s_{-f,i} = s^*_{-f,i} + \beta_{f,i}(p^*_i, s^*_{-f,i})(p_{f,i} - p^*_i), \tag{R1}$$

where $(p^*_i, s^*_{-f,i})$ is an equilibrium (price, sales) pair that is exogenous to the firm f's profit maximization problem but endogenous to the market. Considering the function $\beta_{f,i}(x, y)$ is a positive constant $\beta_{f,i}$, we have

$$p_{f,i} = \left(p^*_i - \frac{s^*_{-f,i}}{\beta_{f,i}} \right) + \frac{1}{\beta_{f,i}} s_{-f,i}$$

By substituting (R1) into the demand function, we get

$$p_{f,i} = P_{i0} - \frac{P_{i0}}{Q_{i0}} \left[s_{f,i} + s^*_{-f,i} + \beta_{f,i}(p_{f,i} - p^*_i) \right]$$

which yields

$$p_{f,i} = \frac{P_{i0} - \dfrac{P_{i0}}{Q_{i0}}\left[s_{f,i} + s^*_{-f,i} - \beta_{f,i}p^*_i\right]}{1 + \dfrac{P_{i0}}{Q_{i0}}\beta_{f,i}},$$

$$p_{f,i} = p_{f,i}(s_{f,i}, s^*_{-f,i}, p^*_i).$$

Hence, the objective function of the firm f's profit becomes

$$\sum\nolimits_i \left[p_{f,i}(s_{f,i}, s^*_{-f,i}, p^*_i) - w^*_i\right] s_{f,i} - \sum\nolimits_{i,h} (C_{f,i,h} - w^*_i) x_{f,i,h} - w^*_l (l_f - l_{f0}).$$

6.3 *Other Transmission Representations*

In our model, the competition among firms on grid is based on linearized DC load flow approximations. To have more realistic representations of transmission costs and constraints, Hobbs et al. [14] studied three different formulation representations: (1) quadratic resistance losses, (2) phase shifters, and (3) controllable DC lines. From their simulation results, they showed that improved transmission representations can change the results of oligopoly market models. On the other hand, Cheng and Overbye [2] proposed a scheme to determine the equivalent reactance of interfaces of a reduced system based on the zonal power transfer distribution factors of the original system. It is believed that a more realistic PTDF-based system is possible and has yet to be devised. Decomposition of multi-regional (zonal) model (Chung et al. [4]) may be one of the effective modeling approaches to develop this realistic PTDF-based system. It is because the submodels of zonal firms can be modeled separately from the PTDF-based system.

6.4 *With Other Non-Electric Power Participants of CO_2 Trading*

It should be noted that there are other players in CO_2 emission permits market, including energy-intensive industries such as aluminum, chemical, paper, and steel. However, most of the profit functions of these players are nonlinear, unlike the linear function in our model. Properties of the nonlinear functions would affect the convergence of the abovementioned algorithms. In general, if the nonlinear demand function is strictly positive on some bounded interval, on which it is continuously twice differentiable, and strictly decreasing, we may follow the work of Murphy et al. [20] to show that the existence of equilibrium and the convergence of Newton's method are guaranteed. Moreover, due to these players, the model may become a multi-commodity market equilibrium model which may be too big to be solved in a single PC. Indeed, individual commodity market equilibrium models can be joined as a multi-commodity model and solved by decomposition methods, see Chung and Fuller [3] for details.

6.5 Equilibrium Models with Continuous and Binary Variables

To resolve the problem of power plant expansion, some binary variables in the equilibrium models need to be handled. Fuller [9] provides some algorithms and a general definition of equilibrium in a mathematical programming framework that allows for prices related to binary variables.

6.6 Multi-Period Model

One of the most important and controversial features is the initial allocation of licenses. For instance, should the power industry be given a free initial assignment of licenses? Or should the government auction the initial licenses? If the "free initial allocation" based on a "grandfathering mechanism" depends on present or future generator decisions, should incentives to deviate from least-cost investment mixes and operation be introduced? If a polluting generator is shut down, it would lose its licenses allocation. As a result, this provides an incentive to keep non-economic capacity in operation. Moreover, should new investment be allocated free initial licenses? On the other hand, can or should the initial licenses be banked for a certain period? Jaehn and Letmathe [17] discussed the unpredicted volatility and crash of license prices in 2005, which led to the hypothesis that electricity producers might use their market power to influence the prices of licenses. They argued that banking is one of the main factors leading to such a price crash if market players act rationally.

A multi-period version of **[VI-Market Equilibrium]** is preferred to resolve the power plant expansion and the problems of banking emissions permits and of initial license allocation (either by auction or grandfathering).

7 Summary and Conclusion

In this paper, we presented a variational inequality framework for the formulation and computation of equilibria in oligopolistic electricity models with marketable CO_2 emissions licenses. The model explicitly combines the electricity and CO_2

licenses markets. Moreover, we describe some model properties and some algorithms. A simple example is used to illustrate the model solved by Newton's method of VI with numerical results. The topics discussed in Sect. 6 require further investigations and may be the subject of future studies.

Appendix: A Brief Introduction to Variational Inequalities

Some key concepts of variational inequality representations are provided to readers who are familiar with constrained convex optimization models. More complete discussions can be found in various sources – e.g., Nagurney [21], Harker and Pang [12], or Facchinei and Pang [8].

Let K be a closed convex set in $\mathbf{R}^n$, and f be a continuously differentiable, convex function defined on K. Consider the optimization problem

$$\mathbf{OPT}(f, \mathbf{K}) : minimize f(z)$$

$$\text{subject to: } z \in K$$

Consider also the variational inequality problem $VI(\nabla f, K)$: find $z^* \in K$ such that $\nabla f(z^*)^T(z - z^*) \geq 0$ for all $z \in K$. Nagurney [21] shows (Propositions 1, 2 and 1.3) that **OPT(*f*, K)** and $VI(\nabla f, K)$ are equivalent, i.e., that z^* is a solution to **OPT(*f*, K)** if and only if z^* is a solution to $VI(\nabla f, K)$. This can be understood intuitively by examining the condition for z^* to be a solution to $VI(\nabla f, K)$, i.e., $\nabla f(z^*)^T(z - z^*) \geq 0$ for all $z \in K$.

This condition says that, to a first order approximation, using the gradient of f at z^*, any change, $(z - z^*)$ away from z^* causes either an increase or no change in f.

The general form of a variational inequality may be written *VI(G, K)*: find $z^* \in K$ such that $G(z^*)^T(z - z^*) \geq 0$ for all $z \in K$. In the general case, it is not required that the vector function G be the gradient of a single valued function, which means that it may not be possible to rewrite *VI(G, K)* as an optimization model in the straightforward manner above.

References

1. Chen YH, Hobbs BF (2005) An oligopolistic electricity market model with tradable NOx permits. IEEE Trans Power Syst 20:119–129
2. Cheng X, Overbye TJ (2005) PTDF-based power system equivalents. IEEE Trans Power Syst 20:1868–1876
3. Chung W, Fuller JD (2010) Subproblem approximation in Dantzig-Wolfe decomposition of variational inequality models with an application to a multicommodity economic equilibrium model. Oper Res 58:1318–1327

4. Chung W, Fuller JD, Wu YJ (2006) A new decomposition method for multi-regional economic equilibrium models. Oper Res 54:643–655
5. Crocker TD (1966) The structuring of atmospheric pollution control systems. In: Wolozin H (ed) The economics of air pollution. WW Norton, New York
6. Dafermos S (1983) An iterative scheme for variational inequalities. Math Progr 26:40–47
7. Day CJ, Hobbs BF, Pang J-S (2002) Oligopolistic competition in power networks: a conjectured supply function approach. IEEE Trans Power Syst 17:597–607
8. Facchinei F, Pang J-S (2003) Finite-dimensional variational inequalities and complementarity problems, vol 2, Springer series in operations research. Springer, New York
9. Fuller JD (2008) Market equilibrium models with continuous and binary variables. Working paper, University of Waterloo, Canada
10. Geogopoulou E, Sarafidis Y, Mirasgedis S, Lalas D (2006) Next allocation phase of the EU emissions trading scheme: How tough will the future be? Energy Policy 34:4002–4023
11. Harker PT (1984) A variational inequality approach for determination of oligopolistic market equilibrium. Math Progr 30:105–111
12. Harker PT, Pang J-S (1990) Finite-dimensional variational inequality and nonlinear complementarity problems: a survey of theory, algorithms and applications. Math Progr 48:161–220
13. Hobbs BF (2001) Linear complementarity models of Nash-Cournot competition in bilateral and POOLCO power markets. IEEE Trans Power Syst 16:194–202
14. Hobbs BF, Drayton G, Fisher EB, Lise W (2008) Improved transmission representations in oligopolistic market models: quadratic losses, phase shifters, and DC lines. IEEE Trans Power Syst 23:1018–1029
15. Hobbs BF, Pang J-S (2004) Spatial oligopolistic equilibria with arbitrage, shared resources, and price function conjectures. Math Progr B 101:57–94
16. Huntington HG, Weyant JP (2004) Modeling energy markets and climate change policy. In: Cleveland CJ et al (eds) Encyclopedia of energy. Academic/Elsevier, Amsterdam, pp 41–53
17. Jaehn F, Letmathe P (2010) The emissions trading paradox. Eur J Oper Res 202:248–254
18. Jensen SG, Skytte K (2003) Simultaneous attainment of energy goals by means of green certificates and emissions permits. Energy Policy 31:63–71
19. Claire Breidenich, Daniel Magraw, Anne Rowley, James W. Rubin (1998) The Kyoto Protocol to the United NationsFramework Convention on Climate Change. The American Journal of International Law 92:315–331
20. Murphy F, Sherali H, Soyster A (1982) A mathematical programming approach for determining oligopolistic market equilibria. Math Progr 24:92–106
21. Nagurney A (1999) Network economics: a variational inequality approach, Rev. 2nd edn. Kluwer Academic Publishers, Dordrecht
22. Oren SS (1997) Economic inefficiency of passive transmission rights in congested electricity systems with competitive generation. Energy J 18:63–83
23. Pang JS, Chan D (1982) Iterative methods for variational and complementarity problems. Math Progr 24:284–313
24. Schweppe FC, Caramanis MC, Tabors RD, Bohn RE (1988) Spot pricing of electricity. Kluwer, Boston
25. Shaidehpour M, Yamin H, Li Z (2002) Market operations in electric power systems: forecasting, scheduling, and risk management. Wiley, New York
26. Unger T, Ahlgren EO (2006) Impacts of a common green certificate market on electricity and CO_2-emission markets in the Nordic countries. Energy Policy 33:2152–2163
27. Ventosa M, Baillo A, Ramos A, Rivier M (2005) Electricity market modeling trends. Energy Policy 33:897–913

Effect of Power Generation Mix and Carbon Emissions Tax on Investment Timing

Ryuta Takashima and Junichiro Oda

Abstract Electricity production accounts for around 40% of global energy-related CO_2 emissions and it is expected that the electricity demand increases to twice the current level in 2050. Therefore it is necessary to invest in low-carbon thermal power plants, nuclear and renewable energy for realizing low-carbon economy. These policies may require a large amount of investment costs, and additionally, the uncertainty increases in a situation surrounding power generation projects and their investments. On the other hand, environmental policy for encouraging use of low carbon emission generation power includes an internalization of the externality for CO_2 emissions such as carbon-emissions tax. In this study, we develop a real option model of power generation investments allowing for two uncertainties of the market risk and the introduction of the policy. We analyze the effect of the uncertainties on the power generation mix and the investment timing.

Keywords Environmental policy • Electricity market • CO_2 emission • Uncertainty • Real options

1 Introduction

Climate change is a global environmental issue and introduction expansion of climate policy is officially discussed in many countries for addressing climate change. Especially, power industry accounts for around 40% of global energy

---

R. Takashima (✉)
Department of Risk Science in Finance and Management, Chiba Institute of Technology, 2-17-1 Tsudanuma, Narashino-shi, Chiba 275-0016, Japan
e-mail: takashima@sun.it-chiba.ac.jp

J. Oda
Systems Analysis Group, Research Institute of Innovative Technology for the Earth, 9-2, Kizugawadai, Kizugawa-Shi, Kyoto 619-0292, Japan

Q.P. Zheng et al. (eds.), *Handbook of CO_2 in Power Systems*, Energy Systems, DOI 10.1007/978-3-642-27431-2_2, © Springer-Verlag Berlin Heidelberg 2012

related CO_2 emissions [9], climate change is a key issue in power industry. Some power companies are charged for explicit carbon cost. Other companies feel implicit carbon cost at least. These climate policies are very uncertain for power companies because climate policy can be changed over sociopolitical trend in the short term compared with long-lived power plants. On the other hand, companies also face an uncertainty of electricity market. Under the circumstance, it is very important to evaluate the value of power plant investment such as expansion and new construction under uncertainties of climate policy and electricity market.

For one of economic analysis methods for investment projects under uncertainties, real options analysis has recently attracted growing attention. Real options analysis, which is pioneered by Brennan and Schwartz [3] and McDonald and Siegel [10] and is summarized in Dixit and Pindyck [4], has been widely used for problems of power plants investments such as problems of modularity [7, 12, 14], capacity sizing [2], technology choice [15, 17], and problems of replacement and refurbishment [11, 16]. Furthermore, the evaluation of the power generation investment under policy uncertainty in real options framework includes Blyth et al. [1], Fuss et al. [5], Yang et al. [18], and Fuss et al. [6]. Blyth et al. [1] analyze the investment options of coal- and gas-fired power plants, and the power plants associated with carbon capture and storage (CCS) technologies taken into accounts uncertain future climate policy. Fuss et al. [5] examine the investment decisions of coal-fired power plant, the plant including a CCS module[1], and the existing plant with a CCS module when investors face uncertainty from climate change policy as well as from volatile prices in the markets. Yang et al. [18] analyze the effects of government climate policy uncertainty on gas, coal and nuclear power investment. Fuss et al. [6] analyze the impact of the frequency of policy changes on investment decisions for low CO_2-emitting electricity generation technologies such as integrated gasification combined cycle plant, the plant including a CCS module, and wind power.

The policy uncertainty presented in these previous works is represented by the dynamics of CO_2 price such as the jump. Our modeling of the policy uncertainty is different from these papers. We focus on uncertain adoption time of environmental policy. Specifically, our modeling setup with respect to environmental policy uncertainty follows Hassett and Metcalf [8] that investigate the effect of tax policy uncertainty on the investment decisions in which tax incentives reduce the capital cost. In Hassett and Metcalf [8], the uncertainty of the policy adoption is assumed to follow a Poisson process. Likewise the model in this paper, suppose that the uncertainty of environmental policy adoption follows the Poisson process.

In this chapter, we develop a real option model of power generation investments allowing for two uncertainties of the market risk such as future price changes and the introduction of the policy. The numerical results show how the power generation mix and the tax rate influence the investment timing.

[1] The plant including a CCS module means the plant to be invested.

The remainder of this chapter is organized as follows. In Sect. 2, we present the basic model for analyzing the investment of capacity expansion for power generations and the model of environmental policy uncertainty. Section 3 provides some results of numerical analysis in which uncertainty of environmental policy adoption affects the investment timing of power generation. Finally, Sect. 4 concludes the paper.

2 The Model

2.1 Model Setup

In this section, we model a profit flow that is obtained from operations of power generations.

Suppose that the firm is a price taker, and, its actions have no influence on the dynamics of the electricity price. Thus, for a straightforward description of uncertainty, we assume that the electricity price at time t, P_t follows the geometric Brownian motion:

$$dP_t = \mu P_t dt + \sigma P_t dW_t, \qquad P_0 = p, \tag{1}$$

where μ is the instantaneous expected growth rate of P_t, and σ is the instantaneous volatility of P_t. W_t is a standard Brownian motion.

Consequently, the profit flow from plant operating at time t, π_t^i can be represented by the following equation,

$$\pi_t^i = Q^i P_t - \sum_{j=1}^{n} q_j^i c_j, \tag{2}$$

where $i = \{0, 1\}$ denote the states before and after the investment[2], respectively, $Q^i = \sum_{j=1}^{n} q_j^i$ is a total capacity of n power generations for any state, q_j^i is a capacity for power generation j, and c_j is the operating cost that is composed of the fuel cost as well as operating and maintenance costs for power generation j. If an internalization of the externality for CO_2 emissions such as carbon-emissions tax is introduced, the profit flow can be rerepresented as follows:

$$\pi_t^i = Q^i P_t - \sum_{j=1}^{n} q_j^i c_j - \tau \sum_{j=1}^{n} q_j^i \eta_j, \tag{3}$$

[2] In this paper, we do not consider the investment in the module retrofitting, but that in the generation expansion.

where τ is a tax rate for carbon-emission, and η_j is a emission basic unit for power generation j.

2.2 Capacity Expansion Investment

In this section, we describe the model that derive the investment timing of the capacity expansion and the change of the generation mix, and its project value. We consider that a firm operates power generations at the present time, and has the investment options of the capacity expansion and the change of the generation mix. Suppose that the firm can determine the investment timing of power generations with a fixed output, Q. The value of the investment opportunity is:

$$F(p) \equiv \sup_T \mathbb{E}\left[\int_0^T e^{-\rho t}\pi_t^0 dt - e^{-\rho T} I(q_j^i) + \int_T^\infty e^{-\rho t}\pi_t^1 dt\right], \tag{4}$$

where T is the investment time, $\rho>0$ is an arbitrary discount rate, $I(q_j^i) = \sum_{j=1}^n \delta_j \max(q_j^1 - q_j^0, 0)$ is the total investment cost for capacity expansion, and δ_j is the investment cost per kW for power generation j.

Prior to determining the investment threshold p^* and $F(p)$, we calculate the now-or-never expected NPV, $V(p)$, of a power generation mix after the investment:

$$\begin{aligned} V(p) &= \mathbb{E}\left[\int_0^\infty e^{-\rho t}\pi_t^1 dt - I(q_j^i)\right] \\ &= \frac{Q^1 p}{\rho - \mu} - \frac{\sum_{j=1}^n q_j^1 c_j}{\rho} - \sum_{j=1}^n \delta_j \max(q_j^1 - q_j^0, 0). \end{aligned} \tag{5}$$

Following standard arguments as in [4], the value of the investment option satisfies the following differential equation

$$\frac{1}{2}\sigma^2 p^2 F''(p) + \mu p F'(p) - \rho F(p) + Q^0 p - \sum_{j=1}^n q_j^0 c_j = 0, \tag{6}$$

where the primes denote derivatives, that is, $F'(p) = \frac{dF(p)}{dp}$ and $F''(p) = \frac{d^2F(p)}{dp^2}$. The general solution of (6) is given by the following equation:

$$F(p) = A_1 p^{\beta_1} + A_2 p^{\beta_2} + \frac{Q^0 p}{\rho - \mu} - \frac{\sum_{j=1}^n q_j^0 c_j}{\rho}, \tag{7}$$

where A_1 and A_2 are constants, and $\beta_1>1$ and $\beta_2<0$ are the positive and the negative root of the characteristic equation $\frac{1}{2}\sigma^2\beta(\beta-1)+\mu\beta-r=0$, respectively. The unknown constants A_1 and A_2 together with the investment threshold of the capacity expansion and the change of the generation mix, p^* are determined by the following boundary conditions,

$$\lim_{p\to 0}(A_1 p^{\beta_1}+A_2 p^{\beta_2})=0, \tag{8}$$

$$F(p^*)=V(p^*), \tag{9}$$

$$F'(p^*)=V'(p^*). \tag{10}$$

Condition (8) requires that the investment option becomes zero if the price level is close to zero. Thus, $A_2=0$. Conditions (9) and (10) are the value-matching and smooth-pasting conditions, respectively. The value-matching condition means that when the level of P_t is p^*, the firm exercises the investment option, and then can obtain the net value of $V(p^*)$. Additionally, the smooth-pasting condition means that if the capacity expansion and the change of the generation mix at p^* is indeed optimal, the differentiation of the value function must be continuous at p^*. From these conditions, we can obtain the investment threshold as follows:

$$p^*=\frac{\beta_1}{\beta_1-1}\frac{\rho-\mu}{Q^1-Q^0}\left[\frac{\sum_{j=1}^{n}(q_j^1-q_j^0)c_j}{\rho}+I(q_j^i)\right]. \tag{11}$$

Furthermore, the unknown constant A_1 is given by,

$$A_1=\frac{1}{\beta_1}\frac{Q^1-Q^0}{\rho-\mu}p^{*1-\beta_1}. \tag{12}$$

When the policy is implemented, it is necessary to consider the CO_2 emissions cost. Thus for the case, the term of the emission cost is embedded in (4), and then, the investment threshold (11) can be rerepresented as follows:

$$p^*=\frac{\beta_1}{\beta_1-1}\frac{\rho-\mu}{Q^1-Q^0}\left[\frac{\sum_{j=1}^{n}(q_j^1-q_j^0)c_j}{\rho}+I(q_j^i)+\frac{\tau\sum_{j=1}^{n}(q_j^1-q_j^0)\eta_j}{\rho}\right]. \tag{13}$$

When the relatively high emission power generation increases, the emission cost becomes large, and then the investment threshold increases, whereas when the low or zero emission power generation increases, the investment threshold decreases.

2.3 Investment Decision under Environmental Policy Uncertainty

In this section we consider the policy instrument that is a carbon tax at a given rate τ. If the policy is implemented, the CO_2 emissions cost as in (3) is incurred. The government can switch from a present regime to policy regime, one in which the emission cost is not imposed, and the other in which it is. The switches from present state to policy regime follow Poisson process with an constant intensity λ. The probability that the policy will be implemented in the next short interval of time dt is λdt. Suppose that p_0 and p_1 are investment thresholds for the present state and the policy regime, respectively. As shown in Fig. 1, over an interval of $0 \leq p < p_1$, the firm will not invest regardless of whether the environmental policy is adopted. Over an interval of $p_1 \leq p < p_0$, if the environmental policy is not adopted, the firm will postpone the investment with the expectation not only that the price increases but also that the environmental policy is adopted, whereas if the environmental policy is adopted the firm will invest. In $p_0 \leq p$, the firm will invest irrespective of the policy adoption.

The value of the investment option for $p<p_1$ in adoption region, $F_1(p)$ is given by

$$F_1(p) = B_1 p^{\beta_1} + \frac{Q^0 p}{\rho - \mu} - \frac{\sum_{j=1}^{n} q_j^0 c_j}{\rho} - \frac{\tau \sum_{j=1}^{n} q_j^0 \eta_j}{\rho}, \tag{14}$$

where B_1 are a unknown constant, The expected NPV for $p_1 \leq p$ in adoption region, $V_1(p)$ is

$$V_1(p) = \frac{Q^1 p}{\rho - \mu} - \frac{\sum_{j=1}^{n} q_j^1 c_j}{\rho} - \frac{\tau \sum_{j=1}^{n} q_j^1 \eta_j}{\rho}. \tag{15}$$

The unknown constant, B_1 and the investment threshold of the capacity, p_1 are determined by the following value-matching and smooth-pasting conditions,

$$F_1(p_1) = V_1(p_1), \tag{16}$$

$$F_1'(p_1) = V_1'(p_1). \tag{17}$$

From these conditions, we can obtain the investment threshold as follows:

$$p_1 = \frac{\beta_1}{\beta_1 - 1} \frac{\rho - \mu}{Q^1 - Q^0} \left[\frac{(C^1 - C^0) + (C_\tau^1 - C_\tau^0)}{\rho} + I(q_j^i) \right], \tag{18}$$

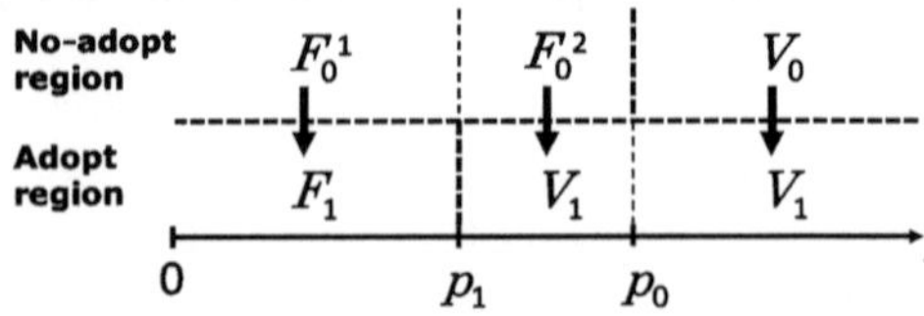

Fig. 1 Value functions and investment thresholds for each region

where $C^i = \sum_{j=1}^{n} q_j^i c_j$, and $C_\tau^i = \tau \sum_{j=1}^{n} q_j^i \eta_j$ $(i = 0, 1)$. Furthermore, the unknown constant B_1 is given by,

$$B_1 = \frac{1}{\beta_1} \frac{Q^1 - Q^0}{\rho - \mu} p^{*1-\beta_1}. \tag{19}$$

For $p<p_1$ in no adoption region, the Bellman equation is given by

$$F_0^1(p) = \left(Q^0 p - \sum_{j=1}^{n} q_j^0 c_j\right) dt + e^{-\rho t}\left[\lambda \mathbb{E}[F_1(P + dP)] + (1 - \lambda)\mathbb{E}[F_0^1(P + dP)]\right]. \tag{20}$$

Following standard arguments, the following differential equation, which is satisfied by the value of the investment option, is derived from the Bellman equation,

$$\frac{1}{2}\sigma^2 p^2 F_0^{1\prime\prime}(p) + \mu p F_0^{1\prime}(p) - \rho F_0^1(p) - \lambda(F_0^1(p) - F_1(p)) + Q^0 p - \sum_{j=1}^{n} q_j^0 c_j = 0, \tag{21}$$

The general solution of (21) is given by the following equation:

$$F_0^1(p) = B_2 p^{\beta_1} + B_3 p^{\gamma_1} + \frac{Q^0 p}{\rho - \mu} - \frac{C^0}{\rho} - \frac{\lambda C_\tau^0}{(\rho + \lambda)\rho}, \tag{22}$$

where B_2 and B_3 are unknown constants, and $\gamma_1 > 1$ is the positive root of the characteristic equation $\frac{1}{2}\sigma^2\gamma(\gamma - 1) + \mu\gamma - (r + \lambda) = 0$. Following standard arguments as in Hassett and Metcalf [8], $B_2 = B_1$. For $p_1 \le p < p_0$ in no adoption region, the investment opportunity value $F_0^2(p)$ is given by

$$F_0^2(p) = B_4 p^{\beta_1} + B_5 p^{\gamma_2} + \frac{(\rho - \mu)Q^0 + \lambda Q^1}{(\rho - \mu)(\rho + \lambda - \mu)} p - \frac{\rho C^0 + \lambda C^1}{\rho(\rho + \lambda)} - \frac{\lambda C_\tau^1}{(\rho + \lambda)\rho} - \frac{\lambda I(q_j^i)}{\rho + \lambda}, \tag{23}$$

where B_4 and B_5 are unknown constants, and $\gamma_2<0$ is the negative root of the characteristic equation $\frac{1}{2}\sigma^2\gamma(\gamma - 1) + \mu\gamma - (r + \lambda) = 0$. The expected NPV for $p_0 \le p$ in no adoption region, $V_0(p)$ is given by

$$V_0(p) = \frac{Q^1 p}{\rho - \mu} - \frac{\sum_{j=1}^{n} q_j^1 c_j}{\rho} - \frac{\lambda C_\tau^1}{(\rho + \lambda)\rho}. \tag{24}$$

The investment threshold in no adoption region, p_0, along with endogenous constants, B_3, B_4 and B_5, are determined via the following boundary conditions:

$$F_0^1(p_1) = F_0^2(p_1), \quad (25)$$

$$F'^1_0(p_1) = F'^2_0(p_1), \quad (26)$$

$$F_0^2(p_0) = V_0(p_0), \quad (27)$$

$$F'^2_0(p_0) = V'_0(p_0). \quad (28)$$

Conditions (25) and (26) are the continuity and high-contact conditions for p_1, respectively. Conditions (27) and (28) are the value-matching and smooth-pasting conditions for p_0, respectively. Since these four equations are highly non-linear, it is not possible to find an analytical solution to the system. However, numerical solutions may be obtained for specific parameters.

3 Numerical Analysis

In the previous section, we presented a model that can analyze the effect of the generation mix change and environmental policy uncertainty on the investment timing. In this section, using this model we present the numerical analysis of various scenarios and the effect of the possibility of policy adoption on the investment thresholds for each region. The solutions for the model can be obtained by means of a numerical calculation method such as the Newton method.

We solve the model numerically using the following base case parameter values: the expected growth rate of the electricity price, μ is 0, the volatility of the electricity price, σ is 20% per year, and the discount rate, ρ is 5% per year. For the parameter values regarding the power generations, we use the data as shown in Table 1 based on OECD/NEA [13]. In this table, the difference of fuel cost between coal-based plants is derived from the assumed power generation efficiency. Suppose that at the present time power generation mix of 3,000 MW is composed of 1,000 MW of aged coal-fired generation, 1,000 MW of gas-fired generation, and 1,000 MW of nuclear power. For present condition, total operating cost, which is composed of the fuel cost as well as operating and maintenance costs, is \$25.9/MWh (579 million\$/year), and CO_2 emissions of present generation mix is 0.465 t-CO_2/MWh (10,390 kt-CO_2/year). We consider the investment projects of the capacity expansion of 5,000 MW for the following four cases: For the case i, the power generation mix is composed of 2,500 MW of gas-fired generation and 2,500 MW of nuclear power. For the case ii, the power generation mix is composed of 2,500 MW of coal-fired generation with CCS and 2,500 MW of nuclear power. For the case iii, the power generation mix is composed of 1,250 MW of coal-fired generation with CCS, 1,250 MW of gas-fired generation, and 2,500 MW of nuclear power. For the case iv, the power generation mix is composed of 625 MW

Table 1 Parameter values with respect to power generations

	Initial cost	O&M cost	Fuel cost	Load factor	Emission intensity
	(US$/kW)	(US$/kWh)	(US$/kWh)	(%)	($t\text{-}CO_2$/MWh)
Coal	1,300	5.0	15.8	85.0	0.830
Coal with CCS	1,600	23.0	18.0	85.0	0.095
Aged coal	–	11.0	20.3	85.0	1.064
Gas	600	2.4	20.3	85.0	0.331
Nuclear	1,750	6.5	5.2	85.0	0

Table 2 Investment cost, operating cost, and CO_2 emissions data and investment thresholds for each case

Case	i	ii	iii	iv
Coal	0	0	0	625
Coal with CCS	0	2,500	1,250	625
Aged coal	0	0	0	0
Gas	2,500	0	1,250	1,250
Nuclear	2,500	2,500	2,500	2,500
Investment cost ($/MW)	3,525	6,625	4,775	4,588
Operating cost ($/MWh (million$/year))	23.3 (867)	26.4 (981)	24.8 (924)	22.3 (830)
CO_2 emissions ($t\text{-}CO_2$/MWh ($kt\text{-}CO_2$/year))	0.166 (6,167)	0.047 (1,761)	0.106 (3,964)	0.198 (7,389)
p_0 ($/MWh)	54.329	84.202	67.393	57.385
p_1 ($/MWh)	50.135	75.510	60.947	54.447

of coal-fired generation, 625 MW of coal-fired generation with CCS, 1,250 MW of gas-fired generation, and 2,500 MW of nuclear power.

In Table 2, the investment cost per MW, the operating cost per MWh, CO_2 emissions per MWh and the investment thresholds for each case are shown. In this numerical example, we use a tax rate of \$15/$t\text{-}CO_2$. The threshold of the investment for case i is the lowest one due to low investment cost. Therefore case i have the highest incentive of the investment decisions, however the reduction of CO_2 emissions in $kt\text{-}CO_2$/year for case i leads to about 60% of CO_2 emissions prior to the investment. On the other hand, although the threshold for case ii is the highest one because of high investment and operating costs of coal-fired generation with CCS technology, the reduction of CO_2 emissions in $kt\text{-}CO_2$/year for case ii induces less than 20% of CO_2 emissions prior to the investment. As the fraction of coal-fired generation with CCS technology in the generation mix decreases as in case iii and then case iv, the thresholds considerably decreases. It can be seen from this table that for cases ii and iii, the difference between p_0 and p_1 are relatively large. Thus this shows that there exists a high possibility that the firm invests at once if the environmental policy is adopted.

Table 3 shows the effect of the tax rate on the investment thresholds for cases iii and iv. In each case, the differences between p_0 and p_1 increases as the tax rate becomes high. This is because the incentive of the investment for power generations of low-carbon emissions becomes large as the tax rate increases. For low tax rate, the threshold, p_1 for case iii is larger than that for case iv, whereas for high tax rate, p_1 for case iii is smaller than that for case iv. It is found that when

Table 3 Effect of the tax rate on the investment thresholds for cases iii and iv

		Tax rate ($/t-$CO_2$)				
		15	20	25	30	35
Case iii	p_0 ($/kWh)	67.393	65.702	64.121	62.666	61.349
	p_1 ($/kWh)	60.947	56.929	52.911	48.893	44.875
Case iv	p_0 ($/kWh)	57.385	56.529	55.696	54.888	54.107
	p_1 ($/kWh)	54.447	52.571	50.694	48.818	46.941

the tax rate is relatively high, the incentive of the investment for power generations of low-carbon emissions such as CCS technology becomes large even if the construction cost is somewhat high.

4 Conclusions

In this chapter we have developed real options models to evaluate the investment decisions of power generations under environmental policy uncertainty. We show the threshold of the power generations investments in each case for various costs and CO_2 emissions degrees. The effect of tax rate on optimal decision rules is indicated.

The model that is resented in this chapter considers uncertain adoption time of environmental policy. However, realistically, uncertainty over tax rate must be considered as a problem that concerns the entire investment decision of power generations under environmental policy uncertainty. Therefore, extension of this chapter's model towards uncertain tax rate would be warranted. Other directions for future work focus on uncertainty of CO_2 prices in the emissions trading market, capacity choice, and sequential investments.

Acknowledgements The authors would like to thank Hiroshi Takamori, Hidetaka Nakaoka, Shoji Nakada, and two anonymous referees for their helpful comments and suggestions. This study was supported in part by the Grant-in-Aid for Young Scientists (B) (grant No. 21710145) from Japan Society for the Promotion of Science.

Appendix A. List of Symbols

P_t	Electricity price at time t
p	Initial electricity price
μ	Instantaneous expected growth rate of electricity price
σ	Instantaneous volatility of electricity price
π_t	Profit flow from plant operating at time t
$I = \{0, 1\}$	States before and after the investment
j	Type of power generation
q_j^i	Capacity for power generation j
Q^i	Total capacity of power generations
c_j	Operating cost for power generation j
C^i	Total operating cost

(continued)

τ	Tax rate for carbon-emission
η_j	Emission basic unit for power generation j
ρ	Arbitrary discount rate
δ_j	Investment cost per kW for power generation j
$I(q_j^i)$	Total investment cost for capacity expansion
p^*	Investment threshold for the case without policy uncertainty
λ	Poisson intensity for the state transition
p_0	Investment threshold for the present state
p_1	Investment threshold for the policy regime

References

1. Blyth W, Bradley R, Bunn D, Clarke C, Wilson T, Yang M (2007) Investment risks under uncertain climate change policy. Energy Policy 35:5766–5773
2. Bøckman T, Fleten S-E, Juliussen E, Langhammer H, Revdal J (2008) Investment timing and optimal capacity choice for small hydropower projects. Eur J Oper Res 190:255–267
3. Brennan MJ, Schwartz ES (1985) Evaluating natural resource investments. J Bus 58:135–157
4. Dixit AK, Pindyck RS (1994) Investment under uncertainty. Princeton University Press, Princeton
5. Fuss S, Szolgayova J, Obersteiner M, Gusti M (2008) Investment under market and climate policy uncertainty. App Energy 85:708–721
6. Fuss S, Johansson DJA, Szolgayova J, Obersteiner M (2009) Impact of climate policy uncertainty on the adoption of electricity generating technologies. Energy Policy 37:733–743
7. Gollier C, Proult D, Thais F, Walgenwitz G (2005) Choice of nuclear power investments under price uncertainty: valuing modularity. Energy Econ 27:667–685
8. Hassett KA, Metcalf GE (1999) Investment with uncertain tax policy: does random tax policy discourage investment? Econ J 109:372–393
9. IEA (2010) Energy technology perspectives. IEA, Paris
10. McDonald R, Siegel D (1986) The value of waiting to invest. Q J Econ 101:707–727
11. Naito Y, Takashima R, Kimura H, Madarame H (2010) Evaluating replacement project of nuclear power plants under uncertainty. Energy Policy 38:1321–1329
12. Näsäkkälä E, Fleten S-E (2005) Flexibility and technology choice in gas fired power plant investments. Rev Fin Econ 14:371–393
13. OECD/NEA (Nuclear Energy Agency) (2005) Projected costs of generating electricity: 2005 update. OECD/NEA, Paris
14. Siddiqui AS, Maribu K (2009) Investment and upgrade in distributed generation under uncertainty. Energy Econ 31:25–37
15. Siddiqui AS, Fleten S-E (2010) How to proceed with competing alternative energy technologies: a real options analysis. Energy Econ 32:817–830
16. Takashima R, Naito Y, Kimura H, Madarame H (2007) Decommissioning and equipment replacement of nuclear power plants under uncertainty. J Nucl Sci Technol 44:1347–1355
17. Wickart M, Madlener R (2007) Optimal technology choice and investment timing: a stochastic model of industrial cogeneration vs. heat-only production. Energy Econ 29:934–952
18. Yang M, Blyth W, Bradley R, Bunn D, Clarke C, Wilson T (2008) Evaluating the power investment options with uncertainty in climate policy. Energy Econ 30:1933–1950

T	Lifetime of generation expansion
[illegible]	Emission basic unit for power generation
ρ	Arbitrary discount rate
[illegible]	Investment cost per kW for power generation plant
I(Q)	Total investment cost for capacity expansion
[illegible]	Investment threshold for the case without policy uncertainty
λ	Poisson intensity for the state transition
[illegible]	Investment threshold for [illegible]
[illegible]	Investment threshold for the policy [illegible]

References

1. Blyth W, Bradley R, Bunn D, Clarke C, Wilson T, Yang M (2007) Investment risks under uncertain climate change policy. Energy Policy 35:5766–5773
2. Bøckman T, Fleten S-E, Juliussen E, Langhammer HJ, Revdal I (2008) Investment timing and optimal capacity choice for small hydropower projects. Eur J Oper Res 190:255–267
3. Brennan MJ, Schwartz ES (1985) Evaluating natural resource investments. J Bus 58:135–157
4. Dixit AK, Pindyck RS (1994) Investment under uncertainty. Princeton University Press, Princeton
5. Fleten S-E, Maribu KM, Wangensteen I (2007) Optimal investment strategies in decentralized renewable power generation under uncertainty. Energy 32:803–815
6. Fuss S, Johansson DJA, Szolgayova J, Obersteiner M (2009) Impact of climate policy uncertainty on the adoption of electricity generating technologies. Energy Policy 37:733–743
7. Fuss S, Szolgayova J, Obersteiner M, Gusti M (2008) Investment under market and climate policy uncertainty. Appl Energy 85:708–721
8. Hassett KA, Metcalf GE (1999) Investment with uncertain tax policy: does random tax policy discourage investment. Econ J 109:372–393
9. IEA (2010) Energy technology perspectives. IEA, Paris
10. McDonald R, Siegel D (1986) The value of waiting to invest. Q J Econ 101:707–728
11. Naito Y, Takashima R, Kimura H, Madarame H (2010) Evaluating replacement project in nuclear power plants under uncertainty. Energy Policy 38:1378–1385
12. Näsäkkälä E, Fleten S-E (2005) Flexibility and technology choice in gas fired power plant investments. Rev Financ Econ 14:371–393
13. OECD/IEA/NEA (2010) Projected costs of generating electricity: 2010 edition. OECD/NEA, Paris
14. Siddiqui AS, Maribu K (2009) Investment and upgrade in distributed generation under uncertainty. Energy Econ 31:25–37
15. Siddiqui AS, Fleten S-E (2010) How to proceed with competing alternative energy technologies: a real options analysis. Energy Econ 32:817–830
16. Takashima R, Naito Y, Kimura H, Madarame H (2007) Decommissioning and equipment replacement of nuclear power plants under uncertainty. J Nucl Sci Technol 44:1347–1355
17. Wickart M, Madlener R (2007) Optimal technology choice and investment timing: a stochastic model of industrial cogeneration vs. heat-only production. Energy Econ 29:934–952
18. Yang M, Blyth W, Bradley R, Bunn D, Clarke C, Wilson T (2008) Evaluating the power investment options with uncertainty in climate policy. Energy Econ 30:1933–1950

Greenhouse Gas Emissions Trading in the Electricity Sector: Model Formulation and Case Studies

Yihsu Chen, Wietze Lise, Jos Sijm, and Benjamin F. Hobbs

Abstract Models formulated as complementarity problems have been applied previously to assess the potential for market power and costs of environmental regulation in transmission-constrained electricity markets. One emerging use of these models is to study the impacts of cap-and-trade (C&T) policies on electricity markets. In this chapter, we first summarize the theoretical background on the choice of environmental instruments to regulate emissions from the power sector. The chapter then presents a mathematical formulation of a power market that incorporates a carbon dioxide C&T program. We illustrate the capability of the model by presenting the results from two analyses. The first analysis examines the impact of the European Union Emissions Trading Scheme (EU ETS) on the northwestern European electricity market. The second study investigates the energy and emissions implications of Maryland's decision to join the Regional Greenhouse Gas Initiative (RGGI) by nesting a regional power sector model within a national

Y. Chen (✉)
School of Social Sciences, Humanities and Arts and School of Engineering, Sierra Nevada Research Institute, University of California, Merced, 5200 N. Lake Rd., Merced, CA 95340, USA
e-mail: yihsu.chen@ucmerced.edu

W. Lise
ECORYS Research and Consulting, Buklum Sokak No: 63/73, Kavaklidere 06680, Ankara, Turkey
e-mail: wietze.lise@ecorys.com.tr

J. Sijm
Energy Research Centre of the Netherlands (ECN), Radarweg 60, 1043 NT, Amsterdam, Netherlands
e-mail: sijm@ecn.nl

B.F. Hobbs
Department of Geography and Environmental Engineering, Whiting School of Engineering, The Johns Hopkins University, 3400 N. Charles St., Baltimore, MD 21218, USA
e-mail: bhobbs@jhu.edu

Q.P. Zheng et al. (eds.), *Handbook of CO_2 in Power Systems*, Energy Systems, DOI 10.1007/978-3-642-27431-2_3, © Springer-Verlag Berlin Heidelberg 2012

model. In both cases, the larger firms in the electric market are modeled as Cournot oligopolists who pursue a quantity strategy while at the same time they act competitively in the C&T programs and transmission markets. We demonstrate how complementarity-based power market models can be easily modified to incorporate details of alternative policy designs.

Keywords Cap-and-trade • Market power • Oligopoly • Carbon cost passthrough • Windfall profits • EU ETS and RGGI

1 Introduction

Process-based models based on bottom-up principles have been used extensively to study the energy sector's response to restructuring and proposed public policies [10]. For example, models formulated as complementarity problems have been applied previously to assess the potential for market power in transmission-constrained electricity markets. In contrast to top-down models, process-based models represent supply curves using step functions, with each step corresponds to the marginal production cost of an individual technology or generating unit. When applying process-based models to studying environmental policy, the marginal emission rate, which differs by each unit/technology and is likely non-monotonic and non-differentiable if generating units are arrayed in order of production costs, can be appropriately represented by this approach. There are at least two strengths of process-based modeling: *explicitness* and *flexibility*. The *explicitness* of the process-based approach allows the changes in technology, policies, input prices or objectives to be modeled by altering decision variables, objective function coefficients or constraints. The transparency of the formulation and inputs of process-based models facilitates review of model assumptions, and allows policy parameters to be readily changed in order to tailor the models for new applications.

In general, the objectives of policy analyses using process-based modeling could be *descriptive* and/or *normative*. When a model is used for descriptive purposes, it shows detailed cost, emission or technology portfolio impacts as well as changes in market prices and consumer welfare due to policy choices. If its purpose is instead normative, it attempts to identify improved solutions in terms of the objectives of the users, and shows tradeoffs among policy objectives and choices. What we illustrate in this chapter is a set of applications that use process-based models in the descriptive way. In particular, we show how to build a process-based power market model that includes a C&T (cap-and-trade) program and illustrate the results from two case studies. Those models show how electricity prices, social welfare, and technology choices can change when introducing a new C&T greenhouse gas policy to power sectors in the US mid-Atlantic region and the European Union.

2 Background

This section gives some background information on the two regional electricity markets that we analyze in our case studies in Sect. 2.1, along with background on the cap-and-trade programs in these two case studies in Sect. 2.2.

2.1 Electricity Markets

Electricity markets around the world have experienced dramatic changes from traditional vertical integrated monopolies with rate-of-return regulation to competitive but regulated markets. In these markets, electricity prices are driven by marginal costs. These changes were initiated by the legal or functional separation of generation, transmission and distribution together with divestiture of generation ownership to enhance competition. By granting nondiscriminatory access to transmission, generation companies can compete for customers. This competitive environment allows for the use of market models based on the economic profit-maximization principle to stimulate competition among electricity producers. This chapter illustrates the capability of process-based power market models in Sect. 3 to analyze the interactions of C&T (cap-and-trade) policies and two deregulated electricity markets: the Pennsylvania-New Jersey-Maryland Interconnection (PJM) and the European Union markets. We provide brief background on these two markets in the next two sections.

2.1.1 Pennsylvania-New Jersey-Maryland Interconnection

PJM (Pennsylvania-New Jersey-Maryland Interconnection) is a regional transmission organization (RTO) that coordinates bulk sales of electricity in more than ten states and the District of Columbia. It began operating as an ISO (independent system operator) in 1998 with a day-ahead and hourly-ahead energy and operating (spinning) reserve market, and an installed capacity market was introduced in 2000. Its coverage has evolved over time and now serves more than 51 million people. We focus on a subregion of PJM that is based on its 2000 footprint, including its neighboring regions of West Virginia, Virginia, western Pennsylvania, Kentucky, North Carolina and South Carolina in order to account for bulk electricity trade (see Fig. 1).

Nuclear and coal plants serve the baseload, accounting for more than 50% of total generation capacity. The capacity shares of oil, gas, and hydro plants were approximately 20%, 18%, and 3%, respectively [25]. In our model, the 2000 PJM system is spatially represented by 14 aggregated nodes (each representing one Power Control Area or a portion thereof) and 18 transmission corridors (Fig. 1). Maryland, the focus of the case analysis in Sect. 4, is separated into four nodes

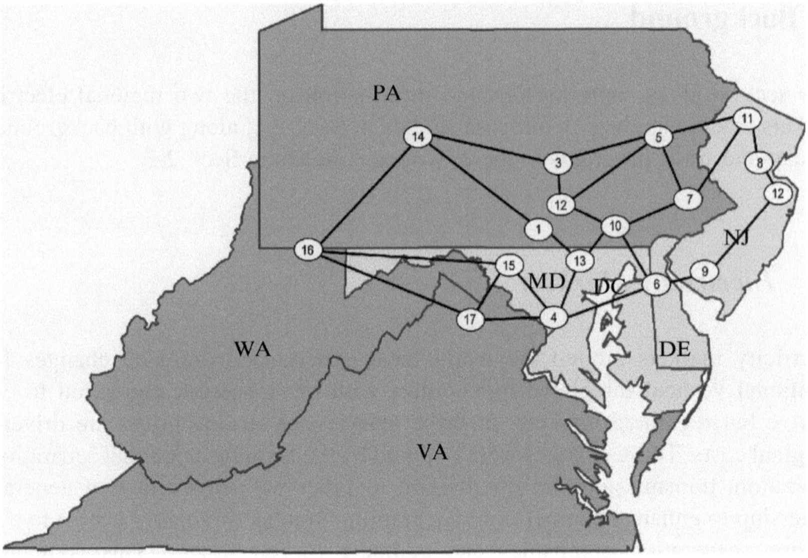

Fig. 1 PJM market area in 2000 updated with 2006 generator and transmission information. *Dark areas* denote states that did not participate in the Regional Greenhouse Gas Initiative

based on flow patterns and network constraints. The PJM energy market during 2006 was moderately concentrated with an hourly average HHI (Herfindahl Hirschman Index) of 1,256, based on shares of the energy market [25]. There are six large generating companies, owning more than 6% apiece of the generating capacity. Although the PJM market monitor reports that prices have generally been near competitive levels, there has apparently been some market power exercised in the installed capacity market. Furthermore, other studies indicate that market concentration is high enough to present a risk of market power [21].

2.1.2 European Electricity Market

The electricity market in Europe is coordinated by the European Network of Transmission System Operators for Electricity (ENTSO-E). Since July 2009, ENTSO-E has represented all electric TSOs (transmission system operators) in the EU and other operators connected to their networks. Forty-two TSOs from 34 countries are members. We focus on a subregion of ENTSO-E that is based on the largest markets of the EU and cover 20 European countries, i.e., Austria, Belgium, the Czech Republic, Denmark, Finland, France, Germany, Hungary, Italy, Luxembourg, the Netherlands, Poland, Portugal, Slovakia, Slovenia, Spain, Switzerland, and the United Kingdom (see Fig. 2).

Baseloaded nuclear and coal plants account for 44% of the regions generation capacity. Oil, gas, hydro and renewable plants make up 9%, 18%, 22% and 7%,

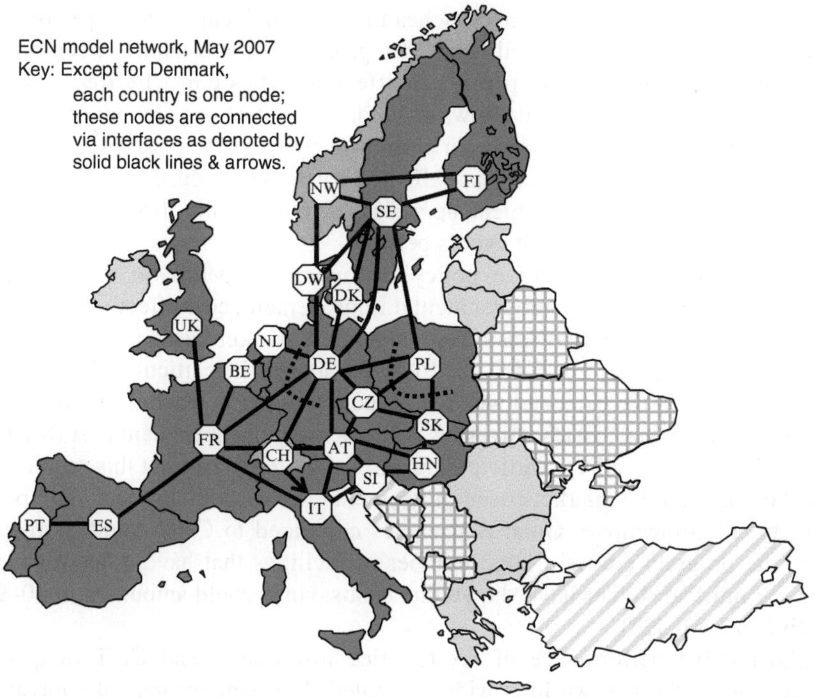

Fig. 2 European market area in 2000 updated with 2006 generator and transmission information

respectively. The ENTSO-E system is approximated by 20 aggregated nodes (one per country, except for Germany and Luxembourg, which are joined into one node, while Denmark is divided into two nodes (East and West Denmark) belonging to two different, non-synchronized networks). The nodes are connected by 34 transmission corridors (Fig. 2). The market is moderately concentrated, with six larger generating companies each owning more than 4% of the regions capacity.

2.2 *Cap-and-Trade Programs*

We briefly summarize the theory of cap-and-trade programs (C&T) in Sect. 2.2.1, along with their institutional settings and actual performance so far for the two C&T programs that are addressed by the case studies. Sections 2.2.2 and 2.2.3 summarize the programs that the eastern PJM and EU regions are subject to.

2.2.1 Theoretical Background

C&T (cap-and-trade) programs, like pollution taxes, are a policy instrument that relies on market incentives to motivate pollution reductions. This is in contrast to command-and-control (C&C) approaches that impose technology or performance requirements on individual facilities. C&T programs first establish a cap on total emissions and then distribute permits to affected parties either by free allocation rules or by auction. One permit allows its holder to emit a fixed amount of pollutant, and polluters are required to surrender permits to cover their emissions. The permits can be traded in secondary markets. Affected parties can reduce pollution through operational changes or equipment upgrades, or buy permits from companies or other permit holders who have excess permits.

An evaluation of alternative policy instruments depends on the evaluation criteria considered, such as environmental improvement, cost-effectiveness, equity considerations and political feasibility. In general, market-based instruments have been found to outperform C&C [8, 9]. This applies in particular to their cost-effectiveness i.e., the extent to which a reduction target is reached at least cost to society. Maximizing cost effectiveness requires marginal abatement cost (MAC) to be equalized among all the participating parties. Theories suggest this requirement will be satisfied by market-based instruments when both product and permit markets are competitive. Great efficiencies compared to C&C could be realized when the program covers a large number of facilities that have a heterogeneous MAC. Empirical studies have shown that cost savings could amounted to 40–90% relative to C&C [23].

The relative performance of tax (a price instrument) and C&T (a quantity instrument) in the real world is widely debated. Yet their theoretical equivalence is well known under some conditions: GHG emissions are the only market imperfection (i.e., the only environmental externality), markets are competitive, and transaction costs are low. Early research argued that this equivalence of price and quantity instruments no longer holds due to uncertainties associated with MAC and marginal damage curve (MD) [35]. In particular, preference between the two approaches hinges on the relative slopes of the MAC and MDC. If MDC curve is flat relative to the MAC, the costs of errors in setting the price (tax) have relatively small consequences for total emissions and social costs compared to getting the quantity (cap) wrong, and taxes are preferable. Conversely, if the reverse is true C&T is preferable [32, 34].

However, as the levels of cap and tax are likely to be adjusted by policymakers when new scientific information about climate change becomes available, it makes such a comparison less relevant [34]. Furthermore, for policy performance in general and cost effectiveness in particular, the choice of instruments is less important than their specific design. In addition to possible provisions relating to price containment (e.g., banking/borrowing, price floor/ceiling, offset), the scope and point of regulation (coverage and upstream/downstream regulation), the rules for allocating auctions are used are also crucial permits, and the way in which the proceeds from permit for a well functioning carbon policy [1]. Other essential elements of effective designs for carbon policies include policy credibility (i.e., belief that the policy will endure and be enforced) and flexibility (i.e., ability to change the policy parameters when new information becomes available) [32].

Finally, besides compliance costs and policy effectiveness, the choice and design of domestic emissions control instruments depend on a variety of other considerations. First among these are transaction costs and recycling of tax revenues and auction proceeds: while some favor taxes for lower transaction costs and better public finance effects, others argue that these effects are rather similar in the case of auctioned allowances [1, 2, 8, 32]. Second, the incidence of carbon costs across different income groups affects political acceptability, and depends on the allocation of policy rents and the recycling of tax/auction revenues [1, 8, 32]. Political feasibility depend on other considerations as well; for instance, C&T seems nowadays less politically controversial than taxes. Third, dynamic efficiency is also important, and depends upon price certainty, economic flexibility, and innovation. Whereas some authors favor taxes for offering price certainty to investors in long-term carbon saving innovations, others prefer C&T for reducing the economic effects of unexpected productivity shocks [6], or for adjusting carbon prices flexibly to unexpected changes in economic or technological developments [33]. A fourth consideration is leakage resulting from competition of home industries with imports from countries without carbon policies. To deal with this problem, policymakers seem to prefer C&T as it offers the opportunity of free allocations to industries vulnerable to outside competition, although similar effects could be reached by inframarginal tax exemptions or emissions trading based on intensity targets [6].

In sum, direct regulatory instruments (i.e., command-and-control) would be preferred if emissions are dominated by a handful of large emitters. Within the group of market-based instruments, the choice between taxes and C&T (cap-and-trade), as well as the effectiveness of these instruments, depend on the design, price certainty, flexibility as well as revenue recycling effects, equity political feasibility. However, mainly for reasons of political feasibility and social acceptability, in developed regions (especially the EU and US) policymakers and industrial stakeholders – as well as some environmental groups – seem generally to opt for C&T rather than carbon taxes.

2.2.2 Regional Greenhouse Gas Initiative

The Regional Greenhouse Gas Initiative (RGGI) is a joint effort by ten northeastern states in US that started in 2009 to control for CO_2 emissions from the electric sector [27]. The program covered coal-fired, oil-fired, and gas-fired electric-generating units with a nameplate capacity of at least 25 MW [16]. The compliance schedule set forth was to maintain annual emissions at 2007 emissions level until 2015, representing roughly 171 million metric tons of CO_2 annually for the entire region and 34 million tons in Maryland [16, 18]. The emissions are then reduced incrementally to achieve a 10% reduction by 2019. Overall, the program is estimated to result approximately in a 35% reduction of power sector CO_2 by 2020 in the original RGGI states (excluding Maryland), compared with the business-as-usual scenario [17] or, equivalently, a 13% reduction from 1990 levels [20].

In terms of allocation, although RGGI left the decisions about ways to allocate permits to each state, a large portion of the permits has been allocated by auctioning [28]. The clearing prices in the first five rounds of auction ranged between 2 and 4 $/t [26].

2.2.3 European Union Emissions Trading Scheme

Member States of the EU decided in 2003 to establish an Emissions Trading Scheme (ETS) to regulate CO_2 emissions. EU ETS is a C&T with phase I as the trial period (2005–2007) and phase II as the compliance period to the Kyoto Protocol. The program covers more than about 11,000 installations which are collectively responsible for approximately 2 Gt of CO_2 emissions (nearly 50% of EU CO_2 emissions or 40% of total GHG emissions). It includes all large installation in the power and heat sector as well as all other combustion plants with a thermal input greater than 20 MW, as well as some energy-intensive industries such as oil refineries, coke ovens and plants making iron and steel, cement, lime, glass, bricks, ceramics, and pulp and paper. The allowance allocation is mostly through grandfathering with some portions reserved for new entries and by auctions (i.e., maximum of 5% in phase I and 10% in phase II). EU allowances (EUAs) can be traded through either organized marketplaces or exchanges such as Powernext or the European Climate Exchange (ECX) or over-the-Counter (OTC) transactions through banks or brokers (e.g., Natsource or Evolution Markets.) EUA price was volatile in the early years in the first phase EU ETS reflecting speculations about permits shortage and has been relatively stable around 15–20 €/t [31].

3 Mode Formulation

This section presents mathematical formulation of power market mode with tradable CO_2 permits. The model is a variant of the authors' early work that accounts for the CO_2 C&T program [11]. When the CO_2 price is endogenously determined, the CO_2 cap program creates an interdependence across periods such that producers have to coordinate their output level over periods to ensure compliance with the annual cap. Reference [22] describes the mathematical properties of the model, including uniqueness and existence of equilibria. In the remainder of this section, we first explicitly write out the individual market participant optimization problems (i.e., generator, grid operator and consumer), including objective function and constraints. Second, we derive the first-order (Karush-Kuhn-Tucker, KKT) conditions associated with each variable. Third, collecting KKT conditions together with market clearing conditions will define the market equilibrium conditions as a linear complementarity problem, which can be solved using PATH or other complementarity solvers. The model presented herein is the price-taker model, in which generators do not attempt to manipulate price. We also add some modifications to the models to the case of Cournot oligopoly. In this section, we use capital letters to indicate parameters and sets. Lower-case letters refer to variables and indices. Dual variables are designated with greek lower-case letters. The variable in the parentheses to the right of each constraint is its dual variable. (The notation of "$x \perp y$" implies that $x^{\mathrm{T}}y = 0$.)

3.1 Optimization Models for Market Participants

Consumers. Consumers are assumed to have no market power and their willingness-to-pay for electricity is implicitly represented by the inverse linear demand function:

$$p_j^E = P_j^0 - \frac{P_j^0}{Q_j^0}\left(\sum_f s_{fj} + a_j\right), \forall j \tag{1}$$

where the nodal consumption is equal to $\Sigma_f s_{fj} + a_j$. In this presentation, we suppress the subscript t that represents the time period.

Producers. We assume that the bulk sales of power are in the form of bilateral contracts between producers and consumers. The producer f maximizes profits by determining output level (denoted by x_{fih}) of plant h located in node i.

$$\begin{aligned}
\underset{x_{fih}, s_{fi}}{\text{maximize}} & \sum\nolimits_j (p_j^E - w_j)s_{fj} - \sum\nolimits_{i,h \in H(i,f)} (C_{fih} - w_i)x_{fih} \\
& - p^{CO_2}\left(\sum\nolimits_{i,h \in H^{CO2}(i,f)} E_{fih}x_{fih} - N_f\right) \\
& \text{subject to } x_{fih} \le X_{fih}, (\rho_{fih}) \\
& \sum\nolimits_j s_{fj} = \sum\nolimits_{j,h \in H(j,f)} x_{fjh}, (\theta_f) \\
& x_{fih}, s_{fj} \ge 0.
\end{aligned} \tag{2}$$

The term $p_j^E - w_j$ is the per MWh revenue of price-taking producers by selling s_{fj} MWh of electricity to consumers at j, less the transmission charge paid to the grid operator to bring power to consumers from hub (w_j). The C_{fih}-w_i term is the per MWh cost of producing electricity from plant, where w_i is the price charged by the transmission operator to bring generation to the hub. The term $\sum_{i,h \in H^{CO2}(i,f)} E_{fih}x_{fih} - N_f$ is the number of CO_2 permits purchased (positive) or sold (negative) over the compliance period. N_f is the number of pollution permits possessed by producer f in its account at the beginning of the compliance cycle. Turning to the constraints, in addition to non-negativity restrictions, producers have two types of constraints. In the first constraint, the output level for a generator is no more than its derated capacity. Second, energy generation and sales have to balance during each period.

In the case of oligopoly competition, a few firms with substantial capacity are designated as strategic firms, exercising Cournot (quantity) strategies in the energy market. Under that strategy, those firms adjust their generation and sales as if they believe that rival firms will not react to such output changes. Mathematically, this is accomplished by substituting the inverse demand function (1) for p_j^E in the objective function for the strategic firm in (2).

Power Arbitrager. The equilibrium power prices are assumed to be common knowledge, and arbitrager is assumed to move power from the low-price nodes to the high-price nodes to maximize its profit without exercising market power. So, the prices of energy and transmission are exogenous to the arbitrager. Its decision variable is a_i denoting the power moved from the network hub to node i.

$$\underset{a_i}{\text{maximize}} \sum\nolimits_i (p_i^E - w_i)a_i \tag{3}$$

$$\text{subject to} \sum\nolimits_i a_i = 0.$$

The mass-balance constraint in (3) implies that total sales equal total purchases by the arbitrager.

Grid Operator/Independent System Operator. The operator allocates scarce transmission capacity among demands for transmission services:

$$\underset{y_i}{\text{maximize}} \sum\nolimits_i w_i y_i \tag{4}$$

$$\text{subject to} \sum\nolimits_i PTDF_{ki} y_i \leq T_k, (\lambda_k),$$

We use the DC approximation to derive power transfer distribution factors (PTDFs) to represent load flows in the network [30]. More complicated and realistic formulations such as transmission losses, phase shifters, DC lines and path-based flows can also be considered [12]. The constraint associated with the grid operator are that the total flow has to be no more than the upper bound for interface k based on thermal or other limits. It is common to assume that the grid operator is a regulated entity that allocates transmission capacity efficiently among demands for transmission service, which is equivalent to the price-taking assumption embodied in the objective function. The KKT conditions also relate the price of hub-to-point service w_i to the price of the interface that service requires in (3).

Market Clearing and Consistency Conditions. These two conditions are essential to calculating a market equilibrium. These conditions serve two purpose: for each commodity (e.g., electricity, transmission, CO_2 permits) the clearing condition, first, ensuring balance of the physical system and, second, implicitly generating a price for each commodity by finding the level that forces demand and supply to equilibriate. The first condition in (5) requires that the demand for transmission service from each node equals the quantity supplied by the grid operator in each period. The CO_2 permit clearing condition is the second condition. That is written as a complementary condition such that if there are excess permits in the market, the permit price is zero; otherwise, it is positive.

$$\sum_f s_{fi} + a_i - \sum_{f,h \in H(i,f)} x_{fih} = -y_i, \forall i \tag{5}$$

$$0 \le p^{CO2} \perp \sum_{f,i,h \in H^{CO2}(i,f)} (E_{fih} x_{fih} - N_f) \le 0$$

Market Equilibrium Model. The next step in developing the model is to derive the KKT conditions for each of the market participant optimization problems. The market equilibrium problem is then defined as the collection of all KKT conditions for all the above problems together with the market clearing and consistency conditions (analogous to the derivations in [3]). We summarize the resulting market equilibrium problem below. For firms that are designated as Cournot producers in the oligopoly runs, their KKT condition with respect to the variable s_{fj} is not the first line but is instead the condition within the parenthesis marked by * in (6). Likewise, when plant h is subject to the CO_2 cap, the KKT conditions associated with x_{fih} are also listed within the parenthesis marked by ** in (6).

Producers

$$\begin{aligned}
&0 \le s_{fj} \perp p_j^E - w_j - \theta_f \le 0 \\
&\left(0 \le s_{fj} \perp P_j^0 - \frac{P_j^0}{Q_j^0}\left(\sum_g s_{gj} + a_j\right) - \frac{P_j^0}{Q_j^0} s_{fj} - w_j - \theta_f \le 0\right)^* \\
&0 \le x_{fih} \perp -(C_{fih} - w_i) - \rho_{fih} + \theta_f \le 0 \\
&(0 \le x_{fih} \perp -(C_{fih} - w_i) - E_{fih} p^{CO2} - \rho_{fih} + \theta_f \le 0)^{**} \\
&0 \le \rho_{fih} \perp x_{fih} - X_{fih} \le 0 \\
&\theta_f \text{ free}, \quad \sum_j s_{fj} = \sum_{i,h \in H(i)} x_{fih}
\end{aligned} \tag{6}$$

Arbitrager

$$\begin{aligned}
&a_i \text{ free}, \; p_i^E - w_i - p^H = 0 \\
&p_H \text{ free}, \; \sum_i a_i = 0
\end{aligned} \tag{7}$$

Grid Operator/Independent System Operator

$$\begin{aligned}
&y_i \text{ free}, \quad w_i - \sum_k PTDF_{ki} \lambda_k = 0 \\
&0 \le \lambda_k \perp \sum_i PTDF_{ki} y_i - T_k \le 0
\end{aligned} \tag{8}$$

Together with (1) that defines consumer willingness-to-pay as well as market clearing conditions (5) above, the resulting complementarity problem can then be implemented in GAMS and solved with the complementarity solver PATH [5].

4 Case Studies

We illustrate the capability of the models in Sect. 3 by presenting selected results from two studies that examine the energy and emissions implications of two emissions trading programs on regional electricity markets. We first summarize the background and assumptions of these two studies, and then present their results.

4.1 Regional Greenhouse Gas Initiative

The State of Maryland General Assembly passed the Healthy Air Act (SB 154/HB 189) in 2006, which was then signed by Governor Ehrlich. Among other provisions, the Act commits the State to join the RGGI, subject to the results of a study of the impacts upon power system costs and reliability in Maryland. We summarize in this section a subset of the results from the study that examined how network constraints and the possible exercise of market power might alter the calculated impacts of Maryland joining RGGI. We seek to answer following question: Could the results concerning the impact of Maryland joining RGGI appreciably change under a market power assumption for large electricity generators? We report the selective results concerning production costs, electricity loads and prices, price–cost markup by oligopolists, and CO_2 emissions. More detailed results could be found in the report [7].

4.1.1 Method and Assumptions

To determine its impact, we relied on nesting a short-run regional power market model JHU-OUTEC (The Johns Hopkins University Oligopoly Under Transmission and Emissions Constraints) [3] within a long-run national model Haiku. Haiku simulates interregional trade among regional electricity markets based on market equilibrium concepts [24]. Its solution identifies a market equilibrium for competitive least-cost investment and operation of the electricity system that meets price-responsive demand given a wide set of regulatory institutions. JHU-OUTEC is based on complementarity formulation (see Sect. 3). With plant ownership data, it allows analyses of market power. In contrast to the representation of Maryland within Haiku as a single node, JHU-OUTEC separates Maryland into four nodes based on flow patterns and network constraints [7] (see Fig. 2). In this nesting framework, the information produced by Haiku for future years will be fed into

JHU-OUTEC as boundary conditions: generating capacity, variable costs, electricity loads, allowance prices and import flows.

Another crucial issue in this analysis is the effect of possible additional transmission lines that could increase power transfer capacity between Maryland and neighboring states. By 2015, our assessment concludes that the following two major east–west 500 kV circuits would be added that would provide stronger connections between Maryland and power supplies to the west:

1. The Mid-Atlantic Power Pathway (MAPP), consisting of 500 and 230 kV circuits through PEPCO service territory, connecting PEPCO to Virginia in the west and the Delmarva Peninsula (and ultimately NJ) in the east.
2. A 500 kV corridor proposed by Allegheny Power from the Wylie Ridge substation in northern West Virginia to the Kemptown substation in Maryland.

We simulate two policy alternatives for the years 2010, 2015, 2020 and 2025: (1) Maryland joins RGGI and (2) Maryland does not join RGGI. Four scenarios are simulated in each of the 4 future years. They differ in terms of whether Maryland joins RGGI or not and assumptions concerning market behavior:

- **Competitive Baseline (Run 1).** This case assumes that Maryland does not join RGGI and that all generating companies behave as price-takers (e.g., they are not able to manipulate prices to their advantage by withholding power production).
- **Competitive Maryland Joins RGGI (Run 2).** The same model is run, and it is assumed that Maryland becomes a member of RGGI. This has two main direct effects. First, the opportunity cost of power production will increase, due to the fact that generators must surrender a RGGI CO_2 allowance for every ton of CO_2 emitted. Second, load will decrease, because some of the RGGI allowances will be auctioned and the proceeds will be used to sponsor demand-side management programs such as energy efficiency improvements.
- **Oligopoly Baseline (Run 3).** The ten largest power producers in PJM are assumed to compete as oligopolies, following a Cournot (quantity) strategy.
- **Oligopoly Maryland Joins RGGI (Run 4).** This is the same as Run 3, plus the two direct changes just described (an opportunity cost of CO_2 allowances and changes in loads due to demand-side programs) as a result of Maryland becoming part of RGGI.

5 Results

The presence of market power can alter consumer and producer costs, and emissions. When large firms withhold capacity, it could drive up prices because more expensive supply enters the market either as imports or as production from small producers with inefficient plants. Production costs increase, and levels of emissions can change depending on the relative emissions rates of larger and smaller producers. We contrast the impacts of Maryland joining RGGI under competition (comparing Runs 1 and 2) with the impacts estimated under assumption of oligopoly (comparing Runs 3 and 4). Because the focus of this analysis is on the impacts of Maryland joining RGGI under different market structures, we compare the incremental effects of Maryland joining RGGI under competitive versus oligopoly assumptions, in order to assess whether the conclusions are very different under oligopoly.

Table 1 summarizes the conclusions from potential strategic behavior by generators on the impacts of Maryland joining RGGI. Each entry in Table 1 represents the impact under the oligopoly assumption (Run 4 minus Run 3) as a percentage of the result of the competitive comparison (Run 2 minus Run 1) in which variables of interests are arranged by rows and results of years (e.g., 2010, 2015, 2020 and 2025) by columns. In particular, we take the difference between Runs 1 and 2 as 100% for each of the variables. Then, the value of the variable under the oligopoly result is shown as a percentage of the competitive result for each of the scenario years so, a number less than 100 means that the oligopoly assumption resulted in a smaller impact, while a number greater than 100 means a greater impact. Mathematically, the values in Table 1 are of the form:

$$100 \times \frac{X_{Case4} - X_{Case3}}{X_{Case2} - X_{Case1}} \tag{9}$$

where X is the variable of interest and the subscript corresponds to case. As an example, the projected change in the CO_2 emissions in the JHU-OUTEC region (eastern PJM) of Maryland Joins RGGI is a decrease of 6.6 MT/year in 2025 in the oligopoly simulation ($X_{Case4} - X_{Case3}$). Under competition, however, the projected

Table 1 Impact of Maryland joins RGGI upon cost and loads: oligopoly market as% of competitive market results

	2010	2015	2020	2025
		A. Costs, loads, prices		
MD energy demand	27	51	107	84
MD consumer energy bill	140	107	18	99
Sale-weighted prices of bulk				
Electricity				
Maryland	98	86	311	94
Pennsylvania	63	−13	143	109
NJ/Delaware	94	28	−78	106
Overall JHU-OUTEC market	61	319[a]	−6125	127
		B. Gross margin		
+ Energy and O.R. revenue to				
Generators	122	−12137[a]	229	536
– Variable cost	96	3	−2	63
= Total gross margin	97	35	114	159
		C. CO_2 emissions gross margin		
Maryland	137	265	165	149
Entire JHU-OUTEC region	−77	97	96	117

decrease is 5.6 MT/year in 2025 ($X_{Case2} - X_{Case1}$). Thus, the oligopoly change is $\frac{6.6}{5.6} \times 100\%$ or 117% of the competitive change, which is the value shown in Table 1. As shown in Panel A of Table 1, the effects on Maryland electricity prices as well as consumer bills (price times quantity) are sensitive in magnitude depending on whether the market is competitive or oligopolistic. Effects on prices in other markets are less stable. Take for example Pennsylvania in 2015 where the impact is of the opposite sign. That is, under competition, Maryland Joins RGGI decreases Pennsylvania prices, but they increase when the market structure is oligopolistic. However, this instability is reasonable, as the price changes other than Maryland markets due to Maryland joining RGGI are relatively small, so slight shifts in results can translate into large percentage changes.

Panel B in Table 1 gives the results from an analysis of the generator gross margins. The gross margin is defined as the sum of revenues from energy and spinning reserve markets less variable costs (e.g., fuel and variable O&M costs). Overall, the oligopoly market structure has little impact on the projected effect of Maryland joining RGGI upon suppliers' gross margin outcomes in years 2010 and 2020, but a much greater impact in 2025 (a 59% increase) and actually a decrease in 2015. Finally, Panel C in Table 1 summarizes the outcomes for the impact of market structures on CO_2 emissions for Maryland and entire JHU-OUT region. Maryland's participation in RGGI likely increases the state's CO_2 emissions by 37–165% more under the oligopoly scenarios than under the competitive cases. As for the overall impact on PJM subregion, the CO_2 changes are similar under oligopoly and competition, except for 2010.

5.1 *European Union Emission Trading Scheme*

Power prices in EU countries have increased significantly since the EU emissions trading scheme (ETS) became effective on January 2005 [31]. Besides other factors, these increases in power prices may – at least due to the ETS as a result of producers passing through the costs of EU allowances (EUAs) to power prices. This occurred even though, under the ETS, affected facilities in the power sector have usually received most of their needed allowances for free during the first phase (2005–2007). Therefore, there was public suspicion that the increases in power prices were not caused by CO_2 cost pass-through but instead by the exercise of market power. We summarize in this section a subset of the results from a study that examined the effects of the EU ETS on power prices in Europe. We seek to answer the following question: Could the results concerning the impact of the EU ETS appreciably change under a market power assumption for large electricity generators?

We report selected results concerning changes in production costs, electricity loads and prices, markup and CO_2 emissions. More detailed results can be found in our full report [31].

5.1.1 Method and Assumptions

To determine the impact of the EU ETS upon prices, this section applies the COmprehensive Market Power in Electricity Transmission and Energy Simulator (COMPETES) model developed by authors and maintained by Energy Center of the Netherlands (ECN). Virtually all individual power companies and generation units in the 20 countries – including combined heat and power plants owned by industry or energy suppliers – are covered by the input data of the model. COMPETES allows users to specify which generation companies behave strategically (via Cournot) or competitively (i.e., as price takers). The latter subset of companies are collectedly assigned to a single entity in each node as a "competitive fringe." The basic transmission-constrained Cournot formulation underlying COMPETES was first presented in Hobbs [11]. COMPETES itself, including alternative transmission pricing formulations, is presented and applied in Hobbs et al. [14, 15]. COMPETES has previously been used to analyze issues such as effects of proposed mergers among power companies [29], market coupling [13] and market power, trade and droughts [19] as well as the interactions with the EU ETS [4, 31].

For our purpose, we perform simulation of six scenarios that differ by strategic assumptions (i.e., competitive or Cournot) and levels of allowance prices (0, 20, and 40 €/t):

- **Competitive Fixed Demand with CO_2 = 20 €/t (Run 1).** This case assumes that the carbon price will be 20 €/t CO_2 to meet fixed demand as determined by the model base year 2006 and that all generating companies behave as price-takers (e.g., they are not able to manipulate prices to their advantage by withholding power production). (Note that the price elastic demand curves used in the below runs are specified so that a competitive market with a 20 €/t would yield precisely the same market outcomes as this scenario, so that it does not matter whether we use elastic or perfectly inelastic demand in this scenario.)
- **Competitive Price-Responsive Demand with CO_2 = 0 €/t (Run 2).** The same model is run. However, for this and the remaining scenarios below it will be assumed that demand is elastic with a price elasticity of 0.2 to study the impact changes in carbon prices with respect to the Competitive Baseline.
- **Competitive Price-Responsive Demand with CO_2 = 40 €/t (Run 3).** The same model is run as in Run 2 and it is assumed that the carbon cost goes to 40 to show the impact of a higher carbon price.
- **Oligopoly with CO_2 = 20 €/t (Run 4).** This is the same as Run 1, but in addition to elastic demand, the 22 largest firms in Europe can exercise market power, with one exception. The exception is the largest firm, Energy de France (EdF), which is assumed not to be able exercise market power in France due to regulatory threats. That producer controls 14% of installed capacity in the COMPETES database. All other firms are assumed to behave as price takers.

- **Oligopoly with CO_2 = 0 €/t (Run 5).** This is the same as Run 4, and it is assumed that the carbon cost goes to zero to show the impact of a lower carbon price on the market.
- **Oligopoly with CO_2 = 40 €/t (Run 6).** This is the same as Run 4, and it is assumed that the carbon cost goes to 40 to show the impact of a higher carbon price.

6 Results

When there is market power present, the results can be different wholesale prices, demand, and emissions than under perfect competition. When large firms withhold capacity, it could drive up prices because more expensive supply enters the market either as imports or production from small producers who own less efficient plants.

Wholesale bulk price and levels of emissions can change depending on the relative emission rates of larger suppliers that withhold output and smaller producers the expand their output. If demand is fixed, when larger producers are more polluting, the result of market power is a reduction in emissions. To explore this issue, we contrast the impacts of lower carbon costs under competition (comparing Runs 1 and 2) with the impacts estimated under the assumption of oligopoly (comparing Runs 4 and 5). In addition, we compare the impact of higher carbon costs under competition (comparing Runs 1 and 3) with the impacts estimated under the assumption of oligopoly (comparing Runs 4 and 6). Because the focus of this analysis is on the impacts of different carbon costs under different market structures, we compare the incremental effects of different carbon costs under competitive and oligopoly assumptions, in order to assess whether the conclusions are very different under oligopoly.

Table 2 summarizes how assuming oligopoly rather than competition changes the model results concerning the effect of changing carbon costs on electricity prices and CO_2 emissions in several key countries in the EU market, as well as on aggregate energy demand in the EU market area. Each entry in Table 2 represents the level of impact under oligopoly assumption, relative to the competitive case, in percentage terms. In particular, we take the difference between Runs 1 and 2 as

Table 2 Summary of impact of changes in carbon cost in the six largest markets in Europe: oligopoly market results as percentage of competitive market results

	Bulk electricity prices		CO_2 emissions	
Scenario comparison	20–0	20–40	20–0	20–40
Poland	100	100	107	97
Italy	99	100	83	338
Germany	92	86	68	88
France	109	88	146	42
UK	105	94	52	187
Spain	102	81	187	89
EU20	93	85	76	107
Energy demand	94	83		

100% for each of the variables. Then, the value of the variable under the oligopoly result is shown as a percentage of the competitive result. Mathematically, the values in the table are of the form:

$$100 \times \frac{X_{Case4} - X_{Case5}}{X_{Case1} - X_{Case2}} \tag{10}$$

$$100 \times \frac{X_{Case4} - X_{Case6}}{X_{Case1} - X_{Case3}} \tag{11}$$

where X is the variable of interest and the subscript corresponds to cases. As an example of (10), the projected change in the CO_2 emissions of Poland is a decrease of 12.4 MT/year in 2006 in the oligopoly simulation ($X_{Case4}-X_{Case5}$). Under competition, however, the projected decrease is 11.6 MT/year in 2006 ($X_{Case1}-X_{Case2}$). Thus, the oligopoly change is $\frac{12.4}{11.6}100\%$ or 107% of the competitive change, which is the value shown in Table 2. As shown in Table 2 columns 2 and 3, the effects on wholesale electricity prices are modestly sensitive to assumptions concerning whether the market is competitive or oligopoly.

However, the oligopoly results diverge far more from the competitive results in the case of CO_2 emissions. Sometimes, oligopoly results in a smaller sensitivity of emissions to CO_2 prices, and sometimes in a much larger sensitivity. This points out the importance of particular local market conditions in determining the impact of a policy change upon regional costs and, especially, emissions. These local conditions are readily incorporated in the complementarity-based market models of this chapter.

7 Conclusion

Regulatory policies are becoming increasingly based on market-based principles because of their compatability with liberalized power markets that operate on profit-maximizing objectives. Formulating models based on complementarity is a natural way to study economic equilibrium problems as it allows formulation of various constraints such emissions caps and transmission and generation capacity limits, while capturing the profit seeking behavior of competitive and oligopoly power producers. Advances in computational capacity coupled with efficient algorithms for solving complementarity problems allow large-scale models to be solved in a reasonable time. This chapter presents the basic formulation of a regional electric market that is subject to a C&T (cap-and-trade) program. We illustrate how the models could be useful by two case studies that examine the economic and emissions impacts of proposed C&T policies in the US and EU. Other C&T related designs, such as offsets and allowance price floors or ceilings, are readily incorporated in other existing models.

References

1. Aldy J, Krupnick A, Newell R, Parry I (2009) Designing climate mitigation policy. Technical report RFF DP 08-16, resources for the future, Washington
2. Bennear L, Stavins R (2007) Second-best theory and the use of multiple policy instruments. Environ Res Econ 37:111–129
3. Chen Y, Hobbs BF (2005) An oligopolistic power market model with tradable NOx permits. IEEE Trans Power Syst 20(1):119–129
4. Chen Y, Sijm J, Hobbs BF, Lise W (2008) Implications of CO_2 emissions trading for short-run electricity market outcomes in northwest Europe. J Regul Econ 34(3):252–281
5. Dirkse SP, Ferris MC (1995) The PATH solver: a nonmonotone stabilization scheme for mixed complementarity problems. Optim Method Softw 5(2):123–156
6. Fischer C, Springborn M (2009) Emissions targets and the real business cycle – intensity targets versus caps or taxes. Discussion paper RFF DP 09-47, resources for the future
7. Center for Integrative Environmental Research (2007) Economic and energy impacts from Maryland's potential participation in the regional greenhouse gas initiative. A study commissioned by the Maryland Department of the Environment. Technical report, University of Maryland, College Park. http://cier.umd.edu/RGGI/documents/UMD$_R$GGI$_S$TUDY$_F$INAL.pdf
8. Goulder L, Parry I (2008) Instrument choice in environmental policy. Rev Environ Econ Policy 2(2):152–174
9. Gupta S, Tirpak A, Burger N, Gupta J, Höhne N, Boncheva A, Kanoan G, Kolstad C, Kruger J, Michaelowa A, Murase S, Pershing J, Saijo T, Sari A (2007) Climate change 2007: mitigation, ConPanel on climate change. In: Climate change 2007: mitigation, contribution of working group III to the fourth assessment report of the intergovernmental panel on climate change. Cambridge University Press, Cambridge
10. Hobbs BF (1995) Bottom-up analysis for utility decisions: company, state and regional models. In: Andrews CJ (ed) Regulating regional power systems. Quorum Books, Westport
11. Hobbs BF (2001) Linear complementarity models of Nash-Cournot competition in bilateral and Poolco power markets. IEEE Trans Power Syst 16(2):194–202
12. Hobbs BF, Drayton G, Fisher EB, Lise W (2008) Improved transmission representations in oligopolistic market models: quadratic losses, phase shifters, and DC lines. IEEE Trans Power Syst 23(3):1018–1029
13. Hobbs BF, Rijkers FAM, Boots MG (2005) The more cooperation, the more competition? A Cournot analysis of the benefits of electric market coupling. Energy J 26(4):69–98
14. Hobbs BF, Rijkers FAM, Wals A (2004) Strategic generation with conjectured transmission price responses in a mixed transmission pricing system–part I: formulation. IEEE Trans Power Syst 19(2):707–717
15. Hobbs BF, Rijkers FAM, Wals A (2004) Strategic generation with conjectured transmission price responses in a mixed transmission pricing system–part II: application. IEEE Trans Power Syst 19(2):872–879
16. Regional greenhouse gas initiative. Regional Greenhouse Gas Initiative model rule, final with corrections
17. Regional Greenhouse Gas Initiative. States reach agreement on proposed rules for the nation's first cap and trade program to address climate change
18. Lee CA The nation's first GHG cap and trade program. Carbon finance North American workshop NYC
19. Lise W, Hobbs BF (2008) Market power in the European electricity market – the impacts of dry weather and additional transmission capacity. Energy Policy 36(4):1331–1343
20. Litz F (2010) Regional Greenhouse Gas Initiative: lessons learned from the design of the nation's first carbon cap-and-trade program. In: Presentation at California, climate change center, market advisory committee meeting, Sacramento

21. Mansur ET (1999) Pricing behavior in the initial summer of the restructured PJM wholesale electricity market. Working paper PWP-083, University of California, Berkeley Energy Institute
22. Metzler C, Hobbs B, Pang J-S (2003) Nash-Cournot equilibria in power markets on a linearized DC network with arbitrage: formulations and properties. Netw Spat Econ 3 (2):123–150
23. Newell N, Stavins R (2003) Cost heterogeneity and the potential savings from market-based policies. J Reg Econ 23(1):43–59
24. Paul A, Burtraw D (2002) The RFF Haiku electricity market model. Technical report, Resources for the future. http://www.rff.org/Documents/RFF-RPT-haiku
25. PJM The PJM state of market (2006) Technical report, Pennsylvania-New Jersey-Maryland Interconnection. http://www.monitoringanalytics.com/reports/PJM_State_of_the_Market/2006.shtml
26. Regional Greenhouse Gas Initiative. Fourth RGGI auction yields 104.2 million for investment in the green economy
27. Regional Greenhouse Gas Initiative. Regional Greenhouse Gas Initiative memorandum of understanding
28. Ruth M, Gabriel SA, Palmer KL, Burtraw D, Paul A, Chen Y, Hobbs BF, Irani D, Michael J, Ross KM, Conklin R, Miller J (2008) Economic and energy impacts from participation in the regional greenhouse gas initiative: a case study of the state of Maryland. Energy Policy 36 (6):2279–2289
29. Scheepers MJJ, Rijkers FAM, Wals AF, Boots MG (2003) Verticale concentratie op de nederlandse elektriciteitsmarkt. ECN-CX–03-066 ECN-CX–03-066. Energy research centre of the Netherlands, Petten
30. Schweppe FC, Caramanis MC, Tabors RD, Bohn RE (1988) Spot pricing of electricity. Kluwer Academic Publishers, Boston
31. Sijm J, Hers JS, Lise W, Wetzelaer BJHW (2008) The impact of the EU ETS on electricity prices. final report to DG environment of the European commission. Technical report, Energy research centre of the Netherlands. http://www.ecn.nl/publications/default.aspx?nr=ECN-E–08-007
32. Stern N (2007) The economics of climate change: the stern review. Cambridge University Press, Cambridge, UK
33. Weber T, Neuhoff K (2009) Carbon markets and technological innovation. EPRG working paper no. 916. Electricity Policy Research Group, University of Cambridge
34. Weisbach D (2009) Instrument choice is instrument design. Technical report John M. Olin law and economics working paper no. 490, the law school, The University of Chicago
35. Weitzman M (1974) Prices versus quantities. Rev Econ Stud 41:457–491

Comparing Cap-and-Trade and Carbon Tax Policies in Generation Expansion Planning

Yanyi He, Lizhi Wang, and Jianhui Wang

Abstract Cap-and-trade and carbon taxes are two types of environmental policies that differ in their goals and approaches but both have the potential to reduce greenhouse gas emissions and promote renewable energy. We compare the effectiveness and efficiency of these policies in the context of generation capacity expansion. Game theoretic models are used to integrate the environmental policies in a generation expansion planning framework. The most efficient tax policies and variations are obtained using the inverse equilibrium models. We compare cap-and-trade and carbon taxes with respect to five criteria: carbon price and tax, renewable energy portfolio, total energy generation, generation companies' and grid owner's total profit, and government revenue. Numerical experiments show the relative advantages, disadvantages, similarities, differences, and limitations of the policies.

Keywords Cap-and-trade • Carbon tax • Generation expansion planning • Inverse equilibrium • Subsidy

1 Introduction

With increasing concerns on reducing greenhouse gas (GHG) emissions, which are a major contributor of global warming, renewable energy has been identified as a key solution to a sustainable future of energy supplies. Compared to the mature non-renewable generation technologies, one of the major disadvantages of renewable energy is the capital-intensive construction and manufacturing. In an effort to

Y. He • L. Wang (✉)
Industrial and Manufacturing Systems Engineering, Iowa State University, Ames, IA, USA
e-mail: lzwang@iastate.edu

J. Wang
Decision and Information Sciences Division, Argonne National Laboratory, Argonne, IL, USA
e-mail: jianhui.wang@anl.gov

Q.P. Zheng et al. (eds.), *Handbook of CO_2 in Power Systems*, Energy Systems, DOI 10.1007/978-3-642-27431-2_4, © Springer-Verlag Berlin Heidelberg 2012

offset such disadvantages, environmental policies at the federal and regional levels have been playing an important role in promoting renewable energy for decades. According to the Energy Information Administration [1], renewable energy subsidies in the U.S. increased from $1.4 billion to $4.9 billion between 1999 and 2007. The results of the subsidies were mixed: electricity production from wind increased at a yearly rate of 32%, but the production of all other renewable energy sources (hydroelectric, biomass, geothermal, and solar) failed to increase continuously or significantly during that period.

There are mainly three types of environmental policies: market (such as cap-and-trade), incentive (such as carbon taxes and production tax credits), and mandate (such as renewable portfolio standard). Although they all have the overall effect of promoting renewable energy, they differ in specific policy goals and approaches, which also result in varying effectiveness and efficiency. The cap-and-trade policy, for example, aims at reducing GHG emissions by setting an upper bound, or cap, of the total emissions from a power system. Initial emissions allowances are allocated to individual power suppliers, who can trade their allowances among them but the overall emissions cap will not be exceeded. This policy does not directly penalize non-renewable energy or subsidize renewable energy, but focuses on controlling the total emissions as a system output. An alternative policy that also receives great attention and discussions is carbon tax, which is imposed on GHG emissions from electricity generation. This chapter contributes to the debates on the comparison of these policies from the following perspectives.

First, the comparison is in the context of generation expansion planning (GEP). This is motivated by the fact that once installed, the marginal cost of renewable generation is usually much lower than non-renewable ones; but the major obstacle is that the prohibitive capital investment requirements limit new renewable capacity installment in the first place. The policies should therefore be compared based on their capability to stimulate new investment in renewable energy. Our analysis of the GEP problem is in a restructured electricity market environment, which differs from traditional GEP in various ways [2]. The unbundling of generation and transmission makes the GEP much more competitive for self-interested independent participants in electricity markets. In an old vertically-integrated electricity market, the investment in new generation resources was centrally planned in order to achieve a pre-specified system reliability level, and the cost was covered by ratepayers [3–5]. In a restructured electricity market, however, each generation company (GENCO) can pursue its own maximized profit by strategically investing in and operating its generation units. A non-arbitrage bilateral market equilibrium model presented in Ref. [6] is used in our analysis. This model not only captures major factors of interest but also has the nice property of possessing a unique solution under mild conditions, thus provides a good basis for our extended analysis of GEP under environmental policies.

Second, we propose a novel inverse equilibrium model to determine the optimal carbon taxes. For a given overall emissions cap, the modeling of the cap-and-trade policy is relatively straightforward. However, there could exist multiple and different tax policies that can all achieve the same policy goal. For comparison with cap-and-trade, it is only fair and meaningful to use the *most efficient* tax policy. The inverse equilibrium model we present enables us to determine such optimal tax policy. The carbon tax policies we consider in our models are more general than just taxes; we also consider subsidies to renewable energy, which represent the production tax credits and investment tax credits.

Third, the comparison between cap-and-trade and carbon taxes is based on five criteria: carbon price and tax, renewable energy portfolio, total energy generation, GENCOs' and grid owner's total profit, and government revenue. Our intention is not to declare a winner or loser, but to provide a modeling tool and framework to analyze their advantages, disadvantages, similarities, differences, and limitations from objective and meaningful perspectives.

Environmental policies in electricity market have been a topic of numerous studies. Game theoretic models are used to analyze the equilibria with NO_x limits in Refs. [7, 8]. A case study of the State of Maryland is conducted in Ref. [9] to investigate the economic and energy impacts from participation in a regional cap-and-trade system. Johnson [10] discusses the subsidies, carbon taxes and cap-and-trade policies. The author finds that new-source subsidies will accelerate the de-carbonization of the electricity generation. A comprehensive discussion of key technical challenges for the power industry in face of climate change is provided in Ref. [11]. Tabors et al. [12] analyze the impacts of different policies including carbon taxes for the U.S. electricity industry to reduce both acid rain production and global warming gases. A CO_2 cap-and-trade system is modeled in Ref. [13] to simulate the energy production of some regions in the U.S. However, this research does not consider the long-term generation expansion issues in sufficient details.

There also exist various models for analyzing market power exercise and GEP in an electricity market. Game theoretic models [14–16] are particularly suitable to capture the oligopolistic nature of GENCOs in a restructured electricity market. Wang et al. [17] present a GEP model with uncertain demands and generator availabilities. Coalition and collusion of participants in electricity markets are studied using a cooperative game approach in Refs. [18, 19]. A non-cooperative incomplete information game approach is applied in Ref. [20] to model GENCOs' optimal bidding strategies within a set of discrete bids.

2 Models

In this section, we introduce the energy market models that we use to analyze and compare cap-and-trade and carbon tax policies. Section 2.1 lists the nomenclature used in this chapter. Section 2.2 summarizes the non-arbitrage bilateral market equilibrium model presented in Ref. [6]. This model provides a basis for our analysis of GEP under environmental policies. Section 2.3 extends the bilateral market equilibrium model to include GEP, in which individual GENCOs choose the quantity and type (renewable or non-renewable) of new generation units to compete for maximum profits in a Nash-Cournot manner. The resulting Nash equilibrium is a linear complimentarity problem (LCP). Section 4 incorporates the carbon cap-and-trade policy into the GEP model. By regulating carbon emissions, the regulator intends to encourage more investment in new renewable generation capacity. Section 2.5 presents new inverse equilibrium models for designing optimal carbon tax and subsidy policies for promoting new renewable generation and limiting carbon emissions. The optimal policy design models are linear programs with complementarity constraints (LPCC).

2.1 *Nomenclature*

2.1.1 Sets

$\mathcal{F}$	Set of GENCOs, $f \in \mathcal{F}$.
$\mathcal{I}$	Set of nodes, $i, j \in \mathcal{I}$.
$\mathcal{H}$	Set of power generating types, $h \in \mathcal{H}$
$\mathcal{G}$	Set of generators, $(f, i, h) \in G$.
$\mathcal{K}$	Set of transmission lines, $k \in \mathcal{K}$.

2.1.2 Parameters

X_{fih}	Capacity of generator (f, i, h), which is of type h owned by GENCO f at node i	(MW)
P_i^0, Q_i^0	Parameters of linear demand function $p_i = P_i^0 - (P_i^0/Q_i^0)q_i$, where p_i and q_i are the price and quantity of electricity, respectively.	(\$/MW), (MW)
E_{fih}	Emissions rate of generator (f, i, h).	(t/MW)
H	Power transfer distribution factors matrix.	dimensionless
T_k	Transmission capacity of line k.	(MW)
K_f	Emissions allowance for GENCO f.	(t)
C_{fih}	Operating cost of generator (f, i, h).	(\$/MW)
I_{fih}	Investment cost of generator (f, i, h).	(\$/MW)

2.1.3 GENCOs' Decision Variables

x_{fih}	Power generation of generator (f, i, h).	(MW)
Δx_{fih}	Expanded generation of generator (f, I, h).	(MW)
s_{fj}	Generation quantity sold by GENCO f to consumers at node j.	(MW)
c_f	Amount of sales (positive) or purchase (negative) of emissions certificates at GENCO f.	(t)

2.1.4 Grid Owner's Decision Variables

y_i	Power injection at node i.	(MW)

2.1.5 Dual Variables

ρ_{fih}	Dual variables of constraint (2).	(\$/MW)
θ_f	Dual variables of constraint (3).	(\$/MW)
λ_k^-	Dual variables of constraint (6).	(\$/MW)
λ_k^+	Dual variables of constraint (7).	(\$/MW)
δ_f	Dual variables of constraint (24).	(\$/t)

2.1.6 Incentive Policy Maker's Decision Variables

tax_{fih}	Tax (positive) or subsidy (negative) of generator (f, i, h).	(\$/MW)

2.1.7 Market Clearing Variables

w_i	Payment (charge) for generating (receiving) electricity at node i.	(\$/MW)
p^{co_2}	Carbon price.	(\$/t)

2.2 *Energy Market Equilibrium Model*

In this section, we summarize the non-arbitrage bilateral market equilibrium model presented in Ref. [6]. This model simulates the Nash-Cournot competition of power suppliers in a bilateral market and results in an LCP formulation, which provides a modeling basis for our extended analysis of GEP under environmental policies.

A group of GENCOs $\mathcal{F}$ compete in a Nash-Cournot manner to maximize their own profits by determining the quantity of electricity generation and sales in a bilateral market. Each GENCO $f \in \mathcal{F}$ owns a portfolio of generators $(f, i, h) \in \mathcal{G}$, with marginal cost C_{fih} and generation capacity X_{fih}. The index (f, i, h) indicates the ownership $f \in \mathcal{F}$, location $i \in \mathcal{I}$, and type $h \in \mathcal{H}$ of the generator. The electricity network consists of a group of nodes $\mathcal{I}$, connected by a group of transmission lines $\mathcal{K}$. The linear demand function at node i is assumed to be $p_i = P_i^0 - (P_i^0/Q_i^0)\sum_f s_{fi}$. Here P_i^0 and Q_i^0 are parameters, p_i is the price at node i, and s_{fi} is the sales of GENCO f to node i. A GENCO gets paid (pays) w_i for generating (receiving) power at node i, thus $w_j - w_i$ is the cost of transmitting power from node i to j. A grid owner is assumed to determine the net power injection at each node y_i to maximize revenue $\Sigma_i w_i y_i$ subject to transmission constraints.

GENCO f's profit maximization problem is:

$$\max \sum\nolimits_j [P_j^0 - (P_j^0/Q_j^0)(\sum_g s_{gj}) - w_j]s_{fj} - \sum\nolimits_{i,h}(C_{fih} - w_i)x_{fih} \tag{1}$$

$$\text{s.t.} \quad x_{fih} \le X_{fih} \quad (\rho_{fih}) \quad \forall i,h \tag{2}$$

$$\sum\nolimits_j s_{fj} = \sum_{i,h} x_{fih} \quad (\theta_f) \tag{3}$$

$$s_{fj}, x_{fih} \ge 0 \quad \forall i,j,h. \tag{4}$$

Here, the two terms in the objective function (1) are sales revenue and generation cost, adjusted by the transmission costs. Constraint (2) is the generation capacity constraint, (3) is the balance of production and sales, and (4) requires that all sales and generation be non-negative.

The grid owner's revenue maximization problem is:

$$\max \quad \sum_i w_i y_i \tag{5}$$

$$\text{s.t.} - \sum_i H_{ik} y_i \le T_k \quad (\lambda_k^-) \quad \forall k \tag{6}$$

$$\sum_i H_{ik} y_i \le T_k \quad (\lambda_k^+) \quad \forall k. \tag{7}$$

Here, H is the power transfer distribution factors matrix, which is derived from the Kirchhoff's laws and transmission topology and can be employed to calculate power flows through transmission lines for a given set of nodal power injections. Variable y_i is the power injection into node i. The term $\Sigma_i H_{ik} y_i$ computes the amount of power flow through line k, which is constrained in (6) and (7) to be within T_k in either direction. The injection y_i is a free variable, which can be positive (or negative) when power generation is a surplus (or deficit) at node i.

Assuming Nash competition among GENCOs and the grid owner, we can obtain the market equilibrium solution by solving the following first order conditions:

$$0 \le s_{fj} \perp - P_i^0 + (P_i^0/Q_i^0)\left(2s_{fj} + \sum\nolimits_{g \ne f} s_{gj}\right) + w_j + \theta_f \ge 0 \quad \forall f,j \tag{8}$$

$$0 \le x_{fih} \perp C_{fih} - w_i + \rho_{fih} - \theta_f \ge 0 \quad \forall f, i, h \tag{9}$$

$$0 \le \rho_{fih} \perp X_{fih} - x_{fih} \ge 0 \quad \forall f, i, h \tag{10}$$

$$\sum_j s_{fj} = \sum_{i,h} x_{fih} \quad \forall f \tag{11}$$

$$w_i + \sum_k H_{ik}(\lambda_k^- - \lambda_k^+) = 0 \quad \forall i \tag{12}$$

$$0 \le \lambda_k^- \perp T_k + \sum_i H_{ik} y_i \ge 0 \quad \forall k \tag{13}$$

$$0 \le \lambda_k^+ \perp T_k - \sum_i H_{ik} y_i \ge 0 \quad \forall k \tag{14}$$

$$\sum_f s_{fi} - \sum_{f,h} x_{fih} = y_i \quad \forall i. \tag{15}$$

Here constraints (8)–(14) are the optimality conditions of all GENCOs' and the grid owner's optimization problems, and (15) is the market clearing condition. An underlining assumption behind this formulation is that the GENCOs naively believe that their decisions will not affect transmission prices. It is shown in Ref. [6] that the market equilibrium solution to (8)–(15) uniquely exists.

2.3 Generation Expansion Planning Model

We extend the market equilibrium model to include GEP. Our analysis and comparison of environmental policies are based on their effectiveness and efficiency in promoting new renewable generation capacity and reducing GHG emissions.

We denote the investment cost of expanding the capacity of generator (*f*, *i*, *h*) as I_{fih}, and the new decision of expanded capacity as Δx_{fih}. GENCO *f*'s profit maximization problem under GEP becomes:

$$\max \sum_j \left[P_j^0 - \left(P_j^0/Q_j^0\right)\left(\sum_g s_{gj}\right) - w_j\right] s_{fj} - \sum_{i,h} (C_{fih} - w_i) x_{fih} - \sum_{i,h} I_{fih} \Delta x_{fih} \tag{16}$$

$$\text{s.t.} \quad x_{fih} \le X_{fih} + \Delta x_{fih} \quad (\rho_{fih}) \quad \forall i, g, h \tag{17}$$

$$\sum_j s_{fj} = \sum_{i,h} x_{fih} \quad (\theta_f) \tag{18}$$

$$s_{fj}, x_{fih}, \Delta x_{fih} \ge 0 \quad \forall f, i, j, h \tag{19}$$

The grid owner's problem remains the same. The market equilibrium conditions become:

$$0 \leq \rho_{fih} \perp X_{fih} + \Delta x_{fih} - x_{fih} \geq 0 \quad \forall f, i, h \tag{20}$$

$$0 \leq \Delta x_{fih} \perp I_{fih} - \rho_{fih} \geq 0 \quad \forall f, i, h \tag{21}$$

$$\text{Constraints } (8)-(9) \text{ and } (11)-(15) \tag{22}$$

2.4 Cap-and-Trade Policy Model

In this section, we modify the GEP model from Sect. 3 to include the cap-and-trade policy. The cap-and-trade policy sets a limit, or cap, on a power system. Each GENCO receives an initial amount of allowances, which limit the amount of emissions it is allowed to produce. GENCOs can trade their allowances among themselves so that individual GENCOs' emissions may differ from their initial allowances but the total amount does not exceed the overall cap. Cap-and-trade has been a successful policy tool for limiting SO_x and NO_x emissions in the U.S. A variant of a cap-and-trade plan for GHG emissions is proposed in the American Clean Energy and Security Act of 2009, which was approved by the House of Representatives on June 26, 2009.

For modeling simplicity, we assume that initial allowances are grandfathered to GENCOs rather than auctioned. The market clearing conditions also include the balancing constraint between total sales and total purchases of the emissions allowances, which imposes the pre-determined emissions cap on the entire market. While the target of the cap-and-trade policy is to control emissions, it is expected to also encourage more investment in renewable generation. To evaluate its effectiveness in doing so, we classify GENCOs' generations into four types: existing renewable ($h = 1$), existing non-renewable ($h = 2$), new non-renewable ($h = 3$), and new renewable generations ($h = 4$). The emissions rate of generator (f, i, h) is denoted as E_{fih}. The initial emissions allowances of GENCO f is assumed to be pre-determined and denoted as K_f. The market clearing price for carbon emissions is denoted as p^{CO_2}.

Under the cap-and-trade policy, GENCO f's profit maximization problem is:

$$\begin{aligned} \max \sum\nolimits_j \left[P_j^0 - (P_j^0/Q_j^0)(\sum\nolimits_g s_{gj}) - w_j\right] s_{fj} \\ - \sum\nolimits_{i,h} (C_{fih} - w_i) x_{fih} - \sum\nolimits_{i,h} I_{fih} \Delta x_{fih} + p^{CO_2} c_f \end{aligned} \tag{23}$$

$$\text{s.t.} \quad \sum\nolimits_{ih} E_{fih} x_{fih} + c_f \leq K_f \quad (\delta_f) \tag{24}$$

$$\text{Constraints } (17) - (19). \tag{25}$$

The four terms in the objective function (23) are power sales revenue, generation cost, investment cost, and allowances sales revenue. Constraint (24) requires that the emissions $\Sigma_{ih} E_{fih} x_{fih}$ and allowance sales c_f must not exceed the initial allowance K_f. The variable c_f could also be negative to represent allowance purchases. The grid owner's problem remains the same. The market equilibrium conditions become:

$$0 \leq x_{fih} \perp (C_{fih} - w_i) + \rho_{fih} - \theta_f + \delta_f E_{fih} \geq 0 \quad \forall f, i, h \tag{26}$$

$$0 \leq \delta_f \perp K_f - \sum_{i,h} E_{fih} x_{ih} - c_f \geq 0 \quad \forall f \tag{27}$$

$$\delta_f = p^{\mathrm{CO}_2} \quad \forall f \tag{28}$$

$$\sum_f c_f = 0 \tag{29}$$

$$\text{Constraints}\,(8)-(9),(11)-(15)\text{ and }(20)-(21) \tag{30}$$

A solution that satisfies constraints (26)–(30) yields a Nash equilibrium solution under this model.

2.5 *Optimal Incentive Policy Models*

As alternatives to the cap-and-trade policy, taxes and subsidies also provide incentives for power producers to invest more in renewable energy and less in non-renewable energy. In this section, we construct an optimal incentive policy design model to facilitate the comparison of cap-and-trade and carbon taxes. The incentive policy design problem is inherently a bilevel problem. The lower level is the Nash game among the GENCOs and the grid owner. Each GENCO solves a profit maximization problem under exogenously determined tax and subsidy policies. The upper level's problem is to design the minimal incentive intervention in order to achieve a policy goal. In comparison with cap-and-trade, we choose the policy goal to be the limit on total emissions.

Let tax_{fih} denote the tax (positive) or subsidy (negative) of generator (*f*, *i*, *h*). Under this incentive policy, GENCO *f*'s profit maximization problem is:

$$\begin{aligned} \max \sum_j [P_j^0 - (P_j^0/Q_j^0)(\sum_g s_{gj}) - w_j] s_{fj} \\ - \sum_{i,h} (C_{fih} + tax_{fih} - w_i) x_{fih} - \sum_{i,h} I_{fih} \Delta x_{fih} \end{aligned} \tag{31}$$

$$\text{s.t.} \quad \text{Constraints}\,(17) - (19) \tag{32}$$

The grid owner's problem remains the same. The market equilibrium conditions become:

$$0 \leq x_{fih} \perp (C_{fih} + tax_{fih} - w_i) + \rho_{fih} - \theta_f \geq 0 \quad \forall f, i, h \tag{33}$$

$$\text{Constraints } (8), (11)\text{–}(15) \text{ and } (20)-(21). \tag{34}$$

The upper level's optimal incentive policy design problem is formulated as the following bilevel optimization problem:

$$\min \sum\nolimits_i |tax_{fih}|/|\mathcal{I}| \tag{35}$$

$$\text{s.t.} \quad \sum\nolimits_{fih} E_{fih} x_{fih} \leq \sum\nolimits_f K_f \tag{36}$$

$$\text{Constraints } (33)-(34) \tag{37}$$

The objective (35) is to minimize the average tax or subsidy intervention; Constraint (36) defines the goal of the policy, which could not be achieved without the incentives; Constraint (37) is the set of optimality conditions for the lower level's GEP problem under incentive policy intervention.

The incentive policy design model ((35)–(37)) could be modified to reflect different variations of the policies. In particular, we consider the following variations:

Uniform tax policy (UT): The taxes reflect the carbon taxes for non-renewable energy. Under this policy, no subsidy is provided to renewable energy production and the taxes for non-renewable energy production are the same across all nodes and all GENCOs. The following additional constraints will result in this policy:

$$tax_{fih} = tax_{fjh} \geq 0 \quad \forall f, i, j, h. \tag{38}$$

Uniform subsidy policy (US): The subsidies reflect the production tax credits or investment tax credits for renewable energy. Under this policy, no tax is imposed on non-renewable energy production and the subsidies for renewable energy production are the same across all nodes and all GENCOs. The following additional constraints will result in this policy:

$$tax_{fih} = tax_{fjh} \leq 0 \quad \forall f, i, j, h. \tag{39}$$

Uniform subsidy and tax policy (UST): This is a combination of UT and US. The following additional constraints will result in this policy:

$$tax_{fih} = tax_{fjh} \quad \forall f, i, j, h. \tag{40}$$

Non-uniform tax policy (NT): This policy differs from UT in that the taxes for non-renewable energy production are allowed to differ from node to node. For those nodes where taxes are less effective (due to transmission congestion, for example), this policy has the flexibility to reduce the unnecessarily high taxes. The following additional constraints will result in this policy:

$$tax_{fih} \geq 0 \quad \forall f, i, h. \tag{41}$$

Non-uniform subsidy policy (NS): Similar to NT, this policy is designed with non-uniform subsidies and zero tax. The following additional constraints will result in this policy:

$$tax_{fih} \leq 0 \quad \forall f, i, h. \tag{42}$$

Non-uniform subsidy and tax policy (NST): This is a combination of NS and NT. This policy is the most flexible variation in our analysis and is reflected in (35)–(37).

For notational convenience, in the remainder of the chapter, we will use the abbreviation CT to denote the cap-and-trade policy.

3 Numerical Analysis

A simple 5-bus test example is used to demonstrate our models. We will further discuss the advantages and disadvantages between those policies with respect to the five criteria.

3.1 Data

The 5-bus electricity network is shown in Fig. 1. Nodes B, C, and D are demand nodes, and nodes A, C, D, and E are supply nodes. There are two GENCOs in the market. GENCO 1 owns generators in nodes A and C, and GENCO 2 owns generators in nodes D and E. The demand and supply data and transmission data are given in Tables 1 and 2, respectively. The emissions rates for non-renewable

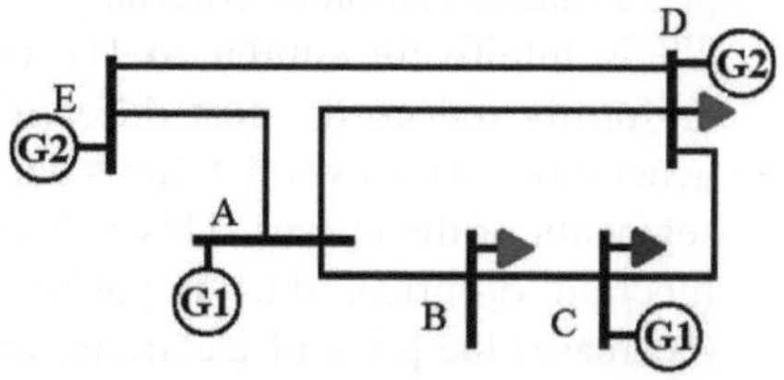

Fig. 1 5-bus test example

Table 1 Demand and supply data

Node	P^0_i	Q^0_i	C_{fi1}	C_{fi2}	C_{fi3}	C_{fi4}	I_{fi3}	I_{fi4}	X_{fi1}	X_{fi2}
A	–	–	1	8	6.5	1	2	15	60	300
B	200	600	–	–	–	–	–	–	–	–
C	100	600	1	8	6.5	1	2	15	50	300
D	200	600	1	5	4.5	1	4	15	70	350
E	–	–	1	5	4.5	1	4	15	55	500

Table 2 Transmission data

Line	A–B	B–C	C–D	D–E	E–A	A–D
T_k	754	154	466	480	720	318

and renewable generators are assumed to be 50 and 0 t/MW, respectively. We also assume that the two GENCOs receive equal initial emissions allowances in the CT policy.

3.2 Results

In this section, we compare cap-and-trade with incentive polices with respect to the five criteria. Our optimal incentive policy design models contain more than 100 complementary constraints, and they can be solved within minutes using a branch and bound algorithm.

3.2.1 Carbon Price and Tax

The carbon price p_2^{CO} (in \$/t) from cap-and-trade and the average carbon tax or subsidy $\frac{\sum_i |tax_{fih}|}{|I|}$ (in \$/MW) are two comparable indicators of policy efficiency, but they have different units. To facilitate the comparison, we convert the carbon price to (\$/MW) by multiplying the emissions rate parameter E_{fih}. Figure 2 shows the tax or subsidy equivalent criteria of different policies as functions of the overall emissions cap. We point out the following findings:

1. UT and CT are equally efficient on this criterion. This is not surprising since in the CT, the carbon price p_2^{CO} is universal for all GENCOs, which is equivalent to a uniform carbon tax. These two policies, however, do show different performances on other criteria.
2. US is intuitively similar to UT (and thus CT) in the sense that they both uniformly reduce the cost differences between renewable and non-renewable generations. However, US turns out to be slightly less efficient than UT in some segments of the curves. This is because the electricity demand is a decreasing function of price, thus imposing taxes (providing subsidies) will increase (decrease) the price of electricity and decrease (increase) the demand and thus

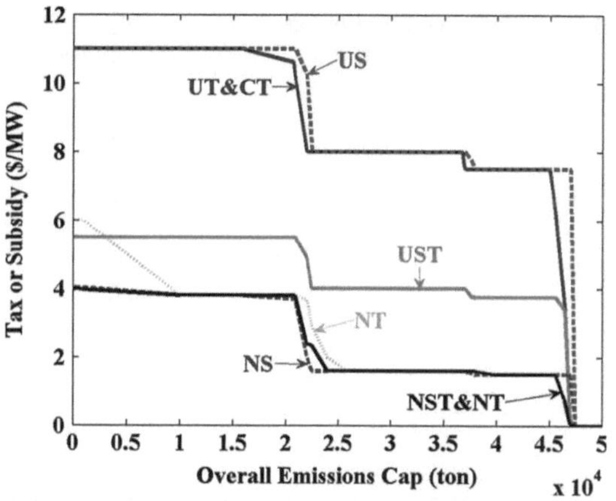

Fig. 2 Policy comparison on carbon price and tax

emissions. Therefore, UT and CT requireless carbon taxes/prices than US requires subsidies to reduce the same amount of emissions.

3. Nonuniform policies (NS, NT, and NST) are generally more efficient than uniform ones (US, UT, and UST) because the former policies have the flexibility to reduce the taxes and subsidies in places where their effectiveness is limited by transmission or generation capacity constraints.
4. When the overall emissions cap is around 46,000 t, the NT shows higher efficiency than the NS for the same reason of demand elasticity. When the cap is around 25,000 t, however, NS appears to be more efficient than NT. This is because when the emissions cap is so tight, significant investment in new renewable generation becomes pivotal, which could be easier to stimulate through subsidies that encourage more electricity demand than through taxes that discourage it. The efficiency of subsidy policies is more likely to be discounted due to lack of transmission capacities to deliver the increased generation.
5. Mixed tax and subsidy policies (UST and NST) are more efficient than tax only (UT and NT) and subsidy only (US and NS) ones.

3.2.2 Renewable Energy Portfolio

Although none of the policies being modeled in Sect. 2 has a specific target on renewable energy portfolio, they are all expected to stimulate more investment and generation of renewable energy and thus increase the renewable energy portfolio. In this section, the criterion for policy comparison is the percentage of renewable energy generation as a function of overall emissions cap. This criterion indicates the policies' effectiveness in stimulating new investment in renewable energy. Figure 3 shows the policy comparison on renewable energy portfolio. We point out the following findings:

Fig. 3 Policy comparison on renewable energy portfolio

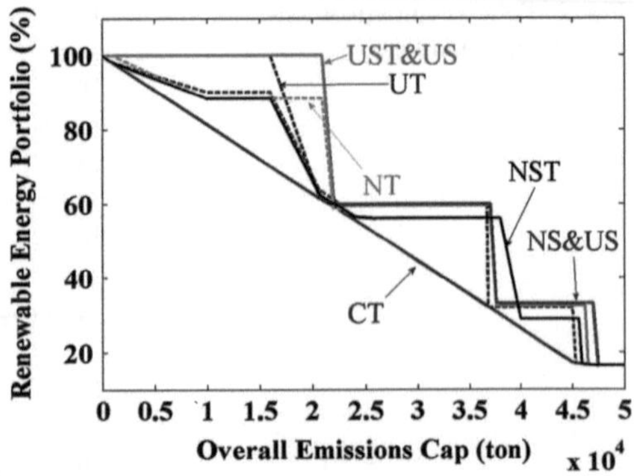

1. As the overall emissions cap reduces from 47,000 t to 0, CT gradually increases the renewable energy portfolio from about 20% to 100% in a straight line. All incentive policies have jumps in their curves, showing lack of flexibility in continuous control. This is due to the nature of incentive policies. Under the pure incentive policy models, GENCOs invest in the least costly generation as much as possible, regardless how close other alternatives are. When the incentives start to make renewable generation even more profitable than non-renewable energy, GENCOs tend to switch to renewable generation all together, resulting in sudden jumps in the curves.
2. Among the incentive policies, NST is the most flexible policy thus its curve looks closest to CT. It also outperforms NS and NT in flexibility, representing the advantage of combining taxes and subsidies over using them alone.
3. The UST overlaps with the US for the most part. Around the 20,000-t cap, the US results in a higher renewable energy portfolio than the UT, since a larger amount of subsidies is paid than taxes collected to achieve the same overall emissions cap, as shown in Fig. 2.

3.2.3 Total Electricity Generation

Total electricity generation is another criterion for comparison, reflecting the policies' impact on electricity prices, since lower generation indicates higher electricity prices. Figure 3 shows the policy comparison on total electricity generation as a function of overall emissions cap. We point out the following findings:

1. US increases electricity generation as the overall emissions cap reduces. This is because a lower cap will lead to more subsidies, lower cost, thus more generation.
2. Opposite to US, the UT and CT policies lead to reduced electricity generation.
3. UST results in a sharp decrease in generation around the 46,000-t cap, which coincides with the sudden increase in renewable portfolio. This is where the incentives, dominatingly taxes, start to be imposed. Below the 45,000-t cap

level, the total generation curve is relatively flat, indicating the overall balance between taxes and subsidies in UST.

4. NS overlaps with US for the most part. When the cap is smaller than around 22,000 t, NS has lower generation because overall it provides less subsidies than US.
5. NST has lower generation than UST, because as demonstrated in Fig. 6, NST is tax-dominant whereas UST is subsidy-dominant. When the cap is smaller than around 10,000 t, NST becomes more subsidy-dominant, which causes its performance to change directions.

3.2.4 GENCOs' and Grid Owner's Total Profit

We also compare the policies' impact on GENCOs' and grid owner's total profit as a function of overall emissions cap in Fig. 5. The total profit is calculated as the summation of the objective values of all GENCOs' and the grid owner's optimization problems under equilibrium. This criterion also indicates the potential perception and reaction from GENCOs and grid owner towards the policies. We point out the following findings:

1. Under CT, the total profit is expected to decrease as the overall emissions cap gets tighter. Counter-intuitively, the opposite trend is observed when the cap is not very tight. This is because the original Nash equilibrium without CT may not be Pareto optimal; when CT introduces additional constraints to GENCOs' optimization problems, it is possible that the new equilibrium solution improves in terms of the total profit. Of course, this phenomenon soon gets offset by the increasingly tight cap, which eventually leads to a gradually decreasing total profit.
2. UT and CT result in different total profits. This is because under UT, the government extracts carbon taxes from the GENCOs, whereas initial allowances in CT are assumed to be grandfathered for free.
3. US provides the most generous subsidies to GENCOs, thus leads to the highest total profit. NS is similar but less so.
4. UT and NT have the opposite effect as US and NS.
5. UST and NST show similar performances as total electricity generation for similar reasons explained in Sect. 3.2.3, since profit is closely related to generation.

3.2.5 Government Revenue

Government revenue, either positive from taxes or negative from subsidies, as a function of overall emissions cap is calculated as another criterion for comparison. Results are shown in Fig. 6 and findings are highlighted as follows:

1. CT is revenue neutral for the government because of our assumption of grandfathered initial allowances.

2. US is the most costly policy to achieve the same level of emissions reduction. NS is similar but less so.
3. UT results in a positive revenue, which nevertheless follows an overall decreasing trend and eventually vanishes as the cap becomes tighter. This is because GENCOs are switching to renewable energy to avoid high tax penalties.
4. UST and NST are generally balanced between tax revenues and subsidy expenses, with occasional dominance of one type over the other.

4 Conclusions

In this chapter, we compare cap-and-trade and carbon tax policies in a GEP context in an electricity market. The contributions of this study include integrated modeling framework for GEP under environmental policies, novel inverse equilibrium approaches for optimal carbon tax policy design, and multiple criteria for policy comparison. It is our expectation that our analysis will shed light on new perspectives and approaches to qualitative and quantitative comparisons of environmental policies.

We make the following summarizing remarks on our findings from Sect. 3:

1. The similarities and differences between cap-and-trade and carbon tax policies are demonstrated in Figs. 2–6. They show similar performances in terms of market-wide uniform carbon taxes and total electricity generation. Cap-and-trade outperforms carbon taxes in continuous and flexible control of renewable energy portfolio and total profit of GENCOs and the grid owner. Carbon taxes create a revenue stream to the government, whereas cap-and-trade does not if initial allowances are grandfathered.
2. Both tax and subsidy policies have their own limitations. Subsidies require expenditure of tax payers money, whereas taxes are politically unpopular. A combination of both appears to be more efficient than either one alone. Subsidies

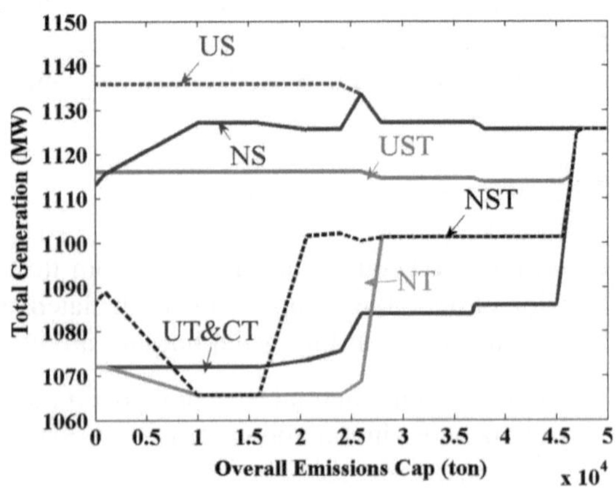

Fig. 4 Policy comparison on total electricity generation

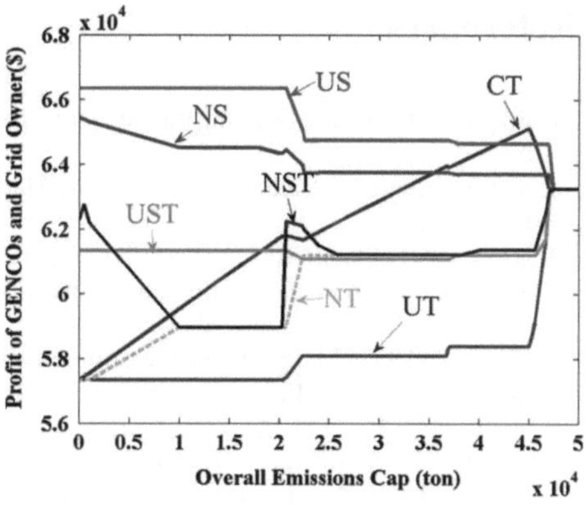

Fig. 5 Policy comparison on GENCOs' and grid owner's total profit

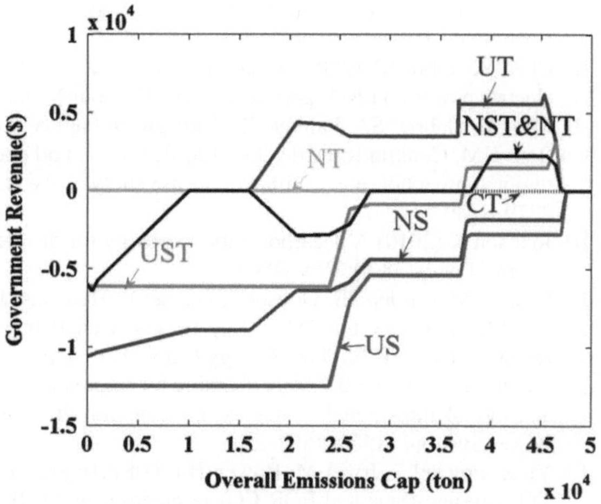

Fig. 6 Policy comparison on government's income

have also shown the tendency of resulting in misleadingly low energy prices, which in turn lead to higher consumptions and thus more emissions.

3. Nonuniform carbon tax policies show superior performances than uniform ones in several criteria.

There are caveats and limitations in our analysis. First, our GEP models are single-period models, which do not provide insights on market dynamics as the emissions cap or carbon taxes change over time. Second, we only consider the grandfathering version of the cap-and-trade policy, thus the allowance auction process is not studied. Third, our optimal incentive policy models are LPCC, the global optimal solutions of which are computationally challenging to obtain. As a result, our case study was conducted on a small 5-bus network system, which could fail to capture some characteristics of real energy markets that are usually much larger. Future research directions should include new modeling approaches to overcome the limitations of our models. Moreover, it would also be revealing to include mandatory environmental policies such as renewable portfolio standards in the comparison.

References

1. Energy Information Administration. http://www.eia.gov/oiaf/servicerpt/subsidy2/pdf/execsum.pdf
2. Shahidehpour M, Yamin H, Li Z (2008) Market operations in electric power systems. Wiley, New York
3. Bjorkvoll T, Fleten S, Nowak M, Tomasgard A, Wallace S (2001) Power generation planning and risk management in a liberalized market. In: IEEE porto power tech proceedings, 1 Sept 2001
4. Bloom J (1982) Long-range generation planning using decomposition and probabilistic simulation. IEEE Trans Power Appar Syst 101(4):797–802
5. Ventosa M, Rivie M, Ramo A, Garcia-Alcalde A (2000) An MCP approach for hydrothermal coordination in restructured power markets. IEEE Power Eng Soc Summer Meet 4:16–20
6. Hobbs BF (2001) Linear complementarity models of Nash-Cournot competition in bilateral and POOLCO power markets. IEEE Trans Power Syst 16(2):194–202
7. Chen Y, Hobbs BF (2005) An oligopolistic power market model with tradable NO_x permits. IEEE Trans Power Syst 20(1):119–129
8. Chen Y, Hobbs BF (2006) Solution of large-scale leader-follower market equilibria problems: electric power and NO_x permit markets. Comput Manag Sci 3(4):307–330
9. Ruth M, Gabriel SA, Palmer KL, Burtraw D, Paul A, Chen Y, Hobbs BF, Irani D, Michael J, Ross KM, Conklin R, Miller J (2008) Economic and energy impacts from participation in the regional greenhouse gas initiative: a case study of the state of Maryland. Energy Policy 36(6): 2279–2289
10. Johnson K (2010) A decarbonization strategy for the electricity sector: new-source subsidies. Energy Policy 38(5):2499–2507
11. Beard LM, Cardell JB, Dobson I, Galvan F, Hawkins, Jewell W, Kezunovic M, Overbye TJ, Sen PK, Tylavsky DJ (2001) Key technical challenges for the electric power industry and climate change. IEEE Tran Energy Convers, to appear 1:1–6
12. Tabors RD, Monroe B (1991) Planning for future uncertainties in electric power generation: an analysis of transitional strategies for reduction of carbon and sulfur emissions. IEEE Trans Power Syst 6(4):1500–1507
13. Yu Z, Preckel P, Irwin M, Bowen B (2008) An enhanced model of capacity expansion with CO_2 cap-and-trade and EOR CO_2 re-emergence. In: IEEE power and energy society general meeting 2008, Pittsburgh, pp 1–7
14. Chuang A, Wu F, Varaiya P (2001) A game-theoretic model for generation expansion planning: problem formulation and numerical comparisons. IEEE Trans Power Syst 16(4): 885–891
15. Keyhani A (2003) Leader-follower framework for control of energy services. IEEE Trans Power Syst 18(2):837–841
16. Li T, Shahidehpour M (2005) Strategic bidding of transmission- constrained GENCOs with incomplete information. IEEE Trans Power Syst 20:437–447
17. Wang J, Shahidehpour M, Li Z (2009) Strategic generation capacity expansion planning with incomplete information. IEEE Trans Power Syst 24(2):1002–1010
18. Bai X, Shahidehpour M, Ramesh V, Yu E (1997) Transmission analysis by Nash game method. IEEE Trans Power Syst 12:1046–1052
19. Ferrero R, Shahidehpour M, Ramesh V (1997) Transaction analysis in deregulated power systems using game theory. IEEE Trans Power Syst 12:1340–1347
20. Ferrero R, Rivera J, Shahidehpour M (1998) Application of games with incomplete information for pricing electricity in deregulated power pools. IEEE Trans Power Syst 13(1):184–189

Cap and Trade Modeling in Electricity Markets Using an Agent-Based Approach

Jianhui Wang, Vladimir Koritarov, and Jin-Ho Kim

Abstract An agent-based approach is proposed in this book chapter to analyze interactions between the emission and electricity markets. A cap-and-trade system is assumed to be in place to regulate emissions from power generation. Generation companies are modeled as adaptive learning agents that can bid strategically into the electricity market by a Q-learning algorithm. These companies also participate in allowances trading in the emission market by adjusting their own allowances positions. In the simulation, generation companies can value their generation capacity and available allowances to maximize their profits. Three different trading strategies are modeled to mimic the decision-making process of generation companies. This modeling framework can help design a sound emission market by simulating market scenarios with different policies. It can also be used to investigate the operation strategies for generation companies in such an environment.

Keywords Cap-and-trade • Power systems • Electricity market • Emission market • Agent-based modeling and simulation • Q-learning • Emission trading • Carbon tax • CO_2 emissions • Emission allowances • CO_2 • Power economics • Bidding • Price • Kirchhoff's laws • Greenhouse gas emission • Climate change • Simulation

1 Introduction

In the past several years, climate change has been drawing increasing attention worldwide. Two methods are usually applied to control greenhouse gas (GHG) emissions. One uses carbon taxes to elevate the cost of producing such emissions. The other one is a so-called "cap-and-trade" system, in which emitters can buy or

J. Wang (✉) • V. Koritarov • J.-H. Kim
Decision and Information Sciences Division, Argonne National Laboratory, Argonne, IL 60439, USA
e-mail: jianhui.wang@anl.gov; koritarov@anl.gov; kimjh@anl.gov1

Q.P. Zheng et al. (eds.), *Handbook of CO_2 in Power Systems*, Energy Systems, DOI 10.1007/978-3-642-27431-2_5, © Springer-Verlag Berlin Heidelberg 2012

sell their emission allowances. Compared with the uncertain emission reductions that carbon taxes can achieve, the cap-and-trade system is widely seen as a highly effective program for reducing emissions of atmospheric pollutants. Europe and the northeastern United States have experimented with the cap-and-trade system with some success. By directly limiting the amount of emissions, the uncertainty in setting emission targets can be reduced. However, one main drawback of a cap-and-trade system is the volatility of allowance prices in the market. Because there are basically no physical constraints associated with allowances trading, it is driven primarily by new information, such as used allowances, policy changes, and anticipated allocation. Allowance price fluctuations can have a tremendous impact on an emitter's operation. Also, proper allocation of allowances is critical in reaching the ultimate reduction target. Excessive allowances may weaken the incentives to pursue more efficient and cleaner generation technologies, while insufficient allocations may have a negative impact on emitters' operations.

As one of the major polluters, the electricity industry is particularly vulnerable to the volatility of allowances prices. Whether in an electricity market or a vertically integrated industry structure, the uncertainty of allowances prices is directly related to the opportunity cost of power generation because the equivalent amount of allowances could have been sold for profit in the emission market. Furthermore, generation companies (GenCos) are exposed to not only allowance price volatility, but also to the volatility of electricity prices resulting from load fluctuation, system component outages, and strategic bidding by market participants. On the other hand, due to the potential arbitrage activities among the electricity and emission markets players can add to the price volatility by breaking the demand–supply equilibrium in either market. Because many electricity markets are already established in places all over the world, modeling cap and trade in the electricity markets is very important for designing a sound emission market and training GenCos to improve their business operations. This book chapter describes a modeling framework for modeling the interplay between the electricity market and the emission market, which can shed light on future policy making and help GenCos optimize their bidding strategies. In this study, we focus on trading CO_2 emissions, which represent the largest emission share generated by the electricity sector. Because other industries, such as the steel and aviation industries, also emit CO_2, we assume that GenCos are price-takers in the CO_2 emission market, which means they have no control over the market clearing prices. However, because of the limited number of participants in the electricity market, GenCos can impact the electricity market prices to a greater or lesser extent by strategically bidding into the market under certain circumstances.

Extensive research has been conducted on the interactions of GenCos in the electricity market. In an electricity market, GenCos submit their bids in the form of available capacity and prices to the independent system operator (ISO) that clears the market. Because of the oligopolistic nature of electricity markets, two dominating research approaches exist: game-theory-based analytical methods [9, 10, 15] and agent-based modeling and simulation (ABMS) [2–4, 13, 14, 18, 19, 23]. Game-theory-based models often demonstrate the difficulty of modeling the practical market in sufficient detail. Simplified assumptions have to be made in the problem formulation in order to make the problem solvable. Additionally, the learning process of GenCos in bidding often results in less-than-optimal (non-equilibrium) bidding strategies. So the absence of dynamic learning modeling for market participants in the game-based models may lead to inaccuracy in the general conclusions.

In comparison, ABMS captures the adaptive learning behaviors of individual GenCos by developing certain learning algorithms for each GenCo agent. The evolutionary process of the interaction among agents can be observed. Moreover, agent-based models are able to simulate the electricity market in greater detail as many assumptions may need to be made in an equilibrium model to describe the equilibrium condition. In an ABMS-based electricity market simulation model, GenCos are assumed to be rational in the sense that they act to maximize their own profit by strategically bidding while interacting with the other agents through the market clearing process conducted by the ISO. Various learning algorithms have been used to represent the learning behaviors by GenCos. In one agent-based electricity market simulation model, Electricity Markets Complex Adaptive System (EMCAS) [18, 19], feedback heuristics are used to define GenCos' learning rules. Botterud et al. [3] discussed the agent-based long-term expansion module in EMCAS. Besides the heuristic learning rules, more sophisticated learning mechanisms, such as reinforcement learning, have been proposed based on the observations in the psychological learning literature [2, 4, 7, 14, 23].

However, in the feedback heuristics or simple reinforcement learning models, GenCos' learning processes are based solely on the retrospective evaluation of their past bidding strategies. The rational forecasting of future market conditions is missing. In fact, GenCos will vary their bidding strategies given different market conditions (e.g., system load, demand, weather). To complement retrospective reinforcement learning algorithms, Q-learning is applied to model GenCos' bidding in more detail. The states (i.e., conditions) that decision makers face and their corresponding actions comprise a matrix of state-action pairs. A Q-value is assigned to each state-action pair. In the context of the electricity market, the states can be interpreted as the system demand, market prices, and so on. The Q-values can be calculated based on GenCos' profits. The learning process is characterized as a reward/reinforcement mechanism to change the Q-value of each state-action pair in a trial-and-error manner. Successful decisions are reinforced by receiving rewards, while unsuccessful decisions are weakened. Tellidou and Bakirtzis [16] employed a new simulated annealing (SA)-Q-learning algorithm [8] to study capacity withholding and the emergence of tacit collusion among market participants in the electricity market. Yu et al. [23] used Q-learning to model suppliers' learning behaviors. Krause et al. [12] analyzed a repeated game in power markets with Q-learning. A conjectural-variation-based bidding strategy in conjunction with Q-learning was proposed [17]. In that study, the other GenCos' reactions to one GenCo's bid were also considered.

However, little research has been done on the emission market side. None of the studies cited above has involved modeling of the emission market in great detail by using ABMS techniques. A game-theory-based model was established to analyze the equilibriums with NO_X limits [5, 6]. The wholesale power trading in Germany was analyzed using ABMS [22]. But no speculation was assumed in the emission market, which means the active participation of GenCos in the emission market was ignored. The portfolio risk management and carbon emissions valuation were investigated by Kleindorfer and Li [11], which is one of the pioneer works that analyzes the emission trading strategies of GenCos in an analytical way. Wang et al. [20] used an agent-based approach to analyze the interactions between the emission market and the electricity market. However, the trading strategies modeled were limited.

In this book chapter, we use ABMS to build a parallel emission market for the trading of emission allowances in addition to the electricity market. The proposed approach takes into account the allocation of emission allowances to GenCos operating in an electricity market, the operating decisions made by GenCos, and the trading of emission allowances among GenCos. GenCos are modeled as individual agents that can apply Q-learning to strategically bid into the electricity market. Trading decisions, in terms of selling or buying allowances in the emission market, can be also made depending on the allowance prices. Three trading strategies including weekly trading, year-end trading and dynamic trading are modeled. This work further expands the study conducted by Wang et al. [20] by more sophisticated dynamic trading strategies.

2 Q-Learning Algorithm

The Q-learning algorithm proposed by Watkins and Dayan is one of the most popular machine learning algorithms [21]. The learning algorithm characterizes the environment in which agents interact with as states $s_i(s_i \in S_i)$ for each GenCo. Each state includes a set of variables that describes the status of the environment. For each state, a group of actions $act_i(act_i \in A_i)$ represents all the possible moves of the agents. A Q-value $Q_{i,t}(s_i, act_i)$ is calculated for each state-action pair at time t. The Q-values are modified during the learning episodes by reinforcing agents' experience for various state-action pairs. The action candidates are prioritized on the basis of past experience (exploitation), while all the actions still have a probability of being selected by certain policy (exploration). Eventually, the most successful action (in terms of the largest Q-value) is identified after the learning process. At time t, the corresponding $Q_{i,t}(s_i, act_i)$ with chosen action $act_{i,t}$ in state s_i is updated by the following formula, in which $\sigma_{i,t}$ is a learning parameter and $r_{i,t}$ is the immediate payoff:

$$Q_{i,t}(s_i, act_i) = \begin{cases} (1 - \sigma_{i,t}(act_i))Q_{i,t-1}(s_i, act_i) + \sigma_{i,t}(act_i)r_{i,t}, & \text{if } act_i = act_{i,t} \\ Q_{i,t-1}(s_i, act_i), & \text{if } act_i \neq act_{i,t} \end{cases} \quad (1)$$

In (1), the discount factor used in the long-run repeated game is not employed. This is because Q-learning in the electricity market game differs from the Markov decision processes in the other learning-related literature in the sense that GenCos' bidding strategies are affected, not only by the current status of the system, but also by their learning experience, rivals' reaction, and future expectations. It has been shown in the literature that the foresight represented by the discount factor is not indispensable [1, 16].

Q-learning is widely used in ABMS because it requires only a proper definition of state-action space and is not limited to one specific kind of problem. However, parameter tuning and balancing exploitation and exploration represent obstacles to further application of Q-learning. Excessive exploration slows down the learning process, while excessive exploitation leads to a locally optimal solution. To better control exploration and exploitation, the SA-Q-learning proposed by Guo et al. [8] borrows the Metropolis criterion from simulated annealing. In SA, the Metropolis criterion is used to make the solution escape from a local optimum by changing corresponding probabilities for exploitation and exploration. The experiments confirm the advantages of using the Metropolis criterion compared with other methods [8].

The Metropolis criterion in SA-Q-learning can be described as follows:

1. Randomly select $act'_{i,t}$
2. Select $act''_{i,t}$ according to the greedy policy, in which the strategy with the highest Q-value is selected:

$$act''_{i,t} = \arg\max_{act_i} Q_{i,t-1}(s_i, act_i)$$

3. Generate a random value $\xi \in (0, 1)$. If $\xi < \exp((Q_{i,t-1}(s_i, act'_{i,t}) - Q_{i,t-1}(s_i, act''_{i,t}))/Temp_{i,t})$, $act_{i,t} = act'_{i,t}$. Otherwise, $act_{i,t} = act''_{i,t}$.

$Temp_{i,t}$ is the temperature, as in the SA algorithm. The geometric scaling factor criterion is used as the temperature-dropping criterion in order to guarantee a slow decay of the temperature factor in the algorithm [8]. φ_i is usually a constant that is close to 1.

$$Temp_{i,t} = \varphi_i Temp_{i,t-1}, t = 0, 1, 2, \ldots \tag{2}$$

In an electricity market, in which the demand varies by time and season, GenCos change their bidding strategies accordingly. They tend to bid higher than their true production cost when load is high because their bids have a higher probability to be accepted to meet the load. In reality, one GenCo's bid is influenced by other system conditions, such as other GenCos' bids, maintenance plans, and system component outages. For simplification, we assume only load as the state space for GenCo agents, which means GenCos define a certain number of load intervals as states of the system. For each interval, GenCos define a set of bidding strategies. A bid price coefficient refers to the coefficient k_i (larger than one) that GenCos impose on their production cost functions to be bid into the market. Once GenCos observe the load published by the ISO, they choose the interval where this load falls. Then GenCos follow the Q-learning algorithm to choose their bidding price coefficients $k_{i,t}$ at t and form their bids for each generator, $\tilde{C}_i(P_{i,t}, k_{i,t})$, to the ISO. The states are defined in this chapter as different load levels, while different bid price coefficients construct the action space according to each state as state-action pairs.

Generators have the same form of quadratic cost function at time t as the following:

$$C_i(P_{i,t}) = a_i P_{i,t}^2 + b_i P_{i,t} + c_i (i = 1, 2, \ldots n) \tag{3}$$

s.t.

$$P_{i,\min} \leq P_{i,t} \leq P_{i,\max}$$

where a_i, b_i, c_i are cost function coefficients, $P_{i,t}$ generation of unit i at t, $P_{i,\max}$ upper generation capacity limit, $P_{i,\min}$ lower generation capacity limit.

With $k_{i,t}$ from Q-learning and the opportunity cost from using the equivalent amount of allowances instead of selling them to the market, we have:

$$\underset{i}{\tilde{C}}(P_{i,t}, k_{i,t}) = a_i k_{i,t} P_{i,t}^2 + b_i k_{i,t} + c_i + P_{i,t} EF_{CO2,i} EAP \tag{4}$$

s.t.

$$P_{i,\min} \leq P_{i,t} \leq P_{i,\max}$$

where EAP is expected allowance price at the end of the year (\$/t), $EF_{CO2,i}$ unit CO_2 emission factor for unit i (t/MWh).

After GenCo i sends its bid to the ISO, the ISO runs the optimal power flow to minimize the supply cost to meet the demand. So the formation of this game changes to a bi-level optimization problem. The upper-level problem is a GenCo's own decision to bid. The lower-level problem is the ISO optimization problem. The market clearing problem will yield the locational marginal prices (LMP) signals to all GenCos to calculate their profits.

For the lower-level problem, the ISO's objective (5) is to minimize the cost to supply the load. Because there are quadratic terms $P_{i,t}^2$ in the objective function, the ISO optimization problem forms a quadratic optimization problem. A DC power flow model is used to represent the load balance (6), line flow (7), and transmission line and generation capacity limits (8)–(9). LMPs can be calculated on the basis of the dual variables of (6) and (7).

$$Min \quad \sum_{i=1}^{NG} \tilde{C}_i(P_{i,t}, k_{i,t}) \tag{5}$$

s.t.

$$\sum_{i=1}^{NG} P_{i,t} = \sum_{m=1}^{M} PD_m \tag{6}$$

$$LF_k(t) = SF_{km} * (P_{i,t} - PD_m) \tag{7}$$

$$|LF_k(t)| \le LF_{k,\max} \tag{8}$$

$$P_{i,\min} \le P_{i,t} \le P_{i,\max} \tag{9}$$

Where k denotes a line, $LF_k(t)$ line flow in line k at t, $LF_{k,\max}$ maximum capacity of line k, i index of generators, m index of buses, M number of buses, NG number of generators, PD_m nodal load, $P_{i,t}$ dispatched generation of unit i at t, SF_{km} shifting factor vector of line flow k to buses.

After the ISO clears the market, it returns the respective generation dispatch $P_{i,t}$ and LMPs $LMP_{i,t}$ to all GenCos. LMPs can be calculated based on the dual variables of (6) and (7).With that information, GenCos can calculate the payoff as (10) and update the Q-values.

$$R_{i,t} = LMP_{i,t}{}^{*}P_{i,t} - C_i(P_{i,t}) \tag{10}$$

Where:
$LMP_{i,t}$ are locational marginal prices at the buses where unit i is located at time t,
$R_{i,t}$ profit for unit i at time t

Also, GenCos update their learning rate $\sigma_{i,t}(act_i)$ at time t as:

$$\sigma_{i,t}(act_i) = \frac{1}{\text{Visited times of } act_i \text{ until } t} \tag{11}$$

3 Cap and Trade Modeling in the Electricity Markets

At the beginning of the simulation period, 1 year in this study, all participating GenCos obtain their allotment of emission allowances for each covered facility. The allowances are grandfathered on the basis of historical emission data or obtained through auctions. Because the internal allocation of actual emissions for GenCos is not necessarily a linear function of time, GenCos may convert their annual emission allowances into periodical limits. Depending on the system load and maintenance schedule, emission limits may have certain seasonality. Also, allocations of emissions by period may be affected by forced outages of generating units. The initial allowance allocation is made by matching the yearly system load profile, which means the allocation is higher when the load is high and lower when the load is low.

During the simulation, GenCos continuously monitor their actual emissions versus the amount of emission allowances that they have available and make appropriate operating decisions. Although our model of the electricity market performs hourly simulations of the market, it is assumed that the emission trading takes place at the end of each week because short-term unit maintenance, scheduling, and other system constraints (such as fuel supply) occur in this time frame. It should be noted that the proposed model can be applied to other trading periods without loss of generality.

At the end of each week, GenCos calculate their actual emissions, compare them with their initial internal emission allowances allocation for the period, and decide on their actions. GenCos can act on a deficit or surplus of allowances. If there is a deficit, GenCos may buy allowances from the market or borrow allowances from the rest of the year based on the allowance price that day. If there is a surplus, GenCos may choose to sell their allowances to the market when the price is high or bank them for future use when the price is not high enough. Accordingly, three trading strategies are modeled as follows:

- Year-end trading

For the year-end trading, GenCos ignore any deficit or surplus of allowances during the year and only trade the allowances at the end of the year to ensure the allowance liability is equal to the allowances allotted. In this way, GenCos bank any surplus or borrow from the future allowances balance if the allowances used are higher than the allocated.

- Weekly trading

For the weekly trading strategy, GenCos try to keep the available allowances to be exactly the same as originally allocated. In other words, at the end of each week, if there is a surplus, GenCos will sell the extra allowances in the emission market while they will buy in some allowances if there is a deficit. This strategy mimics the relatively conservative trading behaviors for GenCos in the sense that they can minimize the risk of trading at the end of the year by not having much deficit or surplus at that time. However, this strategy may not be consistent with the actual market outcome since it is assumed that all the allowances are pre-allotted based on a pre-specified load profile and the generation output also follows the system load pattern.

- Dynamic trading

With dynamic trading, GenCos are allowed to exchange allowances in the emission market by trading in the emission market every week. We define a price threshold for GenCos to make these decisions: GenCos purchase allowances if the allowance price is less than the price threshold; otherwise, they use their available allowances (i.e., borrow from their future allowances). The price threshold for selling works in a similar way: if they have allowances remaining after the week's usage, GenCos sell the balance to the market if the market price is higher than the threshold; otherwise, they keep/bank the allowances because the price incentive is not strong enough for them to sell. The remaining allowances are allocated, again by week, to the rest of the year according to the load. This threshold can be set based on the future price of emission allowances and should vary from time to time. Hence, we model the price movement as an Ornstein-Uhlenbeck process as below:

$$dx_t = \theta(\mu - x_t)dt + \sigma dW_t \tag{12}$$

Where $\theta > 0$ is mean-reverting speed, $\mu > 0$ mean, $\sigma > 0$ volatility, W_t the Wiener process.

At the same time, we assume that GenCos try to even up their annual allowances at the end of a year in order to obtain at least the same level of allowances the next year and avoid the penalty from violating the cap. Also, because of the price volatility in the emission market, the value of the year-of-end allowances is not certain. GenCos try to estimate their total end-of-year allowances by evaluating their available allowances and possible future use. The available allowances include the initial allotment of allowances at the beginning of the year, as well as the allowances purchased so far (minus those sold). Estimate of the future allowances usage follows the initial allocation at the beginning of the year. If an end-of-year shortfall is projected, GenCos need to decide whether they should adjust their generation or increase bid prices to make up the end-of-year expense for buying additional allowances from the market. On the other hand, if an end-of-year surplus is projected, GenCos may decide to decrease their bidding price for generation to increase their share in the market. This adjustment can be realized by a bid adjustment factor (BAF). BAFs, which apply to bidding prices for each generating unit for the next week, are calculated at the end of the previous week. Aiming at balancing out GenCos' weekly allowances positions, BAFs can be computed as a function of expected allowance shortfall, or surplus and remaining time to the end of the year. BAFs are designed to provide financial resources that will be used to purchase additional allowances at the end of the year. Because BAFs increase the bid price, some units may not be competitive anymore in the market, and vice versa. This will affect the CO_2 emissions and, consequently, the shortfall at the end of the year.

The BAF for $t + 1$ for GenCo i can be calculated at the end of time t as follows:

$$BAF_{i,t+1} = (PA_{i,t}/AA_{i,t} - 1) \times EAP \times EF_{CO2} \times N/(N - t) \tag{13}$$

Where

$PA_{i,t}$ Projected allowance need for GenCo i until the end of the year, $AA_{i,t}$ Available allowances for GenCo i until the end of the year, N Number of weeks in the simulation period.

Adding BAFs to the production cost function bid into the market for GenCos, we have

$$\tilde{C}_i(P_{i,t}, k_{i,t}) = a_i k_{i,t} P_{i,t}^2 + b_i k_{i,t} P_{i,t} + c_i + P_{i,t} EP_{CO2,i} EAP + P_{i,t} BAF_{i,t} \tag{14}$$

The overall agent-based market simulation forms as follows:

1. All the GenCos initialize the Q-values and other parameters.
2. In each hour t, GenCos submit their bids to the ISO based on the Q-learning algorithm.
3. ISO clears the market by using (5)–(9), informs GenCos of their generation, and publishes the LMP information.
4. GenCos calculate their profits by using (10) and update $Temp_{i,t}$, the learning rates $\sigma_{i,t}(act_i)$, and Q-values; if it is the end of a week, go to step (5). Otherwise, go to step (2).
5. GenCos calculate the amount of CO_2 emissions for this week and apply one of the following three strategies:
   - Year-end trading
     If there is a surplus, they
     - Bank the surplus.

     If there is a shortfall, they
     - Borrow the balance from the future.
   - Weekly trading
     If there is a surplus, they
     - Sell the surplus to the emission market.

     If there is a shortfall, they
     - Purchase the balance from the emission market.
   - Dynamic trading
     If there is a surplus, they
     - Sell the surplus when the allowance price is higher than or equal to the price threshold.
     - Bank the surplus when the allowance price is lower than the price threshold.

     If there is a shortfall, they
     - Borrow the balance from the future, when the allowance price is higher than or equal to the price threshold.
     - Purchase the balance when the allowance price is lower than the price threshold.

     Reallocate the allowances for the rest of the year and calculate BAFs and apply it to bids for next week.

6. Go to step (2) until the maximum simulation period is reached.

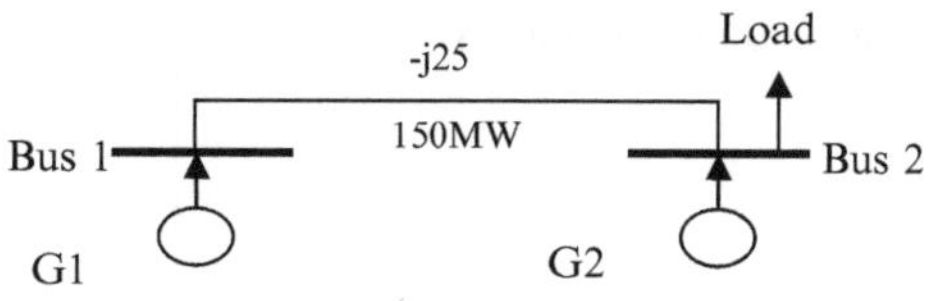

Fig. 1 Two-bus system

Table 1 Unit data

Unit no.	Cost coef. b ($/MWh)	Cost coef. a ($/MW^{2}h)	$P_{i,\max}$ (MW)	$P_{i,\min}$ (MW)
1	10	0.01	300	0
2	10	0.01	200	0

4 Numerical Examples

To illustrate the framework, we use a simple two-bus system to analyze the adaptive learning behaviors of GenCos in detail. The two-bus system is shown in Fig. 1, and the unit data are listed in Table 1. GenCo1 owns unit 1 at bus 1, and GenCo2 owns unit 2 at bus 2. The cost coefficients c of the two units are zero.

Three trading strategies are simulated in this study, i.e., year-end trading, weekly trading and dynamic trading. For the year-end trading, GenCos ignore any deficit or surplus of allowances during the year and only trade the allowance at the end of the year to ensure the allowance liability is equal to the allocated allowances. For the weekly trading strategy, GenCos try to keep the available allowance to be exactly the same as originally allocated every week. In other words, the BAFs are minimized. With dynamic trading, GenCos are allowed to exchange the allowances in the emission market by weekly dynamic trading.

In all three cases, we assume the initial temperatures and decaying factors for Q-learning are $Temp_{1,0} = Temp_{2,0} = 300000$ and $\varphi_1 = \varphi_2 = 0.99$, respectively. The simulation is run on an Intel Core2 Quad CPU (2.83-GHz) computer with 3.25 GB memory. Due to the stochastic nature of the Ornstein-Uhlenbeck process and Q-learning algorithm, each simulation comprises 100 runs; based on these runs, we calculate the average used allowances and profits. The program is coded in Matlab. One run takes only about 15 s to simulate the hourly market clearing for a year. The allowances price profile is assumed based on the data from the European Energy Exchange (EEX). The prices reach their peak during the summer of the year, as shown in Fig. 2.

The mean of the price threshold for selling and buying allowances between the two GenCos is equal to \$4.3/t. The mean-reverting speed, θ, is equal to one with the volatility, σ, also equal to one. The loads are divided into five intervals: [0–150] and the other four between [150–350], with an equal step of 50 MW. To avoid an excessively high or low adjustment on bids, limits are also set for positive BAFs at \$3/MWh and for negative BAFs at −\$3/MWh. Unit emission factors for the two GenCos are equal to 1 t/MWh. The initial allocated allowances for GenCo1 and GenCo2 are 1,150,000 and 900,000 t. The three bidding strategies available for both GenCos to bid into the electricity market are 1.5, 2, and 2.5, respectively.

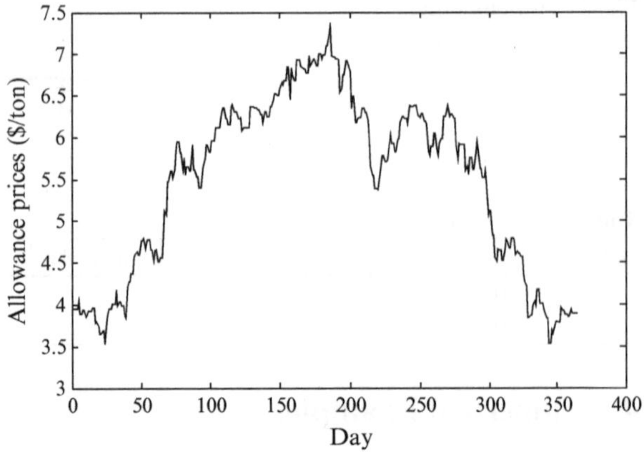

Fig. 2 Daily allowances prices

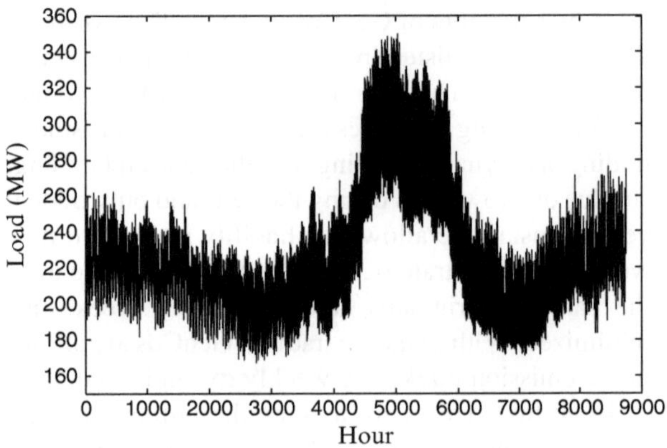

Fig. 3 System load

The system load is shown in Fig. 3, which depicts a summer peak. We also assume that GenCos have a perfect load forecast when they submit their bids to the market.

- **Year-end trading**
  In the first case, we assume that GenCos participate only in the electricity market, which means they simply bank or borrow their own allowances when there is a surplus or shortfall. The average consumed allowances of all the runs in this case are illustrated in Fig. 4. The consumption is basically in line with the system load. During the summer season, GenCos use more allowances because the system load is high. Figure 4 also shows that GenCo1 has a flatter allowances usage curve than GenCo2. This is because GenCo2 is located at bus 2, where the system load is located. The limited transmission line capacity between the two buses prevents GenCo1 from fully competing with GenCo2. So GenCo1 can only supply as much load as the maximum transmission line capacity allows. In this case, the total profit of GenCo1 is $16.34 M and the profit for GenCo2 is $17.03 M. It is interesting to see that GenCo2 can obtain higher profit even though GenCo1 has more generation at most of the time as shown in Fig. 4.

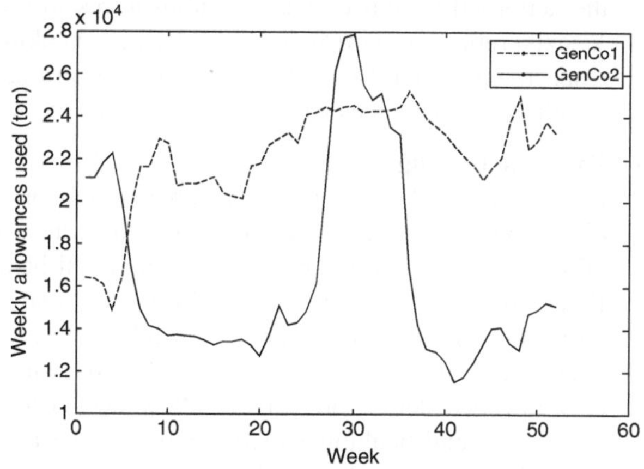

Fig. 4 Weekly allowances used

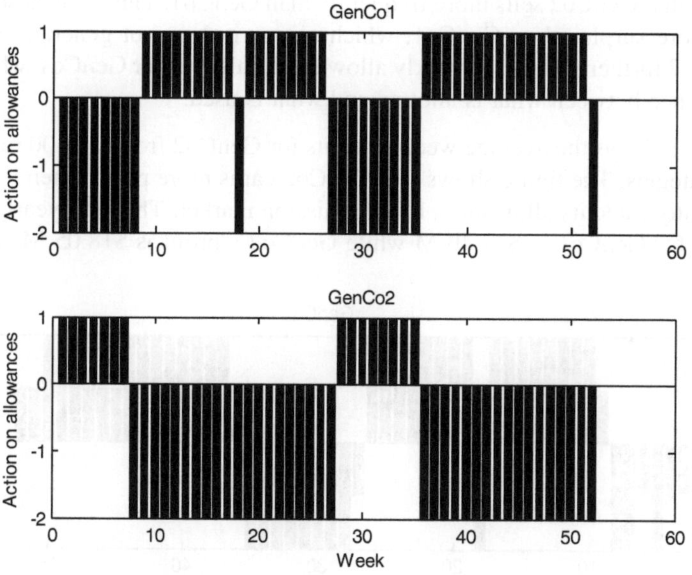

Fig. 5 Actions on allowances in weekly trading

This is again because of the locational advantage that GenCo2 has. The average LMP at bus 1 is \$25.49/MWh as opposed to \$30.80/MWh at bus 2, which leads to a higher profit for GenCo2.

- **Weekly trading**
  With the assumption of weekly trading, GenCos strictly follow the pre-allotted allowances for each week. If the actual used allowances deviate from the allotted, GenCos will try to make up the difference by either buying or selling the allowances in the emission market. In this way, GenCos can avoid trading with the volatility of the allowances price at the end of the year. Figure 5 shows

the actions the GenCos take for allowances in one run. The numbers on the Y axis represent borrowing, purchasing, banking, and selling allowances, respectively, from 2 to −2. As can be seen in Fig. 5, GenCos either purchase or sell their allowances with this trading strategy, with no borrowing or banking.

- **Dynamic trading**
  In dynamic trading, GenCos are able to sell or buy the allowances at their discretion at the end of each week. The transactions occur only when the allowances prices in the market exceed or fall below a certain price threshold. Figure 6 reveals that in the middle of the year, when the system load and corresponding allowances prices are relatively high, GenCos either borrow to avoid the large expense associated with purchasing allowances when they have an allowance deficit or sell the allowances when they have a surplus. The purchasing and banking mostly happen at the beginning and end of the year when the system load and allowance prices are relatively lower. Figure 6 also shows that GenCo2 sells more frequently than GenCo1. This is because GenCo2 has more surplus than GenCo1, which means it does not generate too much. Figure 7 further shows the weekly allowances balances for GenCos, which is the difference between what is allotted and what is used.

Figure 8 shows the average weekly profits for GenCo2 from the 100 runs for the three strategies. The figure shows that GenCo2 earns more profit when it is able to dynamically trade its allowances in the emission market. The total yearly profit in this case for GenCo1 is \$17.40 M while GenCo2's profit is \$18.05 M, which are

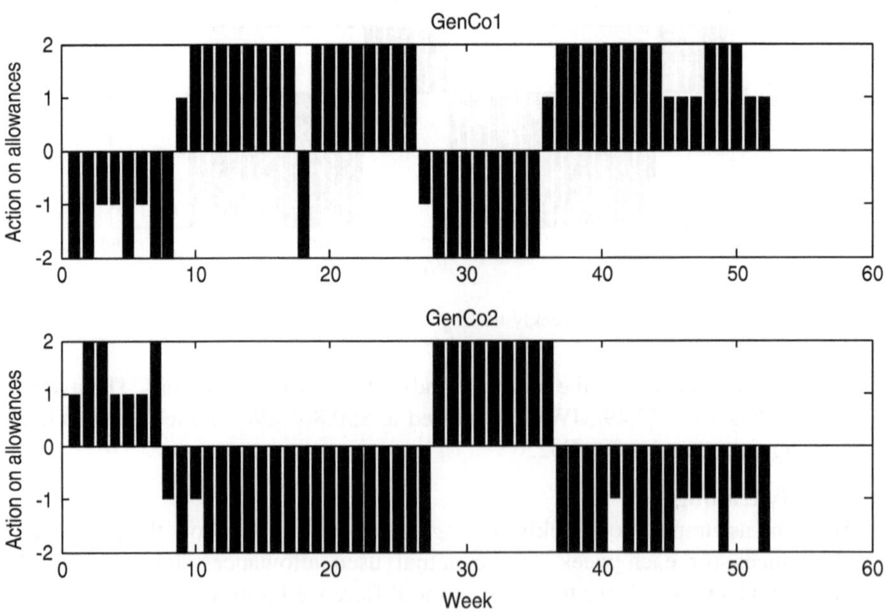

Fig. 6 Actions on allowances in dynamic trading

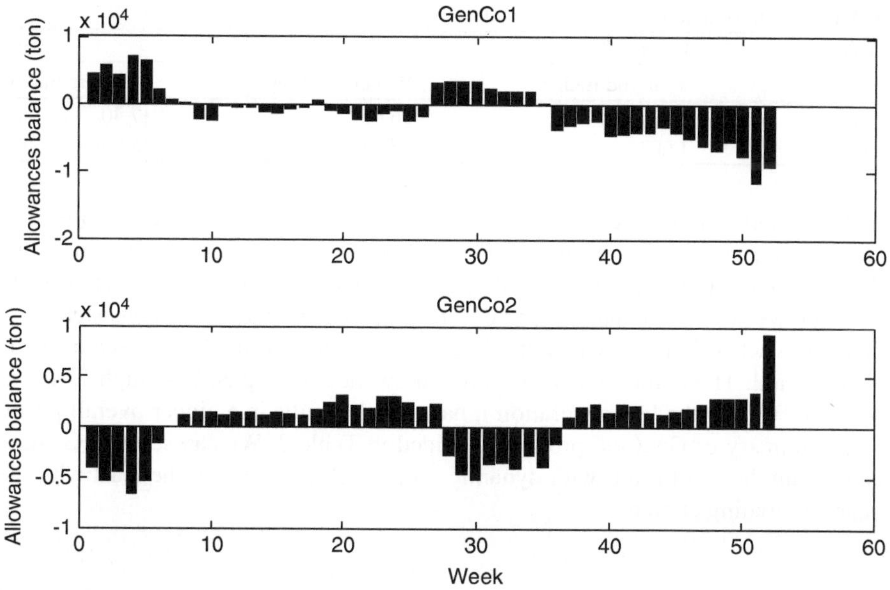

Fig. 7 Weekly allowances balances for GenCos

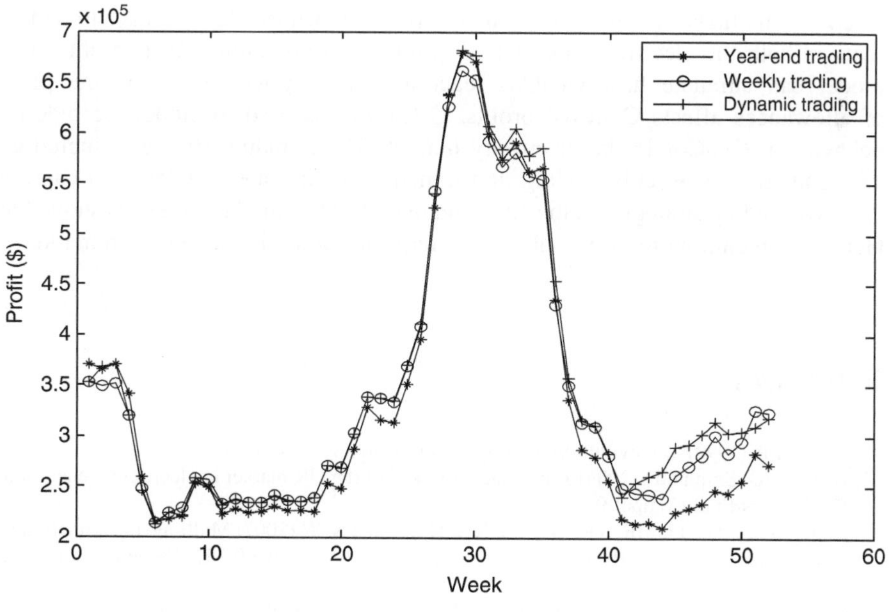

Fig. 8 Average weekly profit for GenCo2

Table 2 Profit summary

GenCo	Profit (M$)		
	Year-end trading	Weekly trading	Dynamic trading
GenCo1	16.34	16.63	17.40
GenCo2	17.03	17.69	18.05

higher than the above two strategies. As can be seen in Fig. 8, the weekly profits of GenCo2 are similar for the period between week 1 and week 40. However, the dynamic trading outperforms the other two alternatives in terms of higher weekly profit between week 40 and the end of the year. This observation is consistent with GenCo2's action in Fig. 6 where it sells most of its additional allowances during the same period. The extra income from the allowance trading and the high LMPs at bus 2 overweigh the less generation it has, which results in a higher overall profit.

A summary of GenCos' profit is provided in Table 2. We can see that GenCos can obtain the most profit with dynamic trading while they earn the least with the year-end trading strategy.

5 Conclusion

This book chapter proposes a simulation framework for modeling cap and trade in electricity markets. An agent-based model is developed to allow GenCos to participate in both the electricity and emission markets to maximize their profits. The model can simulate how GenCos react to market prices and how the price of allowances affects GenCos' profits. Q-learning is used to model the bidding behavior of GenCos in the electricity market. Three trading strategies including year-end trading, weekly trading and dynamic trading are simulated to capture GenCos' trading strategies in the emission market. The simulation results show that GenCos can earn more profit with more frequent trading in the emission market.

References

1. Axelrod R (1990) The evolution of cooperation. Penguin, New York
2. Bagnall AJ, Smith GD (2005) A multiagent model of the UK market in electricity generation. IEEE Trans Evol Comput 9:522–536
3. Botterud A, Mahalik MR, Veselka TD, Ryu H-S, Sohn K-W (2007) Multi-agent simulation of generation expansion in electricity markets. In: Proceedings of the IEEE PES general meeting, Tampa, FL, USA, pp 1–8
4. Bunn DW, Oliveira S (2001) Agent-based simulation: an application to the new electricity trading arrangements of England and Wales. IEEE Trans Evol Comput 5:493–503
5. Chen Y, Hobbs BF (2005) An oligopolistic power market model with tradable NO_X permits. IEEE Trans Power Syst 20(1):119–29

6. Chen Y, Hobbs BF, Leyffer S, Munson T (2006) Solution of large-scale leader-follower market equilibria problems: electric power and NO_X permit markets. Comput Manag Sci 3(4):307–330
7. Erev I, Roth AE (1998) Predicting how people play games with unique, mixed-strategy equilibria. Am Econ Rev 88:848–881
8. Guo M, Liu Y, Malec J (2004) A new Q-learning algorithm based on the metropolis criterion. IEEE Trans Syst Man Cybern B Cybern 34(5):2140–2143
9. Hobbs BF, Metzler C, Pang JS (2000) Strategic gaming analysis for electric power networks: an MPEC approach. IEEE Trans Power Syst 15(2):638–645
10. Kahn E (1998) Numerical techniques for analyzing market power in electricity. Electr J 11(6):34–43
11. Kleindorfer P, Li L (2009) Portfolio risk management and carbon emissions valuation in electric power. In: The 28th annual eastern conference, Skytop
12. Krause T, Andersson G, Ernst D, Beck EV, Cherkaoui R, Germond A (2004) Nash equilibria and reinforcement learning for active decision maker modeling in power markets. In: Proceedings 6th IAEE conference – modeling in energy economics, Zurich
13. Nicolaisen J, Petrov V, Tesfatsion L (2001) Market power and efficiency in a computational electricity market with discriminatory double-auction pricing. IEEE Trans Evol Comput 5:504–523
14. Roth AE, Erev I (1995) Learning in extensive form games: experimental data and simple dynamic models in the intermediate term. Games Econ Behav 8:164–212
15. Smeers Y (1997) Computable equilibrium models and the restructuring of the European electricity and gas markets. Energy J 18(4):1–31
16. Tellidou AC, Bakirtzis AG (2007) Agent-based analysis of capacity withholding and tacit collusion in electricity markets. IEEE Trans Power Syst 22(4):1735–1742
17. Wang J (2009) Conjectural variation-based bidding strategies with Q-learning in electricity markets. In: The 42nd annual Hawaii international conference on system sciences (HICSS 2009), Big Island, Hawaii, USA, pp 1–10
18. Wang J, Botterud A, Conzelmann G, Koritarov V (2008a) Multi-agent system for short- and long-term power market simulations. In: The 16th power system computation conference (PSCC), 14–18 July 2008, pp 1–7
19. Wang J, Botterud A, Conzelmann G, Koritarov V (2008b) Market power analysis in the EEX electricity market: an agent-based simulation approach. In: IEEE PES general meetings, 20–24 July 2008, pp 1–8
20. Wang J, Koritarov V, Kim J (2009) An agent-based approach to modeling interactions between emission market and electricity market. In: IEEE PES general meetings, 26–29 July 2009, pp 1–8
21. Watkins CJCH, Dayan P (1992) Technical note: Q-learning. Mach Learn 8:279–292
22. Weidlich A, Veit D (2008) Analyzing interrelated markets in the electricity sector – the case of wholesale power trading in Germany. In: IEEE PES general meetings, Pittsburgh, PA, USA, pp 1–8
23. Yu N, Liu C, Tesfatsion L (2007) Modeling of suppliers' learning behaviors in an electricity market environment. In: International conference on intelligent systems applications to power systems, Taiwan, China, pp 1–6

A Survey of Carbon Market Mechanisms and Models

Vishnuteja Nanduri and Narges Kazemzadeh

Abstract Climate change is one of the most important challenges of this century. Human activity, particularly that of the electric power sector, has played a dramatic role in exacerbating this scenario. This paper presents a brief survey of some of the policies that are being adopted around the world to tackle the challenge of emissions reductions, with specific focus on the power sector. We discuss some widely used policies such as renewable portfolio standards, feed-in tariffs, carbon taxes, and cap-and-trade, with more emphasis on the latter. We conclude the chapter with a detailed discussion of some of the simulation and mathematical programming models that have been developed to examine economic and environmental impacts of emissions control policies on electricity markets.

Keywords Cap-and-trade • Renewable portfolio standards • Carbon taxes • Banking • Borrowing • Trading • Climate change • Feed-in tariffs • Simulation models • Optimization models • Environmental protection agency • Regional greenhouse gas initiative • Leakage • Offsets

Abbreviations

RGGI	Regional Greenhouse Gas Initiative
EUETS	European Union Emissions Trading Scheme
CRPS	Carbon Pollution Reduction Scheme
GHGs	Greenhouse Gases
MGGA	Midwestern Greenhouse Gas Reduction Accord
WCI	Western Climate Initiative

V. Nanduri (✉) • N. Kazemzadeh
Department of Industrial and Manufacturing Engineering, University of Wisconsin-Milwaukee, 3200 N. Cramer Street, Milwaukee, WI 53211, USA
e-mail: vnanduri@uwm.edu; kazemzad@uwm.edu

Q.P. Zheng et al. (eds.), *Handbook of CO_2 in Power Systems*, Energy Systems, DOI 10.1007/978-3-642-27431-2_6, © Springer-Verlag Berlin Heidelberg 2012

CCE	Chicago Climate Exchange
RPS	Renewable Portfolio Standards
RECs	Renewable Energy Credits
FIT	Feed-in Tariffs
ADAGE	Applied Dynamic Analysis of the Global Economy
IPM	Integrated Planning Model
IGEM	Inter-temporal General Equilibrium Model

1 Climate Change and Its Impact

Climate change refers to long term alterations in the earth's atmosphere caused by both natural events as well as human activities. Natural events such as volcanic eruptions, changes in the earth's orbit, and the amount of energy released from the sun have long influenced our climate. In addition, in the past few decades, human activities, such as burning fossil fuels, have been releasing immense amounts of greenhouse gases into the atmosphere, warming the earth, and thereby leading to climate change [1]. Climate change is a global problem that is impacting life on earth and adversely affecting the health of the people and the environment. Global warming is also causing shifts in seasons, increasing temperatures, and rising sea levels. It is also said to be increasing the risk of drought, fire and floods, and changing natural landscapes. Therefore, mitigating the catastrophic impacts of climate change is one of the most important challenges of the twenty-first century.

Electric power industry is the largest contributor of greenhouse gases, exacerbating the climate change problem. In this paper, we will focus our attention on the electric power industry and the policies to reduce greenhouse gas emissions. Based on the 2010 Annual Energy Outlook [2], electric power and transportation sectors together account for close to three quarters of all the carbon dioxide emissions in the U.S. Projections developed by the Energy Information Administration over the next three decades do not show any major variations in the percentage shares of overall carbon dioxide emissions from electricity and transportation sectors (see Fig. 1).

2 Emissions Control Policies

A variety of policies are being proposed and some are already being used to reduce greenhouse gas emissions responsible for climate change. Some of these policies include cap-and-trade programs, renewable portfolio standards, feed-in tariffs, green certificates, and carbon emissions taxes. In this paper, we will briefly discuss each of these policies.

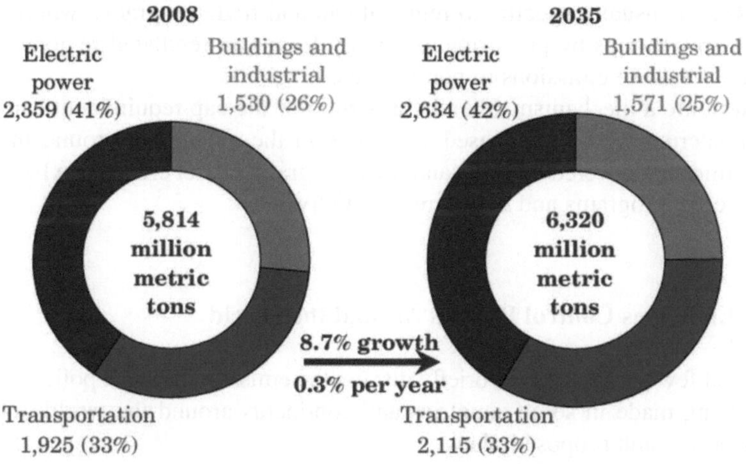

Fig. 1 U.S. energy-related carbon dioxide emissions (Source: Annual Energy Outlook [2])

2.1 Cap-and-Trade

Under a cap-and-trade program for carbon dioxide emissions, policymakers set a limit on the total amount of CO_2 that could be emitted in a given period and issue allowances according to that level of emissions. Entities that are subject to the cap (e.g., power generators) are then required to surrender allowances for their CO_2 emissions. Allowances can be allocated by means of an auction, given away for free (grandfathering), or by using a hybrid mechanism where a percentage are given for free and the rest are auctioned. Each ton of CO_2 is equivalent to one allowance (or permit). Entities that have additional allowances after meeting their requirement, can sell them to those entities in need. Harsh penalties are levied on entities that do no abide by the cap restrictions. Cap-and-trade has been widely discussed as the most effective market-based mechanism to control emissions and mitigate climate change. In fact, the U.S. Environmental Protection Agency had developed a very successful cap-and-trade program in the early 90s for SO_x emissions under the aegis of the Acid Rain Program. The program is said to have achieved its goals within cost and ahead of time [3]. Cap-and-trade programs have a variety of mechanisms to manage the cost of compliance for covered entities such as offsets, banking, and borrowing, which we briefly discuss next. For a comprehensive review of all these parameters, the readers are referred to [3].

Banking is the ability of entities to save unused allowances for future use.

Borrowing is the mechanism using which entities can obtain allowances from a future period (vintage) to meet the current requirements.

Trading is the ability of the compliance entities to sell/buy allowances in/from open markets or from each other.

Reserve prices and price caps refer to the floor and ceiling limits of the auction prices. These are predetermined by policy makers.

Leakage is usually specific to regional cap-and-trade programs, where entities can circumvent caps by procuring electricity from non-regulated regions, thereby causing a 'leak' of emissions to non-regulated regions.

Offsets, are a mechanism whereby a portion of the cap requirement can be met by surrendering credits purchased elsewhere in the country or around the world from entities not covered by a cap and trade program. Offset credits can be obtained from forestry programs and other green initiatives.

2.1.1 Emissions Control Policies Around the World

In the next few paragraphs we briefly discuss the emissions control policy advancements being made in some countries and continents around the world, including some working and proposed plans.

Europe: To meet the ambitious goals set forth by the Kyoto protocol, European Union launched the Emissions Trading Scheme in 2005. European Union Emissions Trading Scheme (EU ETS) is the largest working emissions control program in the world with 30 participating countries. Each country in the EU for the first two phases of the program predominantly gave away allowances for free. This was the subject of numerous criticisms of the program, which resulted in windfall profits to power generators and also increase in costs to consumers. The current goals of the program are to auction 100% of the allowances starting the third phase from 2013 and to reduce the cap by 1.74% each year starting 2013 [4]. The EUETS has served as a good starting point for cap-and-trade program formulations around the world.

Australia: As of June 15th 2010 (last update of the government website related to climate change), the Carbon Pollution Reduction Scheme (CPRS), a cap-and-trade program for Australia set to start in July 2011, did not find bipartisan support and has been postponed. According to [5], the CPRS will impose caps on emissions from multiple sectors including both transportation and electric utilities. The program is expected to have allowance sales and initial free allocation to heavily affected industries like coal plants.

New Zealand: The emissions trading scheme (ETS) which covers all sectors of the economy began in New Zealand in 2008. The government set a time line for entry of different economic sectors into the scheme and intends to not provide any free allocation of allowances to electric power generators. Power generators are expected to buy allowances (New Zealand Units, NZUs) from the market as part of the ETS [6].

Brazil: Brazil passed the National Climate Change Plan (PNMC) to reduce GHG emissions in 2008. Some of the key features of the policy are [7]:

- Increasing energy efficiency in order to decrease electricity consumption,
- Providing a high proportion of electricity supply from renewable sources such as wind, biofuels, and solar, and
- Decreasing deforestation rates.

Russia: The Russian government is in a unique position of having satisfied the emissions reductions obligations outlined in the Kyoto protocol. This, according to [7] has created a surplus of permits that Russia can utilize in the future. The government, following a 2007 Presidential decree is also said to be taking measures to decrease energy intensity of the economy and increase green generation.

India: As one of the largest developing countries in the world, India was pressured during the Copenhagen summit to adopt binding standards of GHGs, which it refused. However, the government has pledged a dramatic increase in renewable energy generation with a goal of reaching 20% by 2020. According to the national tariff policy laid out in 2006, state utilities must purchase a percentage of electricity from green generators. According to the Ministry of New and Renewable Energy [8], India is also poised to spend close to $19 billion on a national solar mission aiming to increase output to 20 GW by 2020.

China: While China has not stated a national target for GHG reductions, it has pledged to cut emissions by enhancing energy efficiency and adopting market based emissions control mechanisms. According to a report by the Information Office of the State Council of the People's Republic of China [9], the government is accelerating the phase out of old/inefficient power plants, initiated feed-in tariff policies, and increased the share of green power in the generation mix [7, 10].

Japan: The Japanese government launched the Kyoto Protocol Target Achievement Plan (KPTAP) in order to deal with the climate change challenges. Japan's current domestic GHG emissions policy is driven by a voluntary emissions trading scheme. The current government also has plans to create a mandatory cap-and-trade program, feed-in tariffs, and also a command and control mechanism for emissions reductions [7].

2.1.2 Federal-Level Policies in the U.S.

A plethora of policies have been languishing in the U.S. Congress over the last few years due to lack of bipartisan support for formulating an economy wide climate policy program. Some of the bills were introduced by Senators Lieberman, Sanders, Feinstein, Kerry, Bingaman, and Representatives Olver and Waxman. A detailed comparison of these bills was presented in [11]. In Table 1, we provide a concise comparison of two of the more recent bills: the Clean Energy and Secuirty Act and the Clean Energy Jobs and American Power Act. Our intention is to provide the readers with a flavor of the type of emissions reduction goals and other critical parameters of these programs. The table is a summary developed based on the detailed comparative analysis by the Congressional Research Service [12].

2.1.3 Regional Cap-and-Trade Policies

Disappointed with the lackluster pace of federal level policies, states (and some cities) have taken matters into their own hands and developed regional initiatives.

Table 1 Comparison of two of the recent federal initiatives, summarized from [12]

Program feature	Clean Energy and Security Act (CESA)	Clean Energy Jobs and American Power Act (APA)
Economy wide GHG Reduction goals	2012 U.S. GHG < 97% of 2005 GHGs 2020 U.S. GHG < 80% of 2005 GHGs 2030 U.S. GHG < 58% of 2005 GHGs 2030 U.S. GHG < 17% of 2005 GHGs	2012 U.S. GHG < 97% of 2005 GHGs 2020 U.S. GHG < 80% of 2005 GHGs 2030 U.S. GHG < 58% of 2005 GHGs 2030 U.S. GHG < 17% of 2005 GHGs
Banking	Unlimited	Unlimited
Borrowing	No interest for immediate year, 15% interest for those allowances upto 5 years into the future	No interest for immediate year, 15% interest for those allowances upto 5 years into the future
Strategic reserve	EPA sets aside pool each year and conducts quarterly auctions	EPA sets aside pool each year and conducts quarterly auctions (pool much larger)
Auction of allowances	Quarterly with reserve price of $10/allowance	Quarterly with reserve price of $11/allowance
Non compliance penalty	Excess emissions multiplied by twice the auction clearing price of most recent auction.	Penalty equals twice the fair market price from registered exchanges
Offsets	By 2050, 66% can be met with offsets half from domestic and from international sources	By 2050, 48% can be met with offsets, but 75% domestic and 25% international sources

Currently, there exist three mandatory regional cap-and-trade initiatives within the United States and a voluntary program. If all these programs become active, over 50% of the U.S. population is expected to be in states with emissions control programs. The three mandatory programs are: (1) Regional Greenhouse Gas Initiative (RGGI, pronounced Reggie), (2) Midwestern Greenhouse Gas Reduction Accord (MGGA), and (3) Western Climate Initiative; and the voluntary program is Chicago Climate Exchange. Figure 2 depicts the regional programs. Out of these three regional programs, RGGI has been fully operational since 2009.

Regional Greenhouse Gas Initiative (RGGI): The Regional Greenhouse Gas Initiative is a mandatory U.S. cap-and-trade program for carbon dioxide. Currently, the following states are participating in RGGI: Connecticut, Delaware, Maine, Maryland, Massachusetts, New Hampshire, New Jersey, New York, Rhode Island, and Vermont [13]. RGGI has rather modest goals of CO_2 reductions, that aim to reduce emissions 2.5% each year starting 2015, after a period of emissions stabilization. A unique feature of the RGGI is that all the participating states have decided to auction almost 100% of their allowances. RGGI participants can buy allowances from the quarterly allowance auctions or via secondary markets such as the Chicago Climate Futures Exchange. A significant percentage of the auction revenue is expected to be recycled back into the economy by investments in clean energy technologies to benefit consumers who are expected to experience increased

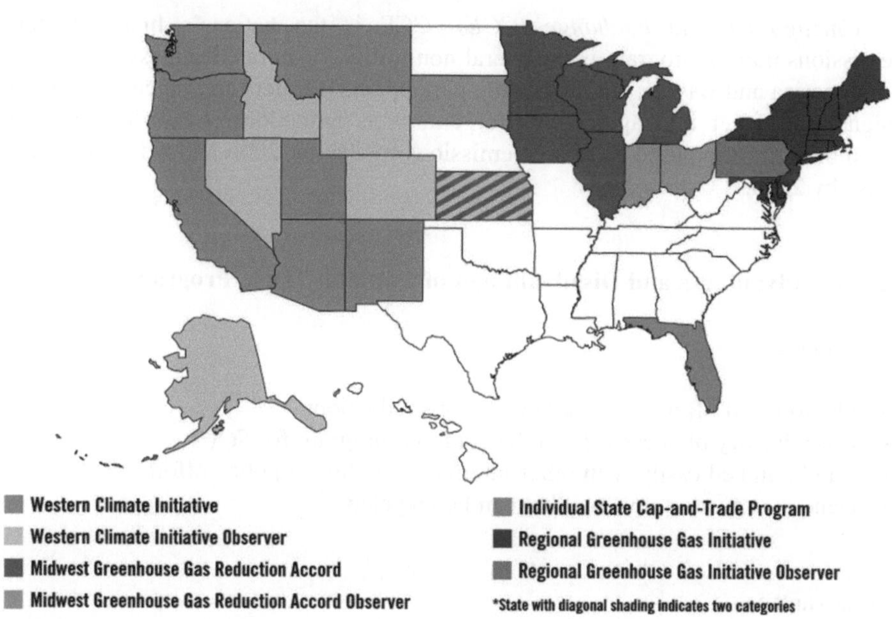

Fig. 2 Proposed and working cap-and-trade programs in the U.S. (Source: Pew Center on Global Climate Change)

prices. RGGI auctions have raised well over half a billion dollars during the last 2 years of operation. In fact, the State of New York borrowed over $90 Million from RGGI auction revenues to meet some of its gaping budget deficits [14].

Midwestern Greenhouse Gas Reduction Accord (MGGA): The MGGA was signed by Governors of six Midwestern states (Illinois, Iowa, Kansas, Michigan, Minnesota, and Wisconsin) and a Canadian province (Manitoba), in Milwaukee in 2007. While the MGGA has not yet started, and does not appear to be heading toward a launch any time soon, it is important for some of the following reasons noted by the World Resources Institute [15]. The MGGA would be the only regional cap-and-trade program including more than one of the top ten GHG emitters in the nation. Furthermore, over 60% of the electricity in the region is obtained from coal [15], which makes an emissions policy in the region a critical necessity. The program had goals to reduce CO_2 emissions by 15–20% by 2020.

Western Climate Initiative (WCI): WCI [16] is a collaborative effort of seven states in the U.S. and four Canadian provinces. They are California, Arizona, New Mexico, Utah, Montana, Washington, Oregon, Manitoba, British Columbia, Ontario, and Quebec. There are several other observer states and provinces. The WCI program is broad in scope covering almost 90% of all the emissions released in the participating regions. A sealed bid uniform price auction, as in RGGI, is expected to be used for auction allocation. As in RGGI, allowances will be sold in lot sizes of a thousand. The program expects to reduce CO_2 emissions 15% below 2005 levels by the year 2020.

Chicago Climate Exchange (CCE): CCE is the nations's first voluntary emissions trading program with several non-utility, corporate business (e.g., Bank of America and Bayer), and individual participants (farmers and ranchers). All the signatories enter into agreements for emissions reductions. According to [17], members were required to reduce emissions by 4% by 2006 and 6% below base case by 2010.

2.1.4 Advantages and Disadvantages of Cap-and-Trade Programs

Advantages

- Flexibility it offers to achieve emissions reductions
- Has a history of success from the acid rain program for SO_X
- Can be linked easily with other international climate policy efforts
- Generates auction revenue that can be recycled

Disadvantages

- Potential for price volatility of allowances
- Administrative burden in creating the trading and tracking platforms
- Potential for speculation by traders

2.2 *Renewable Portfolio Standards*

Renewable Portfolio Standard (RPS) is a policy where utilities are required to produce a percentage of electricity using renewable energy sources by a certain date. Almost 30 states in the U.S. have active RPS policies in place, with a wide array of target renewable percentages, from 33% by 2020 in California, to 20% by 2020 in Kansas. The Database of State Incentives for Renewables & Efficiency (DSIRE) funded by the U.S. Department of Energy hosts comprehensive charts and tables of RPS policies [18]. The following map from the Environmental Protection agency shows the states in the U.S. with renewable portfolio standards. The following technologies are currently eligible under RPS requirements in different states in the U.S.: biofuels, biomass, CHP/waste heat, energy efficiency, fuel cells, geothermal, hydro, landfill gas, municipal waste, photovoltaics, solar thermal, tidal, wave, waste tire, and wind.

RPS can work as follows [19]. RPS could use renewable energy credits (RECs) for compliance. RECs are credits awarded for each MW of renewable energy produced. Power generators obtain RECs for every unit of green energy they generate. Retailers or utilities will need to demonstrate evidence of certain number of RECs for compliance. A regulatory body keeps track of RECs and ensures that

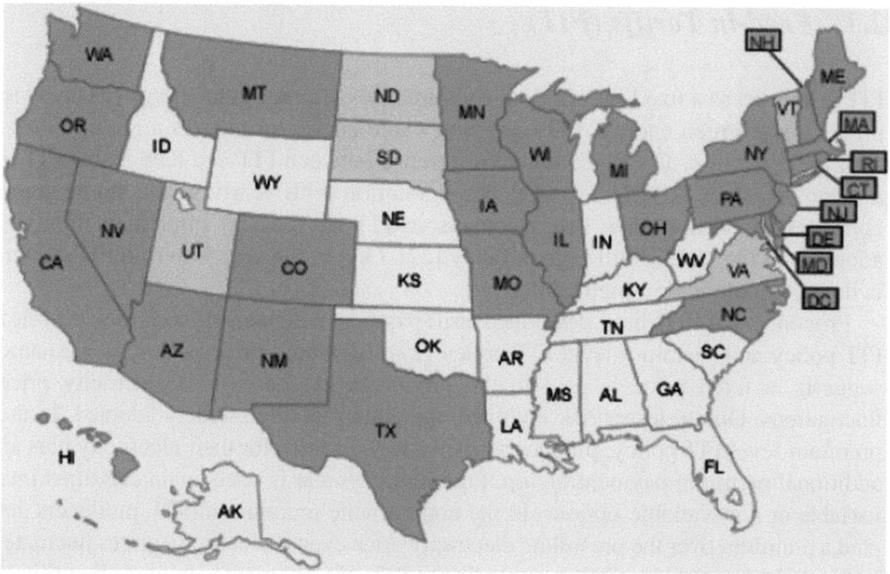

Fig. 3 States with RPS (*blue*) and RPS goals (*Yellow*) (Source: Environmental Protection Agency website on RPS)

transfer and records are maintained. RECs are flexible since they can be purchased either directly from a green generator or from brokers who hold RECs. One of the primary goals of RPS is that green generators receive additional income and therefore will be able to compete on a level playing field with cheaper and dirtier generators. Since it is designed as a market driven program, supply and demand forces are expected to drive the participants to buy/sell RECs as needed to meet the portfolio standards outlined by the state (Fig. 3).

2.2.1 Advantages and Disadvantages of Renewable Portfolio Standards [20]

Advantages

- Direct impact on amount of renewable generation
- Provide some flexibility with RECs

Disadvantages

- Focus on technologies rather than emissions
- Lack of strong enforcement mechanisms may limit its potential

2.3 Feed-in Tariffs (FIT)

FIT is defined as a fixed set of prices for purchases of renewable energy [21] paid to producers of green energy. FITs pay renewable energy producers a premium rate over the retail rate. The fundamental difference between FIT and RPS is that FIT is a mandate, while RPS is a market based solution with yearly goals. While some countries in Europe have been pioneers in FIT design and implementation, its adoption in the U.S. is still in its infancy [22]. One of the core issues in FIT design is the tariff calculation methodology.

Proponents of FITs have developed tariff payment mechanisms such as fixed-price FIT policy and premium-level FIT policy [22]. In a fixed-price policy, as the name suggests, a fixed price is paid to the producers irrespective of electricity price fluctuations. Due to lesser risk involved, this policy is often widely adopted. In the premium-level FIT policy, producers receive market price for their electricity plus an additional premium payment on top. Premium payment is once again classified into variable or non-variable options. In the non-variable premium model, producers get paid a premium over the prevailing electricity price even if electricity prices fluctuate. In the variable premium option, according to [22], the payments are made by utilities to producers in two different streams (while guaranteeing a minimum), where one depends on the spot prices of electricity and the other covers the difference between the guaranteed total and the spot price. This mechanism is being reportedly used in the Netherlands. Cory et al. [22] also note that in some cases, FIT payments mechanisms can be formulated based on some project specific factors such as: type of technology (solar, wind, etc.), project size, quality of resource, and project location.

Mendonca et al. [21] have proposed a novel checklist for designing a FIT in a region. Some of the steps include:

- Correct choice of technologies based on resources that may be available in a country
- Proper design of tariff calculation
- Fixing a duration of payment to encourage long-term investment
- Force utilities to purchase all generated green energy
- Providing an open access for transmission

FIT policies are currently active in over 40 countries with Germany and Spain being some of the most advanced in this green energy policy.

2.3.1 Advantages and Disadvantages of Feed-in Tariffs

Advantages

- Make green technologies cost effective
- Clear market signal for investments
- Mechanism for reaching green targets

Disadvantages

- Setting the optimum tariff is a challenge (too high will result in windfall profits, too low will not spur investments)
- Increased price of electricity for consumers
- May be very expensive if designed inefficiently
- Does not reduce initial installation costs for developers

2.4 Carbon Emissions Taxes

Similar to a cap-and-trade program, carbon tax is a market-based policy instrument to achieve emissions reductions. However, as noted in [3], the fundamental difference lies in the way this reduction is achieved. In a cap-and-trade program, a cap is set on the emissions and price is set by the market in order to adjust to the constraints placed. On the other hand, taxes set a fixed price on emissions and forces the quantities of emissions to adjust accordingly. Stavins notes that unlike cap-and-trade programs, carbon taxes do not guarantee emissions reductions, since it is up to the individuals who tend to reduce emissions only until it costs less to pay the tax than to abate. Some researchers argue that due to the simplicity of adherence, carbon taxes could prove to be successful. Furthermore, just like in a cap-and-trade program, the revenue earned through taxes can be returned to consumers as dividends or reinvested in clean energy technologies. The use of carbon taxes is limited to the Scandinavian countries, Hongkong [23] and most recently being considered in Australia (drawing protests from many citizens). Recent news reports note that Australia intends to slowly transition to an emissions trading scheme from a carbon tax as early as 2015.

Metcalf and Weisbach [23] discuss the key design elements of a carbon tax for the U.S. They note that the main design parameters are: the tax rate, the tax base, and international trade concerns. The tax rate, recommended by economist Arthur Pigou for environmental externalities, must be equal to the social marginal damages from producing an additional unit of emissions. The suggested rates gleaned from surveys by the intergovernmental panel on climate change vary widely from $3 to $95 [23]. Tax base refers to the point of implementation of tax: upstream (on the producers of fossil fuels) or down stream (on users of fossil fuels). International trade concerns deal with the global perspective of emissions, since emissions have global impacts even if the emissions are local. The readers are referred to Metcalf and Weisbach's paper which presents an excellent introduction to the field of carbon taxing and its optimal design and the myriad operational issues.

Debate is ongoing about the better market-based alternative: a carbon tax or carbon trading. Proponents of carbon taxes note that excessive speculation may plague trading programs and that the clear price signal sent by taxing makes it a more stable market policy. On the other hand, proponents of cap-and-trade programs argue that there will be tremendous resistance to new taxes in any political climate, let alone in the current period of fragile economic recovery. Below, we note some of the advantages and disadvantages of carbon taxes discussed in [3, 24].

2.4.1 Advantages and Disadvantages of Carbon Taxes

Advantages

- Simplicity of taxes
- Creates revenues for recycling and redistribution
- No price volatilities
- Stability in revenue

Disadvantages

- Political resistance to taxes
- Not easy to develop international trading programs
- Inflexible in terms of ways to achieve reductions

3 Emissions Policy Models: A Quantitative Perspective

Quantitative research in the area of emissions policy modeling and analysis with respect to electricity markets has progressed along two lines. The first is large-scale simulation-based models [3, 5, 25–32] and the second is mathematical programming based (theoretical) models [33–38]. While there are numerous works that have modeled emissions policies, there still exist very significant opportunities to vertically advance this field. In particular, the large-scale simulation models usually do not consider competitive gaming aspects, stochasticity, and transmission constraints of wholesale electric power markets. The mathematical programming models make some fundamental advancements in the theory of emissions allocation and pricing for small-size (1–5 buses) test networks. However, it has not yet been proven if results from these small theoretical models can translate to large power networks. It is therefore important to vertically advance this field of research in two important ways (1) studying large-scale networks by including all the power network constraints, stochasticity, gaming, and learning behaviors of participants, and (2) harnessing the power of high performance computing to tackle problems of realistic size and complexity in a reasonable amount of time. In the next few paragraphs, we briefly discuss some of the contributions made by researchers in this field.

3.1 Simulation-Based Models

The National Energy Modeling System [26] is an economy-wide simulation model that is often used by the Energy Information Administration to develop forecasts based on energy policies. NEMS consists of several submodules including the electricity market module which is used in developing electricity price forecasts based on detailed econometric simulations. The electricity markets analysis model [39] produced by the Research Triangle Institute is a large-scale integer programming-based simulation model for assessing emissions policy impacts on prices and supply of electricity. The electricity markets analysis model [39] is formulated as a linear programming problem and aims to minimize the total generation cost of the entire U.S. while respecting constraints of transmission, capacity, fuel consumption, emissions, and retrofits. The model assumes to have knowledge of all elements including demand and capacity expansion decisions a priori. The developers note that the electricity markets analysis model has the capability to assess cap-and-trade policies and also forecast electricity prices by linking with the ADAGE [27] model, which is a macroeconomic general equilibrium model of the Environmental Protection Agency. The model, however, does not consider competition among generating companies in wholesale markets, or explicitly study capacity and transmission investment problems, which are complex, standalone research areas by themselves. Nevertheless, the Energy Information Administration of Department of Energy, often uses electricity markets analysis model to make economic forecasts. Other large-scale simulation models include ADAGE [27], IGEM [28], and IPM [30] from the Environmental Protection Agency.

The Haiku model, developed almost a decade ago by Resources for the Future [29], performs econometric simulations of the electricity markets of the entire U.S. Haiku model encompasses all the 13 North American Reliability Council (NERC) regions of the continental U.S. Haiku model, as noted by its developers, uses an iterative algorithm which works by continuously adjusting numerous control variables (e.g., capacity, fuel costs, allowance prices) and achieves an overall market equilibrium. Haiku, however, does not explicitly consider detailed electric power network transmission and security constraints or the gaming and learning behaviors of competing power generators. Palmer and Burtraw [25] perform several numerical simulations using the Haiku model to assess performance of renewable portfolio standards, renewable energy credits (RECs), and cap-and-trade programs with generation based allowance allocation. They note that RPS perform better than RECs and cap-and-trade under generation based allowance allocation. However, note that these are also econometric simulation models and do not consider the intricacies of electrical power networks.

3.2 Mathematical Programming Based Models

There have been some significant contributions on the mathematical programming side of the literature in this area and we discuss some relevant papers here. Chen and Wang [33] present a novel complementarity framework to jointly assess the effectiveness of CT programs, renewable portfolio standards (RPS), and green pricing. In particular, they develop joint optimization models for load serving entities and consumers. This is a significant departure from existing research where typically the generation side is modeled and load serving entities (and/or consumers) are assumed to have inverse demand functions. They demonstrate the methodology on a three bus network. Chen and Hobbs [34] presented one of the pioneering models that captured interactions between a market for NO_X permits and a market for electric power. In [34], the linear complementarity programming model of [40] is extended to incorporate the NO_X market. They demonstrate on a test case that obtaining greater share in the NO_X market has a significant influence over market power in the subsequent electricity market. An interesting feature of the paper is the endogenous representation of the dynamics between energy and NO_X markets.

Zhao et al. [36] develop a set of comprehensive nonlinear complementarity models to assess the impacts of various emissions allocation mechanisms. They assess three rules in particular: (1) capacity-based potential allocation, (2) capacity-based actual allocation, and (3) a sales-based output rule. Since solving non-linear complementarity problems analytically is extremely challenging they develop numerical solutions with different sets of problem parameters. Zhao et al. find that basing free allocation of allowances on sales, in the presence of capacity markets is better than allocation using capacity, if minimizing cost and investment distortions is the objective. The authors do not model banking, borrowing, and offsets of cap-and-trade programs.

In [41], Hobbs et al. compare the upstream versus downstream points of regulation in cap-and-trade programs. The authors develop general equilibrium models and argue that load-based programs, where responsibility of compliance rests with load serving entities, are at best as effective, if not lesser than source-based programs, where compliance responsibility rests with generators. They also note that load-based programs, due to incompatibilities with spot markets as well as with potential federal cap-and-trade programs, may pose some challenges in implementation. The authors also show with the help of their model that load-based programs, even with emissions attributes and credit based proposals, do not surpass source-based programs.

Papavisilou et al. [37] present a Cournot duopoly game in the context of a transmission constrained power network and assess the effectiveness of renewable portfolio standards and taxes. They examine three cases for equilibrium, an unconstrained case, a pollution constrained case, and a transmission constrained case. They find that regulators need to monitor RPS markets closely to ensure that non-polluting generators do not abuse market power.

A bi-level game-theoretic model that links the cap-and-trade allowance market and the electricity market is proposed in [42]. The number of allowances obtained from the cap-and-trade auction affect the bidding strategies and profits of generators in the wholesale electricity markets significantly. Nanduri [42] casts this sequential and competitive gaming process as a bi-level game-theoretic problem. The author demonstrated an application of the approach using data from the Northern Illinois electricity market under the assumption of an RGGI type cap-and-trade program. As expected, it was observed that as the demand for electricity increases and number of allowances become scarcer, the prices of both allowances and electricity show an upward trend.

It can be seen that large scale simulation models of emissions policy often cover a whole range of issues, albeit in a more general fashion, while comprehensive mathematical programming models tend to focus on a subset of emissions policy details to obtain long term and deeper insight. For example, the large scale simulation based models such as Haiku consider a whole range of issues such as capacity expansion, impact of emissions policies, and regulatory regimes in the whole country, while researchers such as Zhao et al., develop comprehensive analytical models for allowance allocation mechanisms and their impact on prices. Similarly, Hobbs et al. [41] focus on upstream versus downstream points of regulation of emissions control policies. Nevertheless, both schools of modeling have played a predominant role in moving the science of emissions policy design forward.

4 Concluding Remarks

This paper discussed some of the carbon market mechanisms that are currently being utilized across the world and also discussed some of the methodologies being used to model and understand them in greater depth. In particular the paper discussed issues such as cap-and-trade programs, renewable portfolio standards, feed-in tariffs, and emissions taxes. Brief discussions of bills in the U.S. congress and some of the regional climate initiatives were also presented. The authors would like to note that this by no means is a comprehensive survey of models available in emissions policy modeling literature. There are a multitude of papers in the literature that deal with modeling various other facets of emissions policies using different approaches. It was our goal to bring to the readers a few of those papers and the references therein.

Given the current state of political uncertainty, it seems unlikely that a comprehensive climate change policy will be adopted by the U.S. However, inaction will only worsen the climate change problem. It may be the case that states will begin to embrace regional initiatives (e.g., RGGI) or may adopt a combination of the policies discussed in this paper. We do believe that pricing carbon will eventually become a necessity to provide some market signals for encouraging investments in green technologies in the near future. The world will be better off with the U.S. taking a leadership role in this fundamental issue of this generation. While, the emissions may be local, the challenge is, indeed, global in scope. Therefore, the largest emerging economies (China, India, Brazil, etc.) would also need to make binding emissions control commitments if this global challenge is to be tackled effectively. We believe that due to the important nature of this area, it must receive much more critical examination by researchers all over the world.

References

1. Climate Change (2011) http://www.epa.gov/climatechange/basicinfo.html. Accessed 5 Feb 2011
2. Annual Energy Outlook 2010 (2010) US Energy Information administration
3. Stavins RN (2007) A U.S. cap-and-trade system to address global climate change. The Hamilton Project, Discussion paper 2007-13, The Brookings Institution, Washington
4. European union emissions trading scheme website (2011) http://ec.europa.eu/clima/policies/ets/index_en.htm. Accessed 5 Feb 2011
5. Betz R, Owen AD (2010) The implications of Australia's carbon pollution reduction scheme for its national electricity market. Energy Policy 38(9):4966–4977
6. Ministry for the Environment and The Treasury (2007) The framework for a New Zealand emissions trading scheme. Ministry for the Environment and The Treasury, New Zealand. www.mfe.govt.nz
7. Overview of greenhouse gas control policies in various countries (2011) http://www.eoearth.org/article/Overview_of_Greenhouse_Gas_Control_Policies_in_Various_Countries. Last accessed 5 Feb 2011
8. Ministry of New and Renewable Energy, Government of India (2009) Government announces Jawaharlal Nehru national solar mission. Ministry of New and Renewable Energy, Government of India. http://mnre.gov.in
9. Information Office of the State Council of the Peoples Republic of China (2008) Chinas policies and actions for addressing climate change. Information Office of the State Council of the Peoples Republic of China. http://www.ccchina.gov.cn
10. Yen-Chiang chang, Nannan Wang (2010) Environmental regulations and emissions trading in China. Energy Policy 38(7):3356–3364
11. Parker L, Yacobucci BD (2007) CRS report for Congress, greenhouse gas reduction: Cap-and-trade bills in the 110th Congress. Congressional research service, report prepared for members and committees of Congress
12. Yacobucci BD, Ramseur JL, Parker L (2009) Climate change: comparison of the cap-and-trade provisions in h.r. 2454 and s. 1733. Congressional research service report prepared for members and committees of Congress
13. RGGI. Regional Greenhouse Gas Initiative auction results (2010). http://rggi.org/co2-auctions/market_monitor
14. http://www.npr.org/2010/12/22/132265866/states-dip-into-climate-funds-to-shore-up-budgets 2010
15. World Resources Institute. www.wri.org
16. Design for the wci regional program (2010) Western climate initiative. http://www.western-climateinitiative.org
17. https://www.theice.com/ccx.jhtml, 2011
18. Database of state incentives for renewables and efficiency. http://dsireusa.org/rpsdata/index.cfm, 2011.
19. Clemmer S, Nogee A, Brower MC (1999) A powerful opportunity, making renewable energy the standard, appendix b: the renewables portfolio standard, using markets to promote clean power. Union of Concerned Scientists

20. Delmas MA, Montes-Sancho MJ (2011) U.S. state policies for renewable energy: context and effectiveness. Energy Policy 38(5):2273–2288
21. Mendonca M, Jacobs D, Sovacool B (2009) Powering the green economy, the feed-in tarifi handbook. Earthscan, Sterling
22. Cory K, Couture T, Kreycik C (2009) Feed-in tarifi policy: design, implementation, and RPS policy interactions. National Renewable Energy Laboratory, Technical report NREL/TP-6A2-45549
23. Metcalf GE, Weisbach D (2009) The design of a carbon tax. Harvard Environ Law Rev 33(2): 499506
24. Wittneben BBF (2009) Exxon is right: let us re-examine our choice for a cap-and-trade system over a carbon tax. Energy Policy 37(6):2462–2464
25. Palmer K, Burtraw D (2005) Cost-effectiveness of renewable electricity policies. Energy Econ 27(6):873–894
26. NEMS (2009) http://www.eia.doe.gov/oiaf/aeo/overview/electricity.html
27. Ross MT (2008) Applied dynamic analysis of the global economy (ADAGE). Working paper 08 01. Technical report, Research Triangle Institute
28. Goettle RJ, Ho MS, Jorgenson DW, Slesnick DT, Wilcoxen PJ (2007) IGEM, an inter-temporal general equilibrium model of the u.s. economy with emphasis on growth, energy and the environment. Technical report, U.S. Environmental Protection Agency (EPA), Office of Atmospheric Programs, Climate Change Division
29. Paul A, Burtraw D (2002) The RFF haiku electricity market model. Technical report, Resources for the future, 2002
30. Environmental Protection Agency (2010) Integrated Planning Model. http://www.epa.gov/airmarkt/progsregs/epa-ipm/index.html. Accessed 6 Oct 2010
31. Burtraw D, Goeree J, Holt C, Palmer K, Shobe W (2007) Auction design for selling $co2$ emission allowances under the Regional Greenhouse Gas Initiative. Final report produced for RGGI
32. Ellerman D, Buchner B (2007) The European union emissions trading scheme: origins, allocation, and early results. Rev Environ Econ Policy 1:66–87
33. Chen Y, Wang L (2008) A power market model with renewable portfolio standards, green pricing, and GHG emissions trading programs. In: IEEE Energy 2030 conference, Atlanta, 17–18 Nov 2008
34. Chen Y, Hobbs BF (2005) An oligopolistic power market model with tradable NOx permits. IEEE Trans Power Syst 20(1):119129
35. Sijm JPM, Hers SJ, Lise W, Wetzealer BJHW (2008) The impact of the EU ETS on electricity prices final report to DG environment of the European commission. Technical report, Energy Research Center of Netherlands
36. Zhao J, Hobbs BF, Pang JS (2010) Long-run equilibrium modeling of emissions allowance allocation systems in electric power markets. Oper Res 58(3):529–548
37. Papavasiliou A, Chen Y, Oren S (2009) Environmental regulation in transmission- constrained electricity markets. In: IEEE conference proceedings, Calgary, 26–30 July 2009
38. Sijm J, Neuhofi K, Chen Y (2006) CO_2 cost pass through and windfall profits in the power sector. Cambridge working Papers in Economics 0639, Faculty of Economics, University of Cambridge, May 2006
39. Ross MT (2008) Documentation of the electricity markets analysis (EMA) model. Technical report, Research Triangle Institute Working Paper 08_01
40. Hobbs BF (2001) Linear complementarity models of Nash-Cournot competition in bilateral and Poolco power markets. IEEE Trans Power Syst 16(2):194–202
41. Hobbs BF, Bushnell J, Wolak FA (2010) Upstream vs. downstream co2 trading: a comparison for the electricity context. Energy Policy 38(7):3632–3643
42. Nanduri V (2010) Equilibrium bidding strategies in electricity and co_2 cap-and-trade markets. Working paper, Department of Industrial Engineering, University of Wisconsin-Milwaukee

Carbon Offset Markets: A Viable Instrument?

K. Kathy Dhanda and L. Hartman

Abstract In this article, we explore the realm of carbon offset markets that have been set up to enable consumers to offset their share of carbon emissions. Though the market is a relatively new one, it has quickly spawned countless offset providers under both regulated and voluntary schemes. Our research points out that the market is widely unregulated and, furthermore, there is no common quality or certification structure for the offset providers. At this point in the evolution of the market, only a handful of offset provider ratings schemes exist; and even these schemes leave a vast void for consumers.

The article begins with a discussion on the state of the carbon markets including the mandate from the Kyoto Protocol. Next, the article will examine the concept of carbon offsets and provide examples of the market. Our purpose is to present the logic and the working of carbon markets, especially in the space of market instruments aimed at lowering of carbon emissions. Moreover, we ask how this market developed and where is it headed. The article will outline the standards environment for offset providers in order to illustrate the need for a single set of criteria among providers. It will then explore the differences among the providers and articulate the specific criteria upon which providers may be evaluated. Finally, we share the results of the data collection and highlight early findings. These finding allow us to compare providers effectively and efficiently on a common scale that services both providers as well as consumer stakeholders.

Keywords Carbon neutral • Carbon offset market • Compulsory reductions • Voluntary offsets • Provider standards

K.K. Dhanda (✉) • L. Hartman
Department of Management, DePaul University, 1 E. Jackson Blvd., Ste. 7000, Chicago, IL 60604, USA
e-mail: kdhanda@depaul.edu; lhartman@depaul.edu

Q.P. Zheng et al. (eds.), *Handbook of CO_2 in Power Systems*, Energy Systems,
DOI 10.1007/978-3-642-27431-2_7, © Springer-Verlag Berlin Heidelberg 2012

1 Introduction: Carbon Neutrality

Companies and consumers are pursuing carbon neutral status increasingly, and at a surprisingly fast rate. The term "carbon neutral" was chosen as the word of the year by the *New Oxford American Dictionary* and thousands of new "green companies" have emerged in the last couple years [16]. Yet, beyond the clear social value component, how does this movement relate to company performance and the generation of market value? The movement is ostensibly concerned with lowering carbon waste on the earth. Carbon offsetting builds on this idea by adjusting operations in a company or the activities of individuals to achieve this goal. Yet, it is not clear how well the carbon neutral movement delivers what it promises [1].

There are five basic steps to achieving carbon neutrality. The first step is to assess a carbon footprint. There are numerous web-based calculators that let individuals and corporations estimate their footprint (Appendix A). The second step is to implement emissions reductions measures such as conservation and energy efficiency. Then, the third step is to compute the remaining carbon emissions. At the fourth step, offsets are purchased for the remainder amount in the previous step. Finally, the fifth step is to communicate carbon neutrality to the market. This step is mostly applicable to public companies or corporations [21].

The concept of being carbon neutral offers the participants an opportunity to take personal responsibility for the global warming implications of their lifestyles. Rather than merely discussing scenarios of climate change, it offers an opportunity to be part of a solution. In this context, environmental commodity markets and retail markets for voluntary carbon neutrality provide many solutions for participants. Both markets are still seeking broader public interest. For instance, there are no widely accepted standards of what qualifies as an offset. As an intangible commodity, it is difficult for an environmentally conscious consumer to make a distinction between a high and low quality offset. Next, we present examples of these gaps before presenting our data.

2 Background to the Market

What are the various options available to reduce emissions? Depending upon the regulatory history, these options can fall under the two categories: command and control and market based options.

Command and Control Strategies: In this category, the policies are either mandated by the government or a regulatory agency in form of taxes, subsidies, caps, or targets. In this case, the emissions could be taxed or a subsidy is provided to encourage adoption of clean technology. A carbon tax would be an example whereby emitters would be charged based on the net emissions of carbon.

Market-Based Solutions: The second category is market-based solutions that are primarily driven by industry or consumers and these include emission permits, carbon allowances, pollution offsets, etc. An example would be emission permits that are exchanged between companies in order to meet the quota of emissions. The market drives the exchange whereby a highly polluting company could purchase permits to cover its excess emissions and a clean company could sell the excess permits in the market. Some of the other approaches include allowances and offsets to cover the carbon usage of an individual or a corporation.

Research has shown that market based solutions might be the most efficient method for cutting emissions and achieving sustainability [7]. However, one of most common criticism towards market-based solutions is that these are unethical by nature. The argument goes that these approaches actually allow companies to pay for emissions and get away with their behavior rather than curtailing back and cutting emissions. This is a point of contention for most of the passionate proponents of the environmental movement.

3 Kyoto Protocol

Kyoto includes "flexible mechanisms" which allow Annex I economies to meet their greenhouse gas emission limitation by purchasing GHG emission reductions from elsewhere. These can be bought either from financial exchanges, from projects which reduce emissions in non-Annex I economies under the Clean Development Mechanism (CDM), from other Annex 1 countries under the JI, or from Annex I countries with excess allowances. Only CDM Executive Board-accredited Certified Emission Reductions (CER) can be bought and sold in this manner. Under the aegis of the UN, Kyoto established this Bonn-based Clean Development Mechanism Executive Board to assess and approve projects ("CDM Projects") in Non-Annex I economies prior to awarding CERs. (A similar scheme called "Joint Implementation" or "JI" applies in transitional economies mainly covering the former Soviet Union and Eastern Europe) [17]. http://unfccc.int/kyoto_protocol/mechanisms/items/1673.php

4 Carbon Market: CERS (Certified Emission Reductions) and Voluntary Carbon Offsets

The Kyoto Protocol offers CERs as part of its flexible mechanism. The Kyoto Protocol allows "flexible mechanisms" to Annex I countries to meet limitations by purchasing or buying greenhouse gas emission reductions elsewhere. One of these flexible mechanisms is the Clean Development Mechanism (CDM) wherein a Certified Emission Reduction (CER) can be traded by countries in order to meet GHG limitations. There CERS are board certified and have to follow a rigorous process of assessment and approval. These trades are highly bureaucratic and, as a result, the transaction costs tend to escalate.

The carbon market falls into two categories: the cap and trade system and the offset carbon market. The cap and trade system includes countries that have ratified the Kyoto Protocol. The companies receive credits that they can sell when they emit less than the set limit. In the offset space, the emissions reductions can be voluntary of compulsory in nature. The compulsory offsets would be those recognized by Kyoto Protocol and are labeled CERs. The voluntary carbon market is voluntary carbon offsets and the offsets sold are also called gourmet offsets. Clean Air-Cool Planet's (CA-CP)'s "A Consumer's Guide to Retail Carbon Offset Providers" provides a ranking of offsetting companies based on factors like transparency, third party certification, efforts to educate consumers, and proof that they're not selling the same carbon offset more than once. The ranking is an effort to create some standardization.

5 Principal Carbon Offset Types

Offset providers usually offer three types: CO_2 offsets, renewable energy credits (RECs), and cap and trade. CO_2 offsets are sponsored projects usually sold by the ton and designed to reduce or eliminate greenhouse gases in the atmosphere. Common examples of these projects are reforestation projects, carbon sequestration, methane abatement, and the development of energy efficient technologies. Here, the greatest issue is verification of ownership since offsets are saleable numerous times to different parties. Another challenge is one of permanence – whether the change created by the offset will stay intact – especially in the case of carbon sequestration. Even on natural grounds, the carbon absorbed by a tree in its lifetime is released back into the atmosphere once the tree dies [16].

RECs are also referred to as green tags. They represent 1 mega-watt hour (mWh) of renewable energy production. The most common examples are wind farms, solar farms, and biogas generators. The most relevant inefficiencies in this market are verifiable ownership and whether the project would have taken place without the required investment [16]. Cap and trade offsets are known as "emission trading schemes" or "pooled carbon commodities." The participants in these offsets commit to reducing greenhouse gas emissions to a certain level. Upon reaching this level, further reductions can be sold or traded in form of certificates. Active entities in this area include the CCX, Clean Development Mechanism (CDR) under Kyoto, and European Trading Scheme (ETS). These certificates of cap and trade offsets are traded like stocks.

There has been a flurry of activity in the market for carbon offsets. Participation in this market is voluntary in nature and the buyers of offsets range from individuals to non-profit groups to corporate entities. There are also many different sellers of offsets such as firms, charities, NGOs, community groups, and international agencies. Some examples of various offsets are to plant trees, prevent trees from being cut down, and replace fuel-guzzling stoves. The cost of offsets varies as well from some studies stating that it ranges from a few cents to $27 per ton of carbon dioxide [8] to other studies claiming it ranges from $5 to $35 per ton of carbon dioxide [4].

6 Carbon Offsets

Simply put, carbon offsetting occurs when an individual or organization pays a third party to reduce the emissions of greenhouse gases on its behalf. Organizations or individuals as varied as the International Federation of Association Football (FIFA), the musical band the Rolling Stones, and a former U.S. Vice President (Al Gore) have served to increase the demand for offsetting. The specific way the transaction works involves the buyer paying a dollar amount to the third party to support project that sequesters carbon from the atmosphere. The result is a unit reduction for a given amount of carbon.

In Chicago, many corporations that emit carbon have joined the Chicago Climate Exchange (CCX), which requires them to offset or eliminate 6% of their emissions by 2010. As of 2006, there were 210 in the membership, including Motorola, DuPont, and Ford. Some of the heaviest emitters, such as BP and British Airways, have taken the offset model in-house by purchasing offsets and passing the costs to consumers or even encouraging customers to engage in their own offsetting. For instance, the CCX claims that many of its members join to improve their image, to gain insight into a nascent industry, prepare for future regulation, and appease their shareholders, customers and staff [8]. The value to these companies can come in several forms, which are reflected in different kinds of offsetting.

7 Rationale

The concept of carbon neutrality has been around for at least a decade. One of the very first companies to reduce and neutralize greenhouse gas (GHG) footprint was Stonyfield Farms. More recently, there has been a surge of interest in carbon neutrality. Rather than a few individuals or companies doing the right thing, this phenomenon has morphed into an environmental commodity market. There are many companies willing to help one calculate carbon footprints and sell offsets (Clean Air Cool Planet, 2006). *BusinessWeek* estimates that the trade in offsets accounts for more than $100 million a year and it is growing [18].

The goal is emissions reduction via energy efficiency, conservation, technology, re-engineering, and greener buildings. Only when further reductions can no longer take place due to technology or monetary constraints does the option to buy offsets emerge. The goal, therefore, is really to "offset" what could not be *reduced*.

Of course, offsets do not make a difference if reductions are not undertaken, but they also play a role in educating and informing the public about climate change and demonstrating that the issue is ripe for public policy.

8 Carbon Offset Examples

The hospitality industry is gaining popularity for its carbon offsetting. Even among the consumers, travelers can neutralize their trips with carbon offsets. This option involves calculating a carbon footprint for flights and road trips and then purchasing offsets that donate money to projects that promise to produce energy without burning fossil fuels. The reduction equals the amount of carbon the trip created per passenger, with the goal being to have a carbon neutral journey.

Many companies offer carbon offsets to consumers. For example, Travelocity.com offers an option to neutralize the environmental impact of trips by planting trees. REI Adventures offsets its carbon emissions produced by air, water, and ground transportation through a contract with an external provider. In their case, each trip has "green tags" for RECs. The external provider, Bonneville, claims to be selling 500,000 RECs every year [5]. Though it is unclear how much impact these programs have on climate change, travelers see this option as a way to help tackle climate change without having to give up their trips [14]. Silverjet, a luxury liner, claims to be the first carbon neutral airline. They offset 1.2 t of carbon per passenger and put a cash equivalent of $28 into a fund for green projects. In this tradition, other companies (HSBC, Google) are making carbon neutrality part of their brand identity [2].

Beyond individual travel, some companies offset travel expenses. For example, HSBC bought more than $3 million in offsets to achieve carbon neutrality for business travel. Swiss Re aims to reduce business travel by encouraging employees to use telephone and videoconferencing.

Whereas advocates believe offsetting builds awareness for emissions control, critics think it is an easy way for those who do not want to change consumer or operational habits [19]. This area is largely un-regulated, and there is a problem with potential fraud [23]. There is no current certification or monitoring system that can confirm what offsetting companies promised to do. Another major concern is that some of the companies are engaging in "greenwashing," or attempting to appear green without actually changing the underlying behavior [2].

Pros for the market

- Makes economic sense
- Easy to adopt
- Social/individual involvement
- Voluntary market
- Corporate Image
- Builds awareness

Cons on moral grounds

- Easy on sacrifice and big on consumerism
- No need to change lifestyle
- Purchase forgiveness with money
- Morality argument: allowing wealthy individuals or organizations to buy themselves out of responsibility to reduce emissions

Cons on policy side

- Market in climate neutrality can blunt public support for binding limits on emission or a tax on GHG fuels
- Introduction or development of greener and cleaner technologies might be hindered
- Flawed principle: gives the impression that the people in rich countries need not change their lifestyle to reduce global warming

9 Market Performance Criteria

The offset market makes economic sense in that market signals lead to economical offset projects. It is easy market to enter in since all that is required is a computation of the carbon footprint and the amount entailed for the offset. There is also an opportunity for social or individual involvement. Furthermore, it is a voluntary market that builds awareness for a pressing global concern. Companies participate to create a positive corporate image [8, 20].

There are two basic concerns with the market. The first is on moral grounds. It can be reasoned that offsets are "easy on sacrifice and big on consumerism." That is, there is no need to change lifestyle since consumers can "purchase forgiveness with money." In essence, the morality argument is about allowing wealthy individuals or organizations to buy themselves out of responsibility to reduce emissions [6, 9, 10, 18, 19, 20].

The second concern is on policy grounds. A market in climate neutrality can blunt public support for binding limits on emission. It can also tax fuels and hinder the introduction or development of greener and cleaner technologies. In addition, there are concerns that the market is based on a flawed principle in that it gives the impression that the people in rich countries need not change their lifestyle to reduce global warming [18].

As with any new market, there are numerous uncertainties that exist in the functioning. The first glaring point is the price differential of the offsets and the variation can be rather confusing for consumers. The second point is the uncertainty as to whether the emissions reductions in a project are verifiable or not. In a worst case scenario, it is possible to double count the offset wherein a provider could sell a single credit numerous times. Since this market is voluntary in nature, the certifications are done arbitrarily and there is no standardization on these certifications. Buyers and developers of offsets need to set a baseline to be used to predict emissions that would occur in case the project did not go ahead. Finally, some the projects are questionable in nature due to natural decay and destruction, which leads to the issue of insurance guarantees.

Next, we detail some of these uncertainties in the market as they concern the market performance criteria: quality of projects, additionality, certification, ownership, price, and transparency.

10 Quality of Projects

Project quality is crucial in the offset market. For example, tree planting was a popular offset projects since trees act as carbon sinks through the process of carbon sequestration. However, trees are subject to decay and destruction. This happened to the offset project of a mango plantation in India (for the band Coldplay) who had purchased the offset for their on-tour flights. As a result, offset providers are moving away from this option [20]. Whereas, forestry used to account for 100% of *Carbon Neutral's* portfolio 2 years ago, it is now down to 20%. The destruction of ecosystems accounts for about one-fifth of carbon emissions worldwide [18].

11 Additionality

It is important to prove to buyers that a project would not have happened without investment. In other words, the energy savings made should be additional to those that are "business as usual" to count as an offset. Under the CDM of the Kyoto Protocol, there are three criteria accepted as basis for project additionality. The project must not be required by current regulation, the technologies are not common practice, and economic, technological, or investment barriers must offset resources. A baseline can predict emissions that would occur if the project did not go ahead. In the CDM market, for instance, there are about 60 methodologies alone. However, a lack of standardization in these methodologies plagues the offset market [20].

12 Certification and Standardization

There has been an increase in the demand for an independent standard because of diminishing confidence in the burgeoning market. Several standards have been established that can cause confusion. For instance, the Greenhouse Gas (GHG) reduction certification standard, sets the criteria GHG reduction projects must meet to gain a Green-e certificate [11]. Some of the criteria include disclosing detailed information, submitting to periodic independent reviews, and review by independent parties.

This standard has generated concerns that rival international standards will confuse consumers. The aim is to develop a transparent standard about the emissions that are being reduced. However, the risk that too many standards will lead to confusion.

13 Ownership

Ownership of offsets is unchartered territory. The price range exhibits a wide variation, there is uncertainty whether the emissions reductions in a project are verifiable, and it is possible to sell a single credit numerous times leading to double counting. Hence, buyers must be assured of the sole ownership of purchased offsets. There needs to be an assurance from the offset providers that the same offset is not being sold multiple times and to different parties. Evidence that ownership is not an issue is becoming an important market performance criterion.

14 Price Differential

The majority of the offsets purchased worldwide are CERs and the estimated value is $2.7 billion. The CDM market, within which the CERs are traded, is bureaucratic and has high transaction costs. On the other hand, the market in voluntary offsets is fragmented with prices ranging from $1 to $20 per ton of carbon dioxide. The projects supported range from planting trees in Tanzania to building hydroelectricity plants in Bulgaria and the price differential is primarily due to the quality of the offset project.

15 Transparency

It is becoming more important to ask providers to give clear information and transparent prices for offsets. This idea is transcending current CDM standards based on certified carbon credits. Voluntary offsets go towards small scale, local community-based renewable energy projects and high transaction costs of CDM rule such projects out. Thus, voluntary offsets can succeed where the CDM fails but those smaller markets still would require a degree of regulation [20].

16 A Standard among Provider Standards?

> Carbon offsets are an intangible good and, as such, their value and integrity depend entirely on how they are defined, represented, and guaranteed. What the market lacks are common standards for how such representations and guarantees are made and enforced [3].

Before exploring the nature of the provider market, one might ask whether there is any potential for external verification through the use of uniform standards. Indeed, there are numerous standards that lead to an information onslaught for the consumers. There is not one standard but over a dozen standards to verify the legitimacy of an offset provider by seemingly infinite combinations of metrics, and no one seems to be able to agree on a valid combination by which to measure providers. If there remains no agreement on a set of standards to which offset providers should be held, how can we even begin to judge whether a corporation has achieved carbon neutrality by using any of these providers?

There is therefore little disagreement that, in the end, the market will likely demand uniform standards or registries in order to vouch for the legitimacy of the purchased offsets [12]. Moreover, there has been an increase in the demand for a consistent and independent standard because of the practices outlined above that threaten to diminish confidence in the burgeoning market. The most effective standards will be those that are clear and rigorous, and have broad support from a wide spectrum of stakeholders, ranging from carbon offset project developers to offset traders and buyers, from environmental non-governmental organizations (NGOs) to the financial industry. Let us consider the types of standards that currently exist in the VER market before we continue to our analysis of the offset providers that strive to meet these standards.

There are three types of standards and certifications in the voluntary carbon offset market. There are some standards whose purpose is to certify the quality of an offset and the projects that it supports. Examples of this type of standard would include the Voluntary Carbon Standard, the Gold Standard, Plan Vivo, and the Climate, Community, and Biodiversity Standard. A second category of standard would be those whose focus is on the certification of offset sellers, products, services, and their claims relating to carbon neutrality. Standards that include this certification include Green-e for GHG Product Standard, Defra's Guidelines, and the Climate Neutral Network. As additional proposals for *new* standards reach the market, they draw criticism from certain circles. There are concerns that rival international standards will confuse consumers. Jan Hamrin, president of *Center of Resource Solutions*, which proposed the Green-e standard, stated that the aim was to "develop a transparent standard that allows individuals to buy carbon offsets" knowing that emissions of GHG are being reduced. It is not clear that this statement represents a new addition to the market of standards.

A third category of standards has been developed by the offset retailers, themselves, in order to ensure quality within their own portfolios. Examples of these types of standards include the Carbon Neutral Company and MyClimate. While the intent of these retailers may be laudable, there seems to be an apparent conflict of interest inherent in a self-imposed standard structure and, accordingly, these standards are likely to phase out as the market matures [12].

According to a comprehensive analysis conducted by Kollmuss et al. [15], an inclusive, complete and credible carbon offset standard must include the following three components:

1. *Accounting Standards*: To ensure that offsets are "real, additional, and permanent."
2. *Monitoring, Verification and Certification Standards*: To ensure that the projects perform according to project design and to quantify the actual carbon savings that happen once the project is up and running.
3. *Registration and Enforcement Systems*: To ensure that carbon offsets are only sold once, to clarify ownership and to enable trading of offsets. The offsets must include a registry with publicly available information in order to uniquely identify offset projects as well as a system by which to track transparently the ownership of those offsets [15].

We reviewed three reports that detailed the above components: the Kollmuss report, itself, a BSR report [4] and an analysis by Hamilton et al. [13]. The four standards most commonly listed in the three reports reviewed include the Gold Standard, the Voluntary Carbon Standard (VCS), the Voluntary Offset Standard (VOS), and Plan Vivo.

- *The Gold Standard* (GS) requires that each carbon offset project that bears its certification demonstrate social and environmental benefits and have a well-developed stakeholder engagement process. The GS can be applied to voluntary offset projects as well as to Clean Development Market (CDM) Standard projects under the compulsory Kyoto Protocol system.
- *The Voluntary Carbon Standard 07* (VCS 07) focuses only on GHG reduction attributes and, in a significant departure from other schemes, does *not* require projects to have additional environmental or social benefits. The VCS 07 is broadly supported by the carbon offset industry (project developers, large offset buyers, verifiers, projects consultants).
- *The Voluntary Offset Standard* (VOS) is a carbon offset screen that accepts other standards and methodologies using its own specific screening criteria. It currently accepts Gold Standard's VER projects, as well as projects that employ CDM procedures but which are implemented in countries that have not ratified the Kyoto Protocol and are therefore not eligible for CDM.
- *Plan Vivo* is an offset project method for small scale "Land Use, Land-Use Change and Forestry" (LULUCF) projects with a focus on promoting sustainable development and improving rural livelihoods and ecosystems. Plan Vivo works very closely with rural communities, emphasizes participatory design, ongoing stakeholder consultation, and the use of native species.

In addition to standards, there is yet another tool that can track credit ownership and eliminate "double counting" of the offsets, a persistent credibility challenge that will be discussed in detail in the next section. *Offset registries* follow two basic frameworks. One category of registries tracks GHG emissions and/or emission reductions, and a second category of registries comprises a carbon credit accounting system. Examples of tracking registries would include the United States' Department of Energy Voluntary GHG Reporting registry, the Canadian Greenhouse Gas Challenge, and the World Economic Forum Global Greenhouse Gas registry, while the Environmental Resources Trust GHG Registry and the Bank of New York Global Registrar and Custody Service fall into the second category. In some cases, a registry might serve in both capacities such as the California Climate Action Registry [12].

17 Examination of Offset Providers

Current ratings in published studies. In order to best understand the needs of the consumer market in terms of information overloads and gaps, it is vital to engage in a comparative content analysis of published reports that provide information or ratings on offsets. We attempted to access only those reports that were publicly available to consumers and free of charge in order to ensure that what we reviewed was also available to the average person seeking information. Four reports met these criteria:

– *A Consumer's Guide to Retail Carbon Offset Providers*, by Trexler Climate + Energy Services [21]
– *Carbon Offsets in Context*, by the Context Group (2006)
– *Voluntary Offsets for Air-Travel Carbon Emissions*, by Tufts Climate Initiative (2006).
– *Carbon Offset Providers Evaluation Matrix*, by Carbon Concierge (2008).

The first report, above, is written primarily for the average consumer. It provides several criteria by which it evaluates and ranks 30 retail providers. Each retail provider was then assigned a score between one and ten. The *Consumer's Guide* lists eight companies that scored more than 5.0 on their rankings. These eight companies are:

- AgCert/DrivingGreen
- Atmosfair
- Carbon Neutral Company
- Climate Care
- Climate Trust
- CO2balance
- NativeEnergy
- Sustainable Travel/MyClimate

The *Consumer's Guide* identifies the seven evaluative criteria that it used in its ranking, along with the respective weights assigned. These criteria with the weights are found in Table 1.

Table 1 Consumer's guide evaluative ranking criteria and respective weights

Providers' prioritization of offset quality	10.0
Buyers ability to transparently evaluate offset quality	9.4
Transparency in provider operations and offset selection	9.2
Providers' understanding of technical aspects of offset quality	9.0
Priority assigned by provider to educating consumers about global warming and global warming policy	7.8
Ancillary environmental and sustainable development benefits of offset portfolios	5.6
Use of third-party project protocols and certification	3.9

Some of the other criteria that are not used in the ranking process, but which are discussed in the report include the cost of the offset to the consumer, the for-profit versus non-for-profit status of the providers, and the proportion of the funding that goes to the offsets. In addition, there is no discussion of the basis for the weights given above. The *Consumer's Guide* was one of the first reports compiled on carbon offsets. The report analyzes only 30 companies; however, it was completed in 2006 and, since that time, there have been numerous other providers entering the market so the report is unfortunately outdated.

Carbon Offsets in Context identifies 23 organizations. The report shares information gathered from websites and provides information on key activities, role in offset market, type of offset provided, guidelines/verification, and geography of portfolio. In addition, this report also provides information on turnover/total CO_2 offset, year since the company was operational, whether the site has a carbon calculator, the cost per ton of carbon, and a list of external partners. The primary drawback of this report is that there is no rationale provided on the selection of the 23 organizations analyzed and, furthermore, there is no mention of any methodology used. It simply lists the 23 companies (mostly taken from the UK) and the data collected from the respective websites. Hence, it could be viewed as a brief report giving a snapshot on the activities of these 23 companies.

The third report, *Voluntary Offsets for Air-Travel Carbon Emissions*, is specifically directed to consumers interested in companies that offer programs to address the effects of air travel, which severely limits its coverage in terms of providers. The 13 companies that it includes are listed below with the "highest recommended" offered to four of 13:

- Atmosfair (*highest recommendation*)
- Better World Club
- CarbonCounter.org
- Carbonfund.org
- The Carbon Neutral Company
- Cleanairpass
- Climate Care
- Climate Friendly (*highest recommendation*)
- MyClimate (*highest recommendation*)
- NativeEnergy (*highest recommendation*)
- Offsetters

- Solar Electric Light Fund
- Terrapass

This report evaluated its subject companies along six criteria:

1. Overhead
2. Quality of Offsets
3. Standards and Verification
4. Air Travel Emissions Calculator
5. Price per ton of carbon offset
6. Transparency

Finally, the *Carbon Offset Providers Evaluation Matrix* rates offset providers using its "evaluation matrix," which is comprised of eight criteria adopted from six sigma process improvement tools. An offset provider is awarded a score of three (poor), six (good) or nine (excellent) based on an overview of each provider's website, and, where insufficient information is available online, information from the providers themselves. The eight criteria are:

1. Business and Project Transparency
2. Offset Quality
3. Project Location and Offset Traceability
4. Industry Leadership
5. Business Model and Program Services Ratio
6. Third Party Evaluation
7. Education
8. Social Benefit

Results are calculated for 17 providers in North America and the companies rated highest by this report were Native Energy and Climate Trust. Interestingly, on the whole, the rated providers scored highest in the categories of third party evaluation, and for Business and Project Transparency. The lowest scores were found in Project Location, Offset Traceability, and Business Model and Program Services Ratio. While this report is one of the most current, and quite comprehensive in its content, it discusses only 17 North American companies, including 3 wholesalers, an extremely small segment of today's offset provider market.

Based on this overview, it appears that there is great demand, therefore, for current information on a larger number of providers across common criteria where contentions have generally arisen. Our analysis below seeks to respond to those areas of need.

18 Identifying Sample for Current Study

The authors did a detailed study on a sample of 125 companies worldwide, though they were predominantly based in the United States, Europe and Australia. The comparative analysis of the carbon credit providers sought to ascertain whether detailed information on market performance criteria was available via standard consumer inquiry methods. To this end, each company was evaluated based on the following five questions, based on key market performance hurdles identified previously.

Q1: Project Quality. Does the provider's website offer information on the quality of the projects? For example, do the projects involve tree planting or clean energy projects? [*This question was delineated by four component parts.*]

(a) *Type of Projects*: What are the types of projects available and how many different types are available?
(b) *Certification*: Are the offsets or the projects certified? If so, by whom? Is the certification third-party verified?
(c) *Additionality*: Does the website provide information to illustrate or prove that the project would not have happened without the offset investments?
(d) *Transparency*: Does the website provide clear and detailed information about projects, company policies, standards, verifications and the pricing of offsets?

Q2: Calculator. Does the provider's website offer information on calculating how many tons of carbon offsets need to be purchased? Is the service for road trips or air travel?

Q3: Quality of Providers. Does the provider's website offer information on the quality of the offset provider, itself? [*This question was delineated by two component parts.*]

(a) *Single Ownership:* Does the website provide assurance that the offsets are sold only to one entity, that the offsets are not being sold forward?
(b) *Other Benefits:* Does the website provide other information on education, sustainable development, overall climate strategy, advising, consulting, or other benefits?

Q4: Price Per Ton of Carbon Offset: What is the price charged for each ton of CO_2 offset? Is there a range and what is the range?

The results of this study are included in Tables 2 and 3. The results of the complete study are presented in Appendix B.

The key market performance hurdles that comprised the variables included in these scores were satisfied based on attributes identified and shared with the raters. In order to allow consumers and others to apply this scheme to other current and future offset providers, the following example model providers are offered.

Q1: Project Quality. Why did a particular company earn a high score in this category?

(a) *Type of Projects*: **AtmosFair** is an example of a high score provider in this category. The company offers a broad number of projects from which consumers may choose. Within that variety, the company offers many different *types* of projects as well (i.e. solar energy, hydro power, irrigation,

Table 2 Tier one group of companies (with an average combined score greater than 4.0)

3Degrees	4.5
The Carbon Fund	4.43
Terrapass	4.43
EcoSecurities	4.33
Carbon Clear	4.21
Climate Friendly	4.14
Tricorona Green	4.07
Pure	4.07
Climate Care	4.02
Climate Neutral Group	4
My Climate	4

Table 3 Tier two group of companies (with an average score 3.5 or greater)

Action Carbone	3.93
Carbon Balanced	3.93
Carbon Impacts	3.93
Carbon Neutral	3.93
Zero footprint	3.93
AtmosFair	3.93
Carbon Aided	3.92
Native Energy	3.86
LivClean	3.79
Sustainable Travel International	3.79
Carbonfund.org	3.5
Planetair	3.5
TargetNeutral	3.5
Zero GHG	3.5

efficient fuel uses and more). AtmosFair also includes the current status of each project in significant detail, and all information relevant to a consumer's decision: a description of each particular project, what it has accomplished and how it has helped. **Climate Care** also received a high score since the company offers a wide variety of projects from which to choose. The projects are diverse in their types ranging from wind, solar, bio fuels, stoves, and others. Furthermore, the website provided extensive, detailed information on each project, including project reports, detailing the status of the projects, certification, immediate impact and the benefits of the project on the community, as well as video and pictures from the project sites.

(b) *Certification*: **Carbon Impacts** received a high score since it was certified by both the Gold Standard and the Voluntary Carbon Standard. In addition, this company offers details about what those standards represent, what it has accomplished and it provided a link for more information. **3Degrees** also scored high because its website clearly stated each of the certifications that it recognizes and illustrates what kind of certification each projects has. They also give details on the criteria for each kind of certification and provide links to more details. A company usually ranked higher if it had certification from Gold standard, CCX, VCS, CCAR, CDM, Green-e climate, and UNFCC.

(c) *Additionality:* **Carbon Balanced** scored high since its website explained the concept of additionality in detail and provided information on how it achieved additionality, both financially and through delivery. Another example of additionality is offered by **Carbon Clear**, which illustrates on its website that its projects are certified and then explains why it would have been otherwise impossible to start a project in that region and how it has helped the community socially, economically and environmentally.

(d) *Transparency:* **Tricorna Green** received a high score on transparency because it explained where the consumer's money was allocated and how the consumer would be kept informed about activities. It went into detail with transparency and provided an abundance of information on its administrative costs. In addition, after a consumer's money was allocated to a project, Tricorna Green offered to the consumer the option to track their offset. **Carbon Clear** details the date on which a project is started, its certification, the current status of the project and the total emission reduction made possible by the project. The FAQ page also gives information as to the allocation of the money and the allocation of the funds between the project and administrative costs.

Q2: Calculator. **Offset the Rest** received a high score for two reasons. First, its website provides complete detail on how its calculator functions. Second, the website offered six different categories for offsetting – private transportation, commuter transportation, home energy, waste production, along with the details on each category. **Carbon Neutral's** website calculator provided an option of calculating carbon footprint from various activities. It went into details of travels (flights, cars, train), household (electricity, number of people living in the house, type of house, waste management), and the process from which the final computation was derived.

Q3: Quality of Providers. Does the provider's website offer information on the quality of the offset provider, itself?

(a) *Single Ownership:* **LivClean** received a high score because the company's website guaranteed that each of their carbon offsets is measured accurately and that there is no double counting. Also, the company's website explained and verified the retirement of purchased offsets. **3Degrees** directly addresses the question in its FAQ page and states that the projects are registered in the VER registries that guard against double selling and have special security checks in order to prevent double selling.

(b) *Other Benefits:* **EcoSecurities** received a high since the website not only concentrated on carbon offsets but also offered consulting, career opportunities, press information and responsibilities. In addition, it provided a great deal of information about green energy. For instance, the press and media sections provide articles on the current issues on green energy.

Q4: Price Per Ton of Carbon Offset: What is the price given for each ton of CO_2 offset? Is there a range and what is the range? The range in terms of price per CO_2 offset ton was extremely broad, as might be expected given the range of the projects supported. The raw data on price per ton of CO_2 from the offset providers was collected in local currencies and was converted to U.S. dollars on 28 May, 2009.[1] The price per ton of CO_2 from all included providers ranges from $2.50 (**Carbon Neutral**) to $43.80 (**Planet Air**).

19 Conclusion: Implications

In this article, we have offered a basic understanding of carbon neutrality, the carbon market, carbon offsets and the standards surrounding the offset market. Given the large and rising number of offset providers in the highly unregulated and often misunderstood carbon offset industry, this article highlights areas of concern for consumers and other stakeholders. The article offers insight into the standards environment for offset providers since the numerous standards themselves need to be standardized.

This market is also not immune to security attacks. According to a recent news article, about two million tonnes of EU carbon credits were stolen by hackers in an online auction [22]. As long as the purchaser of carbon credits remains naïve about the process, uninformed about standards, confused by conflicting claims and overwhelmed by choice, neither the fundamental values of a right to health and to a sustaining physical environment, nor the benefits of a balanced, equitable and fair global carbon emission standard can truly be realized.

Appendix A: Carbon Calculators

Airplane Travel Emissions

Companies	Websites	Types
Atmosfair	www.atmosfair.de/index.php?id=5&L=3	Location to location detail, with layovers
Climate Care	www.climatecare.org/living/calculator_info/index.cfm	Location to location, as well as house and car emissions
Offsetters	www.offsetters.ca/calculators_fights.htm	Location to location detail

[1] We used www.xe.com for all currency conversions.

Business Emissions Calculators

Companies	Websites	Types
Climate Friendly	www.climatefriendly.com/business.php	One of very few business calculators. Includes factory and office electricity, fleet fuel, and corporate air travel

Car Travel Emissions

Companies	Websites	Types
Certified Clean Car	www.certifiedcleancar.com/menu/cleannow/foryou/index.htm	Input exact car make and model
Target Neutral	www.targetneutral.com/TONIC/carbon.do?method=init	Calculate up to four cars at once
TerraPass	www.terrapass.com/road/carboncalc.php	Input exact car make and model
Clean Air Pass	www.cleanairpass.com/treecanada	Input exact car make and model

Other Notable Calculators

Companies	Websites	Types
Carbon Counter	www.carboncounter.org/test.php?testPath = estimate&nextStep = 1	Calculate "estimated" or "exact" emissions
Sustainable Travel International	www.sustainabletravelinternational.org/offset/index.php?p=hotel	Include hotel emissions
World Land Trust	www.carbonbalanced.org/personal/calculator/calctravel.asp	Includes hotel, boat, flight emissions
Atmos Clear	www.atmosclear.org/calculator_tran.php	Includes household and recreational equipment, from leaf blowers to jet skis

Appendix B

Complete results of average rating evaluations	
3Degrees	4.50
The Carbon Fund	4.43
TerraPass	4.43
EcoSecurities	4.33
Carbon Clear	4.21
Climate Friendly	4.14
Tricorona Green	4.07
Pure	4.07
Climate Care	4.02
Climate Neutral Group	4.00
My Climate	4.00
3Degrees	4.50
The Carbon Fund	4.43
TerraPass	4.43
EcoSecurities	4.33
Carbon Clear	4.21
Climate Friendly	4.14
Tricorona Green	4.07
Pure	4.07
Climate Care	4.02
Climate Neutral Group	4.00
My Climate	4.00
Action Carbone	3.93
Carbon Balanced	3.93
Carbon Impacts	3.93
Carbon Neutral	3.93
Zerofootprint	3.93
AtmosFair	3.93
Carbon Aided	3.92
Native Energy	3.86
LivClean	3.79
Sustainable Travel International	3.79
Carbonfund.org	3.50
Planetair	3.50
TargetNeutral	3.50
Zero GHG	3.50
Conservation Fund's	
Go Zero Program	3.43
Live Climate	3.36
Offset the Rest	3.36
Offsetters	3.36

(continued)

Complete results of average rating evaluations	
Carbon Offsets LTD	3.29
Climate Positive	3.29
CO2Logic	3.29
Neco	3.21
Versus Carbon Neutral	3.21
Climat Mundi	3.17
BEF – Bonneville Environmental FDN	3.14
The Nature Conservancy	3.14
World Land Trust	3.07
Carbon Passport	3.07
Delta Offsets	3.07
CLIMACT	3.00
Carbon Positive	2.93
BeGreen Now	2.93
ClearSky Climate Solutions	2.93
Clear-	2.86
carboNZero	2.79
Origin	2.79
Cleaner Climate	2.71
Greenfleet	2.71
Climate Stewards	2.57
Good Energy Initiative	2.57
Carbon Friendly	2.50
Coolaction	2.50
DriveNeutral	2.50
LiveNeutral	2.50
Carbon Footprint Offsetters	2.43
Carbon Forestry	2.43
Standard Carbon	2.43
EcoNeutral	2.43
Solar Electric Light Fund	2.42
Blue Ventures Carbon Offset	2.36
Carbon Planet	2.36
e-blueHorizons	2.36
My Clean Sky	2.36
Carbon Angel	2.29
C Level	2.29
Go Neutral	2.21
PrimaKlima-Weltweit	2.20
EcoVoom	2.14
Plan Vivo	2.08
Carbon Reduction Institute	2.07
Tist	2.01
CO_2 Australia	2.00
Envirotrade/Plan Vivo	2.00
Flying Forest	2.00

(continued)

Complete results of average rating evaluations	
Carbon Retirement	1.93
GroPower	1.93
ClimateSave	1.83
Tree Canada	1.79
Green Pass	1.76
LiveCooler	1.71
ETA	1.71
Carbon Me	1.64
Treeflights	1.63
Trees, Water and People	1.63
PowerTreeCarbon	1.57
Global Cool	1.50
Canopy	1.36
Reforest the Tropics	1.36
Better World Club	1.29
Enviro Friendly Products	1.24
Pembina	0.86

References

1. Ambec S, Lanoie P (2009) Does it pay to be green? A systemic overview. Acad Manag Perspect 22(4):45–63
2. Braden P (2006) Flying perks turn a shade of green. International Herald Tribune, Sep 21
3. Broekhoff D (2007) Testimony before the House Select Committee on Energy Independence and Global Warming. Voluntary carbon offsets: getting what you pay for. U.S. House of Representatives, July 18. http://pdf.wri.org/20070718_broekhoff_testimony.pdf
4. Business for Social Responsibility (2006) Offsetting emissions: a business brief on the voluntary carbon market. The Ecosystem Marketplace. http://bsr.org/reports/BSR_Voluntary-Carbon-Offsets-2.pdf
5. Conlin J (2007) Going green, one spring break at a time. The New York Times, 25 Feb 2007
6. DePalma A (2006) Gas guzzlers find price of forgiveness. The New York Times, 22 April
7. Dhanda K (1999) A market-based solution to acid rain: the case of sulfur dioxide (SO_2) trading program. J Public Policy Market 18(2):1–15
8. Economist (2006a) Upset about offsets. Economist 380(8489):53–54
9. Economist (2006b) Sins of emission. Economist 380(8489):15
10. Friedman T (2007) It's too late for later. New York Times, 16 Dec 2007. http://www.nytimes.com/2007/12/16/opinion/16friedman.html?ref=opinion
11. The Green-e GHG Emission Reduction Product Certification Program DRAFT, Version 2.0, accessed March, 2012 online at www.green-e.org/.../Green-e%20Greenhouse%20Gas%20Product%20Standard%20Version%201.0.pdf
12. Hamilton K, Bayon R, Turner G, Higgins D (2007) State of the Voluntary Carbon Markets 2007: picking up steam. Ecosystem Marketplace and New Carbon Finance, San Francisco. http://ecosystemmarketplace.com/documents/acrobat/StateoftheVoluntaryCarbonMarket18-July_Final.pdf
13. Hamilton K, Sjardin M, Marcello T, Xu G (2008) Forging a frontier: State of the Voluntary Carbon Markets. Ecosystem Marketplace and New Carbon Finance, San Francisco. http://ecosystemmarketplace.com/documents/cms_documents/2008_StateofVoluntaryCarbonMarket.4.pdf

14. Higgins M (2006) Raising the ante on Eco-tourism. The New York Times, 10 Dec 2006
15. Kollmuss A, Zink H, Polycarp C (2008) Making sense of the voluntary carbon market: a comparison of carbon offset standards, WWF Germany. http://assets.panda.org/downloads/vcm_report_final.pdf
16. Main E (2007) Shifting into neutral. The Green Guide. http://www.thegreenguide.com/doc/119/neutral
17. Mechanisms under Kyoto protocol. Accessed July 2010. http://unfccc.int/kyoto_protocol/mechanisms/items/1673.php
18. Revkin A (2007) Carbon-neutral is hip, but is it green? The New York Times, 29 Apr 2007
19. Richardson A (2006) Carbon credits – paying to pollute? 3rd Degree. http://3degree.cci.ecu.edu.au/articles/view/781
20. Russell J (2007) Are emissions offsets a carbon con? Ethical Corporation. http://www.greenbiz.com/news/reviews_third.cfm?NewsID=34804
21. Trexler Climate + Energy Services, Inc. (2006) A consumer's guide to retail carbon offset providers. Clean Air – Cool Planet, Portsmouth, http://www.cleanair-coolplanet.org/ConsumersGuidetoCarbonOffsets.pdf
22. Vancouver Sun (2011) Two million tonnes of EU carbon credits stolen by hackers. Accessed 21 Feb 2011. http://www.vancouversun.com/technology/million+tonnes+carbon+credits+stolen+hackers/4138352/story.html
23. Velasquez-Manoff M (2007) Do carbon offsets live up to their promise? www.csmonitor.com/2007/0110/p132-sten.htm

Locational Carbon Footprint of the Power Industry: Implications for Operations, Planning and Policy Making

Aleksandr Rudkevich and Pablo A. Ruiz

Abstract Jurisdictions across the globe are implementing CO_2 emissions reduction policies. These policies typically ignore most locational issues, probably because the consequences of greenhouse gas emissions do not depend on the exact emission location. However, the response to emission policies and the costs and effectiveness of emissions reduction in power systems are time-varying and locational in nature. The first part of the paper elaborates on the economic properties of the concept of locational marginal carbon intensity and formulates an allocation of the carbon footprint of the electrical grid to individual generating units, transmission facilities and end users on a real time basis. In the second part, the theory of the marginal carbon footprint is applied to the derivation of the optimal investment policy underlying Renewable Portfolio Standards (RPS). The argument is made that the existing RPS policies are at best sub-optimal in their goal to reduce emissions of Carbon Dioxide and other greenhouse gases. A proposed optimal investment rule could serve to improve the efficiency of RPS policies.

Keywords Electrical grid • Transmission congestion • Marginal carbon intensity • Renewable portfolio standard • Carbon footprint

1 Introduction

The study of the economics of CO_2 emissions is the study of costs and benefits of CO_2 abatement. Carbon reduction economic policies are already a reality affecting operations of power systems in many European countries. The U.S. power industry

A. Rudkevich (✉)
Newton Energy Group
e-mail: arudkevich@negll.com

P.A. Ruiz
Charles River Associates, Boston University, MA, USA
e-mail: paruiz@ieee.org

Q.P. Zheng et al. (eds.), *Handbook of CO_2 in Power Systems*, Energy Systems,
DOI 10.1007/978-3-642-27431-2_8, © Springer-Verlag Berlin Heidelberg 2012

is poised for a national carbon control policy being contemplated by the Federal government and is already subject to certain Regional Greenhouse Gas Initiatives (RGGI) [1], Renewable Portfolio Standards (RPS) [2] and demand reduction programs at the utility and state levels, all claiming CO_2 reduction as a policy centerpiece. The investment and power communities have expressed an increasing interest in the development of renewable generation technologies such as wind and solar on a massive scale, in construction of high voltage transmission lines to deliver renewable energy to the markets and in efficiency programs [3, 4].

The emission of greenhouse gases into the atmosphere is a global problem. When a unit of CO_2 is released into the air, to a large extent the exact geographic location, and to some extent the time, of this event makes little difference on its consequences. In contrast, time and geographical location may make a significant difference in the economic efficiency of actions directly or indirectly focused on reducing CO_2 emissions. This is especially true for the power industry, where a diverse technological and geographical mix of generation technologies and a constrained transmission network make avoidance of CO_2 emissions to be temporally and spatially dependent [5, 6]. In spite of this, most emission reduction programs do not explicitly account for network or time effects. In fact, there seems to be a disconnect between the great level of detail employed in the economic modeling, simulation and analysis of the effects of new policies, programs and investments [7, 8] compared to a relatively poor system of concepts addressing the implications of CO_2 abatement policies for the economics and operation of power systems.

This paper discusses the theory of marginal CO_2 emissions in constrained power networks, as detailed in [9, 10], and its implications for power system operations, planning and policy making.

Section 2 gives the basic concepts introduced in [9], such as marginal carbon intensity (MCI). Section 3 expands the MCI theory toward the analysis of the locational carbon footprint of loads, generators and constrained transmission facilities within a power system. This section provides a formal definition of the carbon footprint of a system element, derives mathematical formulas underlying its calculation and establishes the relations between the total system-wide carbon emissions and carbon footprints of system elements. Section 4 presents illustrative examples of marginal carbon intensity and footprint.

Section 5 provides an application of this theory to the analysis of Renewable Portfolio Standards. Implementation of the RPS policy side-by-side with the nationwide or global carbon regulation offers the power industry a two-prong approach to CO_2 emissions control. Cap-and-trade or carbon tax-based policies directly affect the dispatch order of thermal generating plants and electricity prices, and thereby sends price signals with respect to new entry and retirement decisions on the part of generators and at the same time affects transmission planning decisions. RPS policies create incentives for renewable generating technologies to enter the market by providing investment subsidies to project developers. The intent here is to attract technological innovations to renewable generation technologies and bring them into the market at an accelerated pace. Generally, this has a potential to provide a relatively soft transitional path for the industry by not forcing thermal generators, primarily coal, into retirement through an introduction of a carbon price shock before the alternative renewable generating technology can take their place in the electricity market.

At the same time, as demonstrated in Sect. 5, presently introduced RPS policies are poorly designed and inefficient. This is primarily because the subsidies provided to participants of RPS programs are not well aligned with the objective of carbon reduction. Under existing RPS programs, developers of renewable resources are paid on a per MWh of renewable generation regardless of the amount of CO_2 they actually displace.

To that end, we introduce a theoretical construct underlying a design of the optimal RPS program. In developing this construct, we provide a systematic comparison of two alternative RPS designs, one in which the subsidy provided to developers of renewable resources is used to maximize the total renewable energy and another one, in which the subsidy is used to minimize CO_2 emissions. We demonstrate that these two designs lead to different investment strategies.

2 Marginal Carbon Impact Indicators

This section discusses marginal indicators of carbon emissions in power systems. These indicators are used in the subsequent sections of the paper.

2.1 Marginal Carbon Intensity

Consider an electrical grid, and denote by $\mathfrak{C}(t)$ the total mass of CO_2 emissions produced by all generators interconnected to the electrical grid measured in tons of CO_2 over time period t. Assume that $\mathfrak{C}(t)$ can be measured.

Assume that a market participant finds it economically beneficial to reduce electricity demand by a small amount at a given location on the grid. An important question is how many units of CO_2 emissions will this measure help to avoid? The indicator that provides an answer to this question is the *marginal carbon intensity* (*MCI*), defined as the decrease in CO_2 emissions in the electrical network in response to an infinitesimal decrease in electricity demand and measured in (t/MWh).[1] *MCI* depends on the time and location of the applied demand reduction measure,

$$MCI_k(t) = \frac{\partial \mathfrak{C}(t)}{\partial L_k(t)}, \tag{1}$$

where $L_k(t)$ denotes demand at time t at location k. The larger $MCI_k(t)$ is for a given location and time, the greater is the change in the total carbon emission volume in

[1] Here and elsewhere, t stands for metric tons of greenhouse gas in CO_2 equivalent.

response to the change in electricity demand. A positive value of $MCI_k\ (t)$ implies that at a given location and time an increase/decrease in electricity demand causes increase/decrease in CO_2 emissions in the system. A negative value of $MCI_k\ (t)$ implies that at a given location and time changes in electricity demand and CO_2 emissions move in opposite directions.[2] A zero $MCI_k\ (t)$ implies no change in the total CO_2 emissions in the system for load variations.[3]

To get a better insight into this indicator, consider an unconstrained electrical system dominated by three generating technologies: coal-fired generation, combined cycle gas-fired generation (CCg) and a simple-cycle combustion gas turbine (CTg). Their illustrative characteristics are presented in Table 1.

Parameters presented in this table are typical for these generating technologies. For the purpose of this example, we assume a \$10/t value of CO_2 emissions reflecting a CO_2 control policy in a form of a carbon tax or a price associated with a cap-and-trade program. For generators, this value represents an expenditure which they factor into their dispatch cost along with fuel and variable O&M expenses. The dispatch cost determines a merit order in which generators are deployed while serving system demand. The marginal cost of electricity is set by the cost of the marginal generator—the most expensive generator needed to meet the demand in a given time period. All inframarginal generators, those below the marginal generator in the merit order, are dispatched at their full capacity, while supermarginal generators, those above the marginal generator in the merit order, are not dispatched.

A small change in electricity demand in a given time period causes an equal change in output of the marginal generator (in absence of losses). Therefore, in each time period the marginal carbon intensity is determined by the emission rate of the generator that is marginal during that period. This is illustrated graphically on Fig. 1 which depicts a chronological demand profile for a day, marginal generation technology at each point in time and the resulting dynamics of the *MCI*. As this figure demonstrates, temporal changes in the amount of avoided carbon are significant and could vary by a factor of greater than two when the marginal generating technology switches between conventional coal and combined cycle gas-fired generation.

---

[2] A statistical analysis of demand reduction measures relying on real-time prices reported in Ref. [11] indicates that demand reduction could result in an increase in emissions, in this case NO_X and SO_2.

[3] There could be many reasons zero $MCI_k\ (t)$. One is for cases where renewable generation is on the margin and meets the entirety of the change in demand. Another is due to transmission congestion, and happens when increasing demand requires high emission generators to be replaced by lower emission ones, as is discussed later in the paper. A binding instantaneous CO_2 emissions rate cap would also set $MCI_k\ (t)$ to zero. However, this case is not discussed in the paper, because all implemented and proposed cap-and-trade policies are in terms of total, region- or countrywide emissions over an extended period of time, such as 1 year, rather than on maximum instantaneous emissions rates just for the power industry. This is rightly done, as the effect of CO_2 emissions transcends the power industry and does not depend on the emission rates but on total emissions.

Table 1 Illustrative characteristics of generators

Technology	Heat rate (Btu/kWh)	Fuel price ($/MMbtu)	VO&M ($/MWh)	CO_2 emissions rate (t/MWh)	CO_2 cost ($/t)	Dispatch cost ($/MWh)
Coal-fired	9,500	2.0	1.0	0.9	10	29
Combined cycle	7,000	5.0	3.0	0.4	10	42
Combustion turbine	11,000	5.0	5.0	0.6	10	66

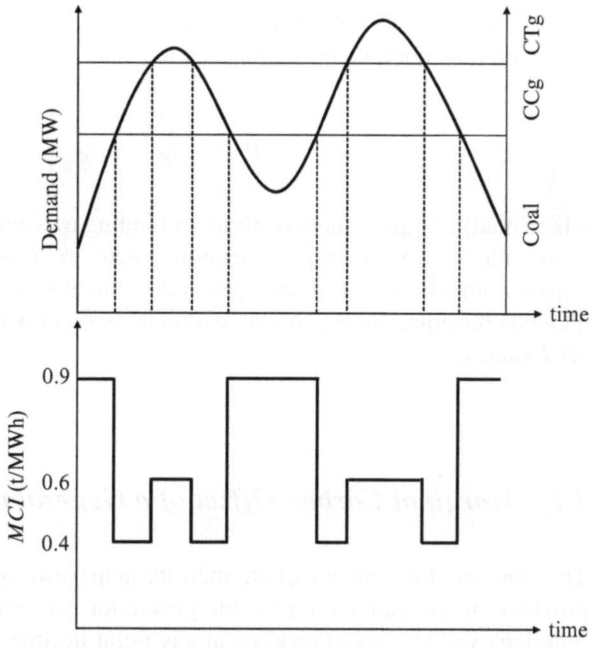

Fig. 1 Electricity demand, marginal generators and marginal carbon intensity over time

The above discussion of the marginal carbon intensity in an unconstrained lossless system could be summarized by

$$MCI_k(t) = \sigma_*(t), \tag{2}$$

i.e., the marginal carbon intensity at all locations is equal to the carbon emission rate of the marginal generator, $\sigma_*(t)$.

In a transmission-constrained power system, there may be multiple marginal generating units,[4] each with a different emission rate. A demand decrease at a given location requires a redispatch of these marginal units. Some marginal units may

[4] Typically, the number of marginal units equals the number of non-redundant binding constraints.

need to decrease their outputs, some may need to increase them, while others may not change it in order to respond to the decremental demand without violating the security of the transmission system. In sum, marginal units have to be moved in tandem, resulting in the following expression for the locational *MCI* for location (node) k:

$$MCI_k(t) = \sum_{j=1}^{m} \beta_{kj}\sigma_j, \tag{3}$$

where m is the number of marginal units, σ_j is the CO_2 emission rate of marginal units j, and β_{kj} are location-specific proportionality coefficients. As detailed in the Appendix, the coefficients β_k for location k are given by the solution of

$$\beta_k = \begin{bmatrix} \mathbf{1}' \\ \tilde{\Psi} \end{bmatrix}^{-1} \begin{bmatrix} 1 \\ \psi_k \end{bmatrix}, \tag{4}$$

where matrix $\tilde{\Psi}$ gives the sensitivity of binding transmission constraints to changes in the injection of marginal generators, vector ψ_k gives the sensitivity of binding transmission constraints to changes in the load at bus k, and $\mathbf{1}'$ is a row vector with all elements equal to one. As the coefficients β_{kj} depend on the location, so do the *MCI* values.

2.2 Marginal Carbon Offset of a Generator

The concept of marginal carbon intensity helps also to determine the carbon offset provided by incremental renewable generation, or more generally, by any generation deployed at a given location at any point in time. The *marginal carbon offset* $\alpha_k^C(t)$ is given by the difference between the *MCI* at the generator's location and the generator's CO_2 emission rate $\sigma_k(t)$.

$$\alpha_k^C(t) = MCI_k(t) - \sigma_k(t). \tag{5}$$

In absence of transmission losses, increasing generator's output by a small amount displaces total output of the marginal generator by the same amount. Marginal carbon offset measures the net impact of this displacement on system-wide carbon emissions.

A positive carbon offset for a generator indicates that if it were possible to increase the generator's capacity (and its optimal output), it would reduce system-wide emissions. A negative carbon offset indicates that increasing this generator's capacity (and its output) increases system-wide emissions.

It is important to note that the carbon offset provided by a generator is dependent on the *MCI* at the generator's location which in turn depends on time. Therefore, the fact that the generator has a low or even zero CO_2 emissions does not necessarily guarantee that it provides a positive carbon offset. As demonstrated in Sect. 4.1 with an example, under certain circumstances, marginal carbon offset of renewable generation could be negative due to transmission constraints. At the same time, depending on system conditions, a non-renewable generator can provide a positive carbon offset as long as its own carbon emission rate is lower than the *MCI* at its location.

2.3 Shadow Carbon Intensity of a Transmission Constraint

The impact of transmission congestion on the locational effectiveness of CO_2 reduction is addressed on a systematic level through *shadow carbon intensities* (*SCI*) of transmission constraints. The *SCI* of a transmission constraint is defined as the reduction in CO_2 emissions in the entire system in response to an infinitesimal increase in the rating of that transmission constraint and measured in t/MWh.

$$SCI_\ell(t) = -\frac{\mathfrak{C}(t)}{\partial F_\ell}, \tag{6}$$

where SCI_ℓ is the shadow carbon intensity of transmission constraint ℓ and F_ℓ is the rating of that constraint. Transmission constraints that do not bind have zero *SCI* values—increasing line ratings for these constraints would make no impact on overall carbon emissions. Binding transmission constraint ℓ has $SCI_\ell(t)$ given by

$$SCI_\ell(t) = \sum_{j=1}^{m} \gamma_{\ell j}\sigma_j, \tag{7}$$

where $\gamma_{\ell j}$ are location-specific proportionality coefficients. As detailed in the Appendix, the coefficients γ_ℓ for binding constraint ℓ are given by the solution of

$$\gamma_\ell = \begin{bmatrix} 1' \\ \tilde{\Psi} \end{bmatrix}^{-1} [0,\ldots,0,1,0\ldots,0]', \tag{8}$$

where the 1 in vector [0,. . .,0,1,0. . .,0] is in the position corresponding to constraint ℓ in $\tilde{\Psi}$. Relieving a constraint with a positive *SCI* value reduces carbon emissions. Relieving a constraint with a negative *SCI* value increases carbon emissions.

Marginal carbon intensities and shadow carbon intensities of transmission constraints are linked by the same fundamental equation as locational marginal prices (LMPs) and shadow prices of binding transmission constraints. As shown in the Appendix,

$$MCI_k(t) = MCI_0(t) - \sum_{\ell=1}^{L} \psi_{\ell k} SCI_\ell, \tag{9}$$

where 0 is the reference bus and $\psi_{\ell k}$ gives the sensitivity of the flow on binding transmission constraint ℓ with respect to a change in load at bus k. The decomposition (9) also suggests an efficient computational scheme for *MCI*. The reference bus MCI_0 and the vector of *SCI* are computed first, which requires the factorization of a matrix of dimension the number of binding constraints, typically very small. Then, *MCI* for all locations of interest are computed with a simple matrix multiplication.

3 Carbon Footprint Theorem

The concept of carbon footprint is widely used in the literature but is loosely defined. A typical definition states that carbon footprint is "the total set of greenhouse gas (GHG) emissions caused by an individual, organization, event or product" [12]. This definition, as well as other similar definitions, implies that this total set of GHG emissions is either known or could be measured. As the above discussion indicates, due to temporal and locational properties of the impact of electricity consumption on carbon emissions, actually measuring carbon footprint of an "individual, organization, event or product" associated with electricity consumption is difficult and ambiguous due to complex network properties of the power system. While the marginal carbon intensity at a given place and time adequately determines the carbon footprint of an incremental (marginal) change in electricity consumption, it is still unclear how to measure the total footprint of electricity end uses for a given time. One way to establish the carbon footprint of each system element is by distributing total emissions among electricity consumers using an allocation rule. However, the potential set of such rules is infinite and there is no clear guidance on why one rule should be preferred over another. For example, computational methods currently used to perform life-cycle assessments associated with electricity consumptions are based on the regional fuel mix of electricity production averaged over some historical period and across all consumers [13]. This simplistic approach is inaccurate, because it does not recognize the temporal and locational impacts of electricity consumption on carbon emissions and therefore provides incorrect signals to electricity market participants. A more accurate approach effectively based on the concept of the marginal carbon intensity is proposed in [14] within the so-called "dispatch data analysis operating margin" methodology. However, this methodology fails to explicitly account for locational properties of the *MCI* and leads to inaccurate estimates in the presence of transmission constraints and parallel flows of power.

A more rigorous and accurate approach to defining the carbon footprint of an element of the power system could be based on the concept of financial responsibility for carbon emissions. Consider, for example, an electricity load *L* at a given location on the grid with locational electricity price P_e with the total cost of serving this load of P_eL expressed, for example, in dollars. Let us further assume that carbon emissions are priced[5] at P_c expressed in dollars per ton of CO_2 and that the cost of carbon emissions are factored into the optimal system dispatch. The carbon footprint associated with this electricity load can be defined as the incremental change in the cost of serving load in response to an infinitesimal increase in carbon price under the assumption that the load is inelastic to price:

$$\mathbf{CF}[L] = \frac{\partial P_e}{\partial P_c} L. \tag{10}$$

Note that since the numerator in (10) is expressed in dollars and the denominator is expressed in dollars per ton of CO_2 the result is expressed in tons of CO_2. For example, if the cost of serving load equals \$1000 and a \$1/t increase in carbon prices causes the cost of serving load to increase by \$20, the carbon footprint of this load is equal to 20 t of CO_2.

Similarly, the carbon footprint could be defined for a generator and for a transmission constraint. The carbon footprint of a generator is the rate at which net revenues of this generator changes in response to the change in carbon price. The carbon footprint of a transmission constraint is the rate at which the congestion rent associated with that constraint changes in response to the change in carbon price.

In order to fully formalize the above definition, consider financial flows for key elements of the power system (e.g., cost of serving load, net generators' revenues and transmission congestion rent) resulting from the optimal environmental dispatch of a power system at carbon price of and corresponding to that dispatch electricity prices. These financial flows could be expressed in the form of a well known identity [15]:

$$Cost(t) = \sum_{n=1}^{N} LMP_n(t)L_n(t) - \sum_{n=1}^{N} GM_n(t)G_n(t) - \sum_{\ell=1}^{L} SP_\ell(t)F_\ell, \tag{11}$$

where N is the number of buses in the network, L is a number of monitored transmission constraints, L_n and $LMP_n(t)$ are electricity demand and locational marginal price (*LMP*), respectively, at location n, $G_n(t)$ and $GM_n(t)$ are the power output and gross margin per unit of output, respectively, for generation at location n, and F_ℓ and $SP_\ell(t)$ are the power flow and shadow price, respectively, of transmission constraint ℓ.

Here, system-wide generation costs are decomposed into loads payments (the first term in (11)), producer surplus (the second term in (11)) and transmission congestion rent (the third term in (11)). By definition, the gross margin of each

[5] In absence of carbon regulation assume that the price equals zero.

dispatched generator is equal to the difference between generator's *LMP* and its short-run marginal dispatch cost,

$$GM_n = LMP_n - c_n. \tag{12}$$

The short-run dispatch cost c_n for the purpose of this analysis could be represented as the sum of a non-carbon component a_n and the product of the emission rate σ_n and carbon price P_c,

$$c_n = a_n + \sigma_n P_c. \tag{13}$$

Definition 1. The carbon footprint $\mathbf{CF}[L_n(t)]$ of the electrical load at a given location n and time t is defined as the change in the cost of serving the load in response to an infinitesimal change in carbon price,

$$\mathbf{CF}[L_n(t)] = \frac{\partial LMP_n(t)}{\partial P_c} L_n(t). \tag{14}$$

Definition 2. The carbon footprint $\mathbf{CF}[G_n(t)]$ of a generator at a given location n and time t is defined as the change in operating revenues accrued to that generator in time t in response to an infinitesimal change in carbon price,

$$\mathbf{CF}[G_n(t)] = \frac{\partial (OM_n(t)G_n(t))}{\partial P_c}. \tag{15}$$

Definition 3. The carbon footprint $\mathbf{CF}[F_\ell(t)]$ of a transmission element ℓ at time t is defined as the change in congestion rent of that element in response to an infinitesimal change in carbon price,

$$\mathbf{CF}[F_\ell(t)] = \frac{\partial SP_\ell(t)}{\partial P_c} F_\ell. \tag{16}$$

Using these definitions, it is possible to establish basic properties of the carbon footprint of each element of the system and of the system as a whole, as the following theorem shows (the proof is provided in [10]):

Theorem 1 *Carbon Footprint Theorem. The carbon footprints of load, generators and transmission are given by*

$$\mathbf{CF}[F_\ell(t)] = MCI_n(t)L_n(t), \tag{17}$$

$$\mathbf{CF}[G_n(t)] = -\alpha_n^C(t)G_n(t), \tag{18}$$

$$\mathbf{CF}[F_\ell(t)] = -SCI_\ell(t)F_\ell. \tag{19}$$

The total carbon footprint of the entire system at a given time moment t could be represented as

$$\mathbf{CF}[System] = \underbrace{\sum_{n=1}^{N} MCI_n(t)L_n(t)}_{\mathbf{CF}[load]} - \underbrace{\sum_{n=1}^{N} \alpha_n^C(t)G_n(t)}_{\mathbf{CF}[generation]} - \underbrace{\sum_{\ell=1}^{L} SCI_\ell(t)F_\ell}_{\mathbf{CF}[transmission]}. \tag{20}$$

The total carbon footprint is always equal to the total mass of carbon emissions of the power system,

$$CF[System] = \mathfrak{C}(t). \tag{21}$$

The Carbon Footprint Theorem offers a straightforward method for allocating the total emission of CO_2 among elements of the power system represented by consumer loads, generators and congested transmission facilities in real time based on system operations.

It is important to compare (20) and

$$\mathfrak{C}(t) = \sum_{n=1}^{N} \sigma_n(t)G_n(t). \tag{22}$$

Equation (22) directly traces carbon emissions to their physical source, i.e. power generation. From that perspective, contribution of power generation to carbon emissions is at best non-negative and in most cases positive. In contrast, (20) allocates the responsibility for carbon emissions among system elements. The emphasis of this equation is on economic activities which cause carbon emissions.

As follows from the theorem, carbon footprints of power system elements depend on time and location on the grid, and could be positive or negative. Elements with a positive carbon footprint could be considered as *virtual* sources of carbon. Elements with negative carbon footprint could be interpreted as *virtual* sinks of carbon. The virtual source does not point to the location on the grid where carbon is being released into the atmosphere, instead it points to the location of economic activity which causes carbon emissions. Similarly, virtual sinks do not physically absorb carbon emissions but point to locations of economic activities that serve to offset carbon emission caused by virtual sources. According to the Carbon Footprint Theorem, virtual sources and sinks are always in balance with the actual CO_2 emissions released by the entire grid.

Virtual sources are loads with positive *MCI* values, generators with negative values of marginal offset (those whose emission rates exceed *MCI* at their locations) and congested transmission elements with negative shadow carbon intensities (those for which congestion relief would increase carbon emissions in the system). Virtual sinks are loads with negative values of *MCI*, generators with positive values of marginal offsets (such as renewable generation at locations with a positive *MCI*) and transmission elements with positive *SCI* (those for which congestion relief would reduce carbon emissions).

According to the Carbon Footprint Theorem, in presence of transmission congestion carbon footprint cannot be fully attributed to loads and generators, some portion of the footprint has to be attributed to congested transmission. This attribution is informative: relieving constraints with positive carbon footprint would increase overall carbon emissions in the system while relieving constraints with negative footprint would reduce overall carbon emissions in the system. This assessment is important for the purpose of transmission planning and assessing its impact on the system-wide carbon emissions.

4 Illustrative Examples of Locational Carbon Footprint

This section presents examples of the optimal environmental dispatch to illustrate the locational properties of the carbon footprint. Three-bus examples based on those in [9, 10] are discussed first, and an example for the NYISO is presented next.

4.1 Three-Bus System

Consider a three-bus example of an electrical network shown in Fig. 2, which includes two generators, $g1$ and $g2$, located at buses 1 and 2, respectively, and one load attached to each bus. Loads at buses 1 and 2 are small, 1 MW each. Bus 3 has the largest load, 50 MW, and no generation attached to it. Each generator is characterized by three parameters: capacity (MW), cost ($/MWh) and CO_2 emission rate (t/MWh), shown in Table 2. The cost reflects the generator's fuel costs, non-fuel variable O&M expenses and the cost of CO_2 emissions computed as a product of the generator's emission rate and the price of carbon, $10/t in this example.

Without transmission constraints and losses, the optimal dispatch utilizes the least expensive resource (generator $g2$) up to its capacity of 30 MW and meets the remaining 22 MW of demand from generator $g1$. Generator $g1$ is the single marginal generating unit that sets the price and defines the marginal carbon intensity for all locations. Thus, the nodal prices at all buses are equal to the cost of generator $g1$, $34/MWh, and the nodal carbon intensity for all buses are equal to the emission rate of generator $g1$, 0.4 t/MWh. In this example, reducing demand by 1 MWh at any location would reduce carbon emission by 0.4 t, and would save $34 in total generation costs, as detailed in Table 3.

Generator $g1$ has CO_2 offset of zero, and so it does not contribute to the carbon footprint. Generator $g2$ has a negative CO_2 offset of −$0.5/MWh, and its system-wide contribution to carbon footprint is 15 t. Loads at buses 1, 2 and 3 have carbon footprint of 0.4, 0.4 and 20 t, respectively. Altogether, generators and loads are responsible for 35.8 t of carbon emissions, which matches the total emissions

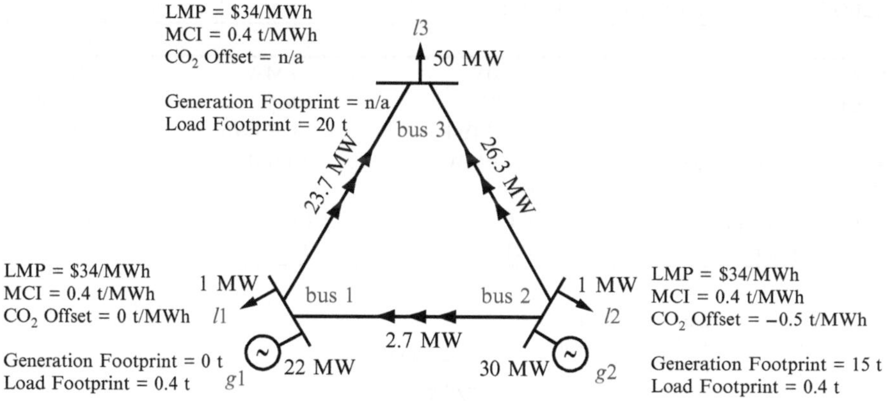

Fig. 2 Unconstrained dispatch in a 3-bus network; all lines have identical impedance

Table 2 3-bus example: generation characterization

Generator	*g*1	*g*2
Fuel + V.O.M. costs	$30/MWh	$20/MWh
Carbon costs	$10/t	$10/t
Carbon emission rate	0.4 t/MWh	0.9 t/MWh
Total variable cost	$34/MWh	$29/MWh
Capacity	50 MW	30 MW

produced by two generators: *g*1 emits 8.8 t (22 MWh times 0.4 t/MWh), *g*2 emits 27 t (30 MWh times 0.9 t/MWh).

Assuming that all lines have the same impedance, 1/3 of the power injected at bus 1 flows to the load at bus 3 along the path 1–2–3, and 2/3 flows along the short path 1–3. For injections at bus 2, 1/3 of the power flows to the load at bus 3 along the path 2–1–3, and 2/3 flows along the short path 2–3. The resulting flows are as shown in Fig. 2.[6]

Consider now the case where the flow on the line 2–3 is limited at 20 MW, shown in Fig. 3. The unconstrained dispatch of Fig. 2 is not feasible, because it results in a flow of 26.3 MW which is above the limit.[7] A redispatch is necessary to accommodate this constraint. The optimal dispatch and corresponding nodal prices are shown on Fig. 3. In this case both generating units are marginal. Nodal prices at their buses are equal to their respective offers of $34/MWh and $29/MWh, respectively. Indeed, the only way to serve an incremental MW of demand at bus 1 is to

[6] Transmission paths were used only for explanation purposes in this example. The dispatch and resulting flows are determined directly, without the need to consider paths, using an OPF formulation such as the one in (49)–(52) in the Appendix.

[7] Note that flows on power systems distribute over the different lines following Kirchoff's laws, and are typically not controllable.

Table 3 3-bus example: unconstrained case

Node	1	2	3
Generation	22 MW	30 MW	n/a
Load	1 MW	1 MW	50 MW
Nodal price	\$34/MWh	\$34/MWh	\$/MWh
Marginal carbon intensity	0.4 t/MWh	0.4 t/MWh	0.4 t/MWh
CO_2 offset	0 t	−0.5 t	n/a
Generation footprint	0 t	15 t	n/a
Load footprint	0.4 t	0.4 t	20 t

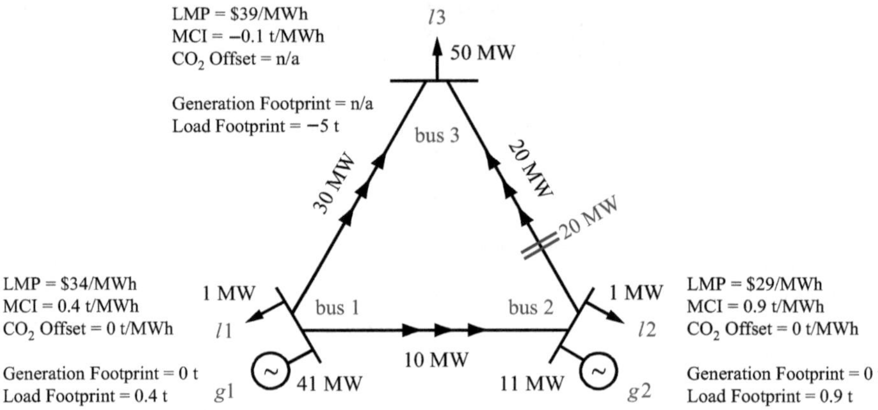

Fig. 3 Constrained dispatch in a 3-bus network; all lines have identical impedance

dispatch generator $g1$, because any attempt to obtain any portion of that power from generator $g2$ would increase the flow over line 2–3 and violate the constraint. The least expensive way to serve an incremental demand at bus 2 is to ramp up unit $g2$. However, to serve an incremental MW at bus 3 at least cost without violating the transmission constraint 2–3, it would be necessary to increase generation at $g1$ by 2 MW and reduce the output of generator $g2$ by 1 MW. Hence the cost of serving that incremental demand is \$39/MWh $=$ 2 $\times$ \$34/MWh−1 $\times$ \$29/MWh.

The same logic is applicable for computing locational marginal carbon emission rates. Since incremental load at bus 1 is served by unit $g1$, the marginal carbon intensity at this bus is 0.4 t/MWh, the emission rate of $g1$. An incremental load at bus 2is served by unit $g2$ and its marginal carbon intensity is 0.9 t/MWh. As explained earlier, to serve an incremental load at bus 3 it is necessary to increase generation at bus 1 by 2 MW and reduce generation at bus 2 by 1 MW. The result of this redispatch action would be an increase of carbon emissions at bus 1 by 2 $\times$ 0.4 t/MWh and a reduction of carbon emissions at bus 2 by 0.9 t/MWh adding up to an overall 0.1 t/MWh reduction in carbon emissions. Thus, the marginal carbon intensity at bus 3 is −0.1 t/MWh (Table 4).

Given these results we conclude that demand reduction at buses 1 and 2 would cause a reduction in carbon emissions but demand reduction at bus 3 would actually increase carbon emissions despite the fact that this location has the highest demand

Table 4 3-bus example: constrained case

Node	1	2	3
Generation	41 MW	11 MW	n/a
Load	1 MW	1 MW	50 MW
Nodal price	\$34/MWh	\$29/MWh	\$39/MWh
Marginal carbon intensity	0.4 t/MWh	0.9 t/MWh	−0.1 t/MWh
CO_2 offset	0 t	0 t	n/a
Generation footprint	0 t	0 t	n/a
Load footprint	0.4 t	0.9 t	−5 t

and highest prices in the system and therefore is the most attractive target of a demand response program. Although a demand response program may make most economic sense, it would not reduce carbon emissions but increase them.

Consider the addition of 1 MW of carbon-free wind generation to the system. Assuming zero operating costs of wind power, this generation addition can be modeled as a 1 MW reduction in demand. Wind addition to buses 1 or 2 would reduce carbon emissions by 0.4 or 0.9 t, respectively. However, the addition of carbon free wind generation at bus 3 would increase total carbon emissions by 0.1 t.

The counter-intuitive results obtained at bus 3 are due to transmission congestion along the line 2–3, which ramps down unit $g2$ with a high carbon emission rate. This congestion appears to be "good" from the emissions point of view and indirect attempts to relieve the constraint by reducing demand or adding supply at bus 3 would reduce dispatch costs but increase emissions. This can be confirmed by the calculation of the shadow carbon intensity of constraint 2–3. Indeed, increasing the rating of this constraint by 1 MW would allow to increase the dispatch of unit $g2$ by 3 MW while reducing by 3 MW the dispatch of unit $g1$. Doing so would reduce dispatch costs by $-\$15 = 3 \times \$29 - 3 \times \$34$ but would increase carbon emissions by $1.5\ \text{t} = 3 \times 0.9\ \text{t} - 3 \times 0.4\ \text{t}$. In other words, the shadow carbon intensity of constraint 2–3 is -1.5 t/MWh.

The distribution of the carbon footprint among system elements in this case is very different from that of the unconstrained example. Since both generators are marginal, their marginal offsets are zero and they make no contribution to the carbon footprint. Loads at buses 1 and 2 have carbon footprint of 0.4 and 0.9 t, respectively, while load at bus 3 has a negative carbon footprint of -5 t. The carbon footprint of binding constraint 2–3 is 30 t. Hence, loads at 1 and 2 and constrained transmission line 2–3 serve as virtual sources of carbon while load at bus 3 serves as a virtual sink. Generators are virtually carbon-neutral. The overall balance matches the system-wide carbon emissions of 26.3 t. The bulk of the carbon footprint is neither attributable to loads, nor to generators, but is concentrated on the congested transmission line 2–3.

Note that the results are not absolute and depend on the underlying price of carbon emissions. Currently, fossil-fueled generation costs are dominated by fuel costs. Thus, the cost minimizing dispatch essentially minimizes the costs of fuel. If carbon costs increase so that they begin to affect the generation merit order, prices and marginal nodal carbon intensities would show a higher correlation.

To illustrate that, consider the case with an increased carbon cost of $30/t, and let the limit on line 1–3 be 32 MW. The total variable costs are then $42/MWh for $g1$ and $47/MWh for $g2$. This change in carbon costs changes the merit order: now $g1$ has lower costs than $g2$, as opposed to the case with $10/t carbon cost. As a result, the dispatch changes such that line 2–3 is no longer constrained, but line 1–3 operates at its maximum rating of 32 MW. The solution is shown in Fig. 4 and Table 5. The price at bus 3 is $52/MWh, since a 1 MW increase in load at bus 3 causes $g1$ to decrease generation by 1 MW and $g2$ to increase generation by 2 MW. Hence, the increase in total emissions due to the load increment at bus 3 is 1.4 t/MWh, compared to −0.1 t/MWh with the low carbon cost case. Note that in this case the increase in the carbon cost made the carbon intensity at node 3 positive. Thus, a change in the carbon price made node 3 the best place for a load reduction program that aims to reduce emissions. Before the price increment, bus 3 was the worst place to conserve energy with the objective of reducing emissions.

In the $30/t carbon case, the carbon footprint of $l3$ is 70 t, well above total carbon emissions in the system. The load carbon footprint is offset by the negative footprint of constrained transmission line 1–3, which serves as a virtual carbon sink with negative footprint of −48 t.

Comparison of the last two examples shows that an increase in the price of carbon could make a significant change in transmission congestion. With shifting of congestion from 2–3 to 1–3, the carbon intensity of bus 3 changes in both the sign and magnitude. As a result, bus 3 is transformed from being a virtual carbon sink in the $10/t carbon case to a virtual carbon source in the $30/t carbon case. In the $10/t carbon case, constraint 2–3 is a virtual carbon source: increasing flow through this constraint by relieving it would increase system-wide emissions. In the $30/t carbon case, constraint 1–3 is a virtual carbon sink: increasing flow through this constraint by relieving it would reduce system-wide emissions.

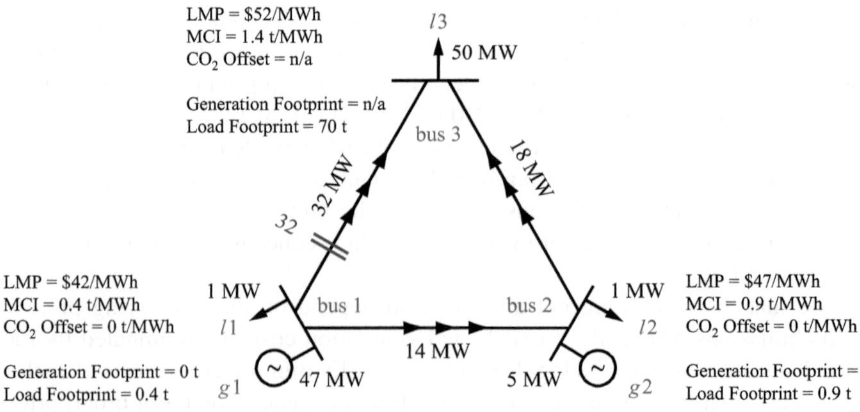

Fig. 4 Constrained dispatch in a 3-bus network; high carbon cost case

Table 5 3-bus example: high carbon cost case

Node	1	2	3
Generation	22 MW	30 MW	n/a
Load	1 MW	1 MW	50 MW
Nodal price	\$42/MWh	\$47/MWh	\$52/MWh
Marginal carbon intensity	0.4 t/MWh	0.9 t/MWh	1.4 t/MWh
CO_2 offset	0 t	0 t	n/a
Generation footprint	0 t	0 t	n/a
Load footprint	0.4 t	0.9 t	70 t

4.2 New York Independent System Operator Example

The *MCI* contour in Fig. 5 is based on an hourly production simulation of the Eastern Interconnection, with a focus on the NYISO system. The hour in question presents binding transmission constraints that create price separation between the eastern portion of New York state and the rest of the state. Effectively, a load increment in eastern New York in this hour is supplied by a combination of combined cycle and gas-fired steam units, and the MCI is above 0.4 t/MWh. Load increments in central and western New York state, on the contrary, are provided by exports from Ontario (combined cycle units) and Pennsylvania (coal-fired steam units). Due to the binding transmission constraints, load increments in the area shown in white require an increase in exports from Ontario and a decrease in exports from Pennsylvania, leading to negative values of MCI. In this rather extensive area, and for this hour, load conservation measures would increase the system wide carbon emissions, as would any increase in renewable generation. In their analyses of Eastern Interconnection simulations, the authors have observed that negative values of MCI occur relatively frequently due to transmission congestion and the significant difference between carbon emission rates for different thermal generation technologies.

5 Theoretical Analysis of Traditional and Carbon Controlling RPS Programs

In this section, we apply the results obtained above to the analysis of the Renewable Portfolio Standards (RPS) policy.

5.1 Supply Curves of Renewable Resources

For the purpose of this analysis, we consider an RPS policy as an investment optimization problem. Assume that there are multiple developers of renewable

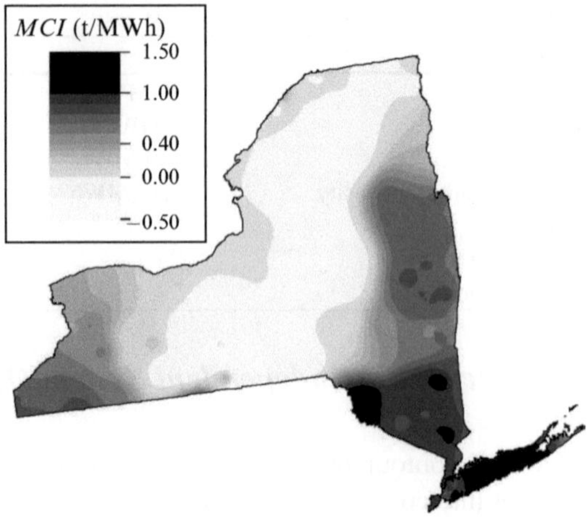

Fig. 5 Marginal carbon footprint in the state of New York for 1 h

resources interested in bringing their projects online. In doing so, they evaluate the physical potential of the resource and the revenue potential in the electricity market (for energy and installed capacity) against costs to develop and finance their projects and on that basis determine locational capacity supply curves for each resource. A supply curve for a given resource indicates the level of subsidy per unit of incremental installed capacity the developer seeks in order to bring the resource online.

Developers then convey these supply curves to the RPS Agency, a hypothetical decision maker responsible for selecting the RPS portfolio. The objective of the RPS Agency is to meet RPS requirements at lowest costs measured as the total subsidy distributed to project developers on the basis of submitted supply curves.

We consider two alternative formulations of this problem: a traditional formulation in which the RPS requirements are set as the desired level of energy generated by renewable resources, and a carbon controlling formulation in which the requirements are set in terms of CO_2 emissions released into the atmosphere by the entire generating system. In both formulations, we assume that the RPS Agency has a perfect knowledge of all future operational details of the power system and therefore is capable of reaching an optimal decision. Decision rules derived from the analysis of these two problems provide valuable theoretical insights that could be used to develop an efficient market mechanism for practical administering of actual RPS programs.

Consider an electrical grid with n potential interconnection points for renewable resources. Assume that each renewable resource is characterized by a specific temporal availability profile $g_j(t)$ such that $0 \leq gj(t) \leq 1$ and that the maximum available generation from that resource is equal to $G_j^{max}(t) = R_{jgj}(t)$ where R_j is the installed capacity of resource j.

Availability profiles for wind and solar generation are determined primarily by weather conditions, whereas biomass profiles are assumed to be driven by forced and planned outages in absence of which the resource is assumed to be available to generate at full capacity. Without loss of generality, we assume that each grid interconnection point is characterized by a unique availability profile and a unique supply curve. In case when multiple resources are interconnected to the same physical point on the grid, we assume multiple replicas of that interconnection point with a unique numerical index assigned to each replica.

For each interconnection point, the RPS Agency aggregates the capacity supply curves provided by the developers. Denote the aggregate by $Z'_j(R_j)$; $j = 1, 2, \ldots, n$, representing the subsidy per MW required to bring the R_jth MW of renewable capacity online at location j. This supply curve is assumed to be a non-negative monotonically non-descending function of cumulative capacity.[8] The RPS Agency then integrates the supply curves $Z'_j(R_j)$ over quantity (R_j) to obtain the cost curves $Z_j(R_j)$; $j = 1, 2, \ldots, n$ representing the total level of capital subsidy required in order to bring online R_j MW of renewable capacity installed at that location. Due to the assumptions on Z'_j, the cost curves represent convex, continuously differentiable, monotonically increasing functions of cumulative capacity.[9]

5.2 *Marginal Impacts of Renewable Generation*

Section 2 studies the impact on carbon emissions of incremental renewable generation added at a given location on the grid which is expressed in the concept of the marginal carbon offset of the generator defined by (5).

As explained in [9], the same mathematics underlying the computations of marginal carbon intensity could be applied to other characteristics of the power system, for example to tracking the locational marginal intensity of renewable generation. To define this concept, consider the total renewable energy dispatched by the power system over a given time period and denote it by R. Clearly, this renewable energy is a function of the renewable installed capacity $R_1, \ldots, R_n$, as well as the corresponding profiles and all parameters impacting the dispatch problem. Assume now an infinitesimal change in electricity demand $dL_j(t)$ at location j and define the locational marginal renewable intensity as

---

[8] Thus we assume that capacity could be added in infinitesimally small increments, a reasonable assumption for wind and solar technology. The assumption that the supply functions are monotonic is more restrictive. Non-monotonic supply functions do not alter the major conclusions of this paper, but create some technical difficulties both from theoretical and market design perspectives which are not discussed in this paper.

[9] For the purpose of this paper we assume smooth and convex cost curves. The introduction of non-smooth, piece-wise differentiable cost curves and capacity constraints would not change the major conclusions. However, while the resulting mathematical problem would not become intractable, it would require the use of a more sophisticated mathematical technique making it more difficult to follow the important concepts of the analysis.

$$\rho j(t) = \frac{\partial \Re}{\partial L_j(t)}. \tag{23}$$

Similarly to the behavior of marginal carbon intensity, marginal renewable intensity equals the weighted average of per unit renewable generation by marginal generating units with exactly the same weighting coefficients that would be used to compute marginal carbon intensities. For a given generating unit, its per unit renewable generation equals 1 for a renewable generator and 0 for a non-renewable generator. If no renewable generators are on the margin, marginal renewable intensity for all locations is zero. Although the dispatch costs of renewable generation such as wind and solar are low, significant penetration of these resources can make them marginal in hours of low demand or local transmission congestion. When renewable generators become marginal, the locational marginal renewable intensity could become non-zero and could be positive or negative depending on location.

Imagine now that we connect a renewable generator to the grid at point j. Each incremental MWh of power produced by that generator will inject 1 MWh of renewable power and emit σ_j tons of CO_2.[10] At the same time, the overall system-wide renewable generation is reduced by $\rho j(t)$ and the system-wide CO_2 emissions are reduced by $MCI_j(t)$. The net result of this would be an increase in system-wide renewable generation by

$$\alpha_j^R(t) = 1 - \rho_j(t), \tag{24}$$

and a reduction in system-wide CO_2 emissions by

$$\alpha_j^C(t) = MCI_j(t) - \sigma_j. \tag{25}$$

Parameters $\alpha_j^R(t)$ and $\alpha_j^C(t)$ represent the *marginal increase* of renewable generation and the marginal offset of carbon emissions, respectively, provided by 1 unit of energy produced by a given renewable resource.

Consider now a build out scenario of renewable resources characterized by the values $R_1, R_2, ..., R_n$, assume the dispatch profile of each renewable resource determined by an economic dispatch of the grid and denote these respective dispatch profiles as $r_1(t), r_2(t), ..., r_n(t)$.

For the purpose of further analysis we introduce average resource impacts over the entire time period $[1, T]$.

Definition 4. The *average marginal renewable increase* is

[10] The CO_2 emission rate σ_j of a renewable resource is typically zero. However, for the sake of generality, we assume that it may deviate from zero (for example if some auxiliary fuel is being used to support the renewable technology).

$$\overline{\alpha}_j^R = \frac{\sum_{t=1}^{T} \alpha_j^R(t) r_j(t)}{TR_j}. \tag{26}$$

It is easy to see that if the resource is never marginal, $\overline{\alpha}_j^R$ is simply equal to the resource capacity factor, if measured in MWh of renewable energy per MW of installed capacity, and equal to 1 measured in MWh of renewable energy per MWh of renewable energy.

Definition 5. The *average marginal carbon offset* is

$$\overline{\alpha}_j^C = \frac{\sum_{t=1}^{T} \alpha_j^C(t) r_j(t)}{TR_j}. \tag{27}$$

5.3 *Optimal RPS Investment Rules*

This section analyzes two RPS formulations: a traditional formulation where the requirement is in terms of renewable generation, and a carbon-controlling formulation where the requirement is in terms of reduction in carbon emissions.

5.3.1 Traditional RPS Problem

The traditional RPS problem could be formulated as a cost-minimization mechanism targeting a desired level of generated renewable energy.

$$\min_R \sum_{j=1}^{n} Z_j(R_j) \tag{28}$$

s.t.

$$\Re(R_1, ..., R_n) \geq \Re_0, \tag{29}$$

where $\Re_0$ is the target level of renewable generation.

A different formulation of the traditional RPS problem would be to maximize renewable generation subject to the budgetary constraint:

$$\max_R \Re(R_1, ..., R_n) \tag{30}$$

s.t.

$$\sum_{j=1}^{n} Z_j(R_j) \leq B_0 \tag{31}$$

where B_0 is the total budget subsidy available.

5.3.2 Carbon-Controlling RPS Problem

The carbon-controlling RPS problem could be formulated as a cost-minimization mechanism targeting a desired level of desired level of CO_2 emissions,

$$\min_R \sum_{j=1}^{n} Z_j(R_j) \tag{32}$$

s.t.

$$\mathfrak{C}(R_1, ..., R_n) \leq \mathfrak{C}_0, \tag{33}$$

where $\mathfrak{C}_0$ is the target level of system-wide carbon emissions.

A different formulation of the traditional RPS problem would be to minimize CO_2 emissions subject to the budgetary constraint:

$$\min_R \mathfrak{C}(R_1, ..., R_n) \tag{34}$$

s.t.

$$\sum_{j=1}^{n} Z_j(R_j) \leq B_0, \tag{35}$$

where B_0 is the total budget subsidy available.

5.3.3 Optimal Investment Rules

Theorem 2 *RPS Theorem. 1a. If the traditional RPS problem* ((28)–(29)) *is feasible, its solution is characterized by the following RPS investment rule:*

$$Z'_j(R_j) = P_R \overline{\alpha}_j^R T, \tag{36}$$

where $\overline{\alpha}_j^R$ *is the resource's generation weighted marginal renewable increase and* P_R *is the RPS energy price (in \$/MWh) such that renewable resource additions satisfying* (36) *at that price meet the target level of renewable generation*:

$$\Re(R1,\ldots,R_n) = \Re_0 \text{ if } P_R > 0 \text{ and } P_R = 0 \text{ if } \Re(R1,\ldots,R_n) > \Re_0. \quad (37)$$

1b. If the budget-constrained problem ((30)–(31)) *is feasible, then the optimal investment rule would satisfy* (36) *and the RPS price* P_R *would be selected such that the investments additions satisfying* (36) *at that price meet the budgetary limitation*

$$\sum_{j=1}^{n} Z_j(R_j) = B_0 \text{ if } P_R > 0 \text{ and } P_R = 0 \text{ if } \sum_{j=1}^{n} Z_j(R_j) < B_0. \quad (38)$$

2a. If the carbon-controlling RPS problem ((32)–(33)) *is feasible, its solution is provided by the following investment rule*

$$Z'_j(R_j) = P_C \overline{\alpha}_j^C T, \quad (39)$$

where $\overline{\alpha}_j^C$ *is this resource's generation-weighted marginal carbon offset and* P_C *is the RPS price of carbon (in \$/t) chosen in such a way that renewable resource additions satisfying* (39) *at that price meet the target level of* CO_2 *emissions,*

$$\mathfrak{C}(R1,\ldots,R_n) = \mathfrak{C}_0 \text{ if } P_R > 0 \text{ and } P_R = 0 \text{ if } \Re(R1,\ldots,R_n) < \Re_0. \quad (40)$$

2b. If the budget constrained problem ((34)–(35)) *is feasible, then the optimal investment rule would satisfy* (39) *and the RPS price* P_C *would be selected in such a way that the investments additions satisfying* (39) *at that price meet the budgetary limitation*

$$\sum_{j=1}^{n} Z_j(R_j) = B_0 \text{ if } P_R > 0 \text{ and } P_R = 0 \text{ if } \sum_{j=1}^{n} Z_j(R_j) < B_0. \quad (41)$$

The proof of this theorem is provided in [10].

The optimal investment rules provided by the two RPS models look mathematically similar, but lead to very different investment policies. The investment rule derived from the traditional RPS problem is based on equalizing supply curves on the basis of the *average marginal renewable increase* of individual resources: Locational Carbon Footprint of the Power Industry

$$\frac{Z'_1(R_1)}{\overline{\alpha}_1^R} = \frac{Z'_2(R_2)}{\overline{\alpha}_2^R} = \ldots = \frac{Z'_n(R_n)}{\overline{\alpha}_n^R} = TP_R. \quad (42)$$

Simply put, this rule means that at optimum an incremental dollar subsidy given to any location should buy the same amount of renewable energy produced by the power system. The subsidy price paid to a renewable generator on top of other revenues they receive from energy and capacity markets would be proportional to its renewable increases and equal to $\overline{\alpha}_j^R TP_R$. This combined with (26) leads to the following subsidy applied on a temporal (e.g., hourly) basis:

$$subsidy_j^R(t) = r_j(t)\alpha_j^R(t)P_R. \tag{43}$$

If we assume that renewable resources are never marginal, then all marginal renewable increases are equal to one (MWh/MWh) and in this case, (43) implies that all renewable resources should receive the same amount of subsidy on a per MWh basis in each hour regardless of their location on the grid. The value of the subsidy is set by the RPS price. If, due to a high penetration of renewables or inadequate transmission capacity renewable resources become marginal, the subsidy should be adjusted to account for that effect and the subsidy rule (42) should provide zero payment to a marginal renewable resource. Note that this subsidy scheme is different to schemes that provide a fixed compensation per MWh, such as production tax credits. In particular, with the subsidy scheme (43) a resource would not have an incentive to produce at maximum available capacity when renewable resources are on the margin, since the subsidy is zero, while production tax credits give incentives to produce at the maximum regardless of what units are on the margin, leading to undesirable effects such as electricity price distortion.

In contrast, the investment rule derived from the carbon controlling RPS problem is based on equalizing supply curves on the basis of *average marginal carbon offsets* of individual resources:

$$\frac{Z'_1(R_1)}{\overline{\alpha}_1^C} = \frac{Z'_2(R_2)}{\overline{\alpha}_2^C} = \ldots = \frac{Z'_n(R_n)}{\overline{\alpha}_n^C} = TP_C. \tag{44}$$

This rule implies that at optimum an incremental dollar subsidy given to any location should buy the same amount of CO_2 emission reduction in the power system. The subsidy price of that carbon reduction paid to a renewable generator on a per ton of CO_2 basis on top of other revenues they receive from energy and capacity markets would be equal to $\overline{\alpha}_j^C TP_R$. This combined with (27), leads to the following subsidy formula applied on a temporal basis:

$$subsidy_j^C(t) = r_j(t)\alpha_j^C(t)P_C. \tag{45}$$

This subsidy formula indicates that the optimal method for providing subsidy to renewable generation is to compensate it on the basis of its carbon footprint. Indeed, according to the Carbon Footprint Theorem, $\mathbf{CF}[Rj](t) = -r_j(t)\alpha_j^C(t)$ and therefore (45) could be restated as

$$subsidy_j^C(t) = -\mathbf{CF}[R_j](t)P_C. \tag{46}$$

Since carbon footprint varies by location and over time, investment decisions resulting from investment rules (42)–(43) and (44)–(45) are structurally and materially different. Between two resources with identical marginal subsidy requirements, the traditional investment rule ((42)–(43)) would favor locations with the higher capacity factor $\overline{\alpha}_j^R$, since they would provide more renewable energy. An investment prescription under the carbon controlling RPS would favor the resource with a higher average marginal carbon offset, since it would offer more contribution to the reduction of CO_2 emissions.

5.4 Comparing Two RPS Programs: An Illustrative Example

Consider a simple example of a two-Zone power system separated by a constrained transmission line, as presented in Fig. 6. In this system, Zone B is always short in energy which it imports from Zone A. Zone A is assumed to contain predominantly coal-fired generation. Zone B is dominated by natural gas-fired generating technology. During on-peak hours, the transmission line connecting these two Zones gets constrained leading to price separation between them.

During on-peak hours, Zone B prices are set by gas-fired technology with average on-peak market heat rate of 9,000 Btu/kWh; Zone A prices are set by coal-fired generation with market coal-based heat rate of 12,000 Btu/kWh. During off-peak hours, transmission is not constrained and the price for the entire system is set by the gas-fired generation in Zone A at market heat rate of 7,000 Btu/kWh.

Both Zones have a significant potential of wind resources, and at both locations technical and cost characteristics of these wind resources are identical. For simplicity, we assume that the addition of wind resources at any Zone makes no impact on market prices and heat rates.

Using the above information we estimate the need for subsidy in order to develop wind resources in each Zone computed as the difference between annual carrying costs associated with building and financing a wind farm and revenues the wind farm is expected to receive in the energy market.

The results of these calculations and underlying market assumptions are presented in Table 6. As shown in that table, we assume a CO_2 controlling

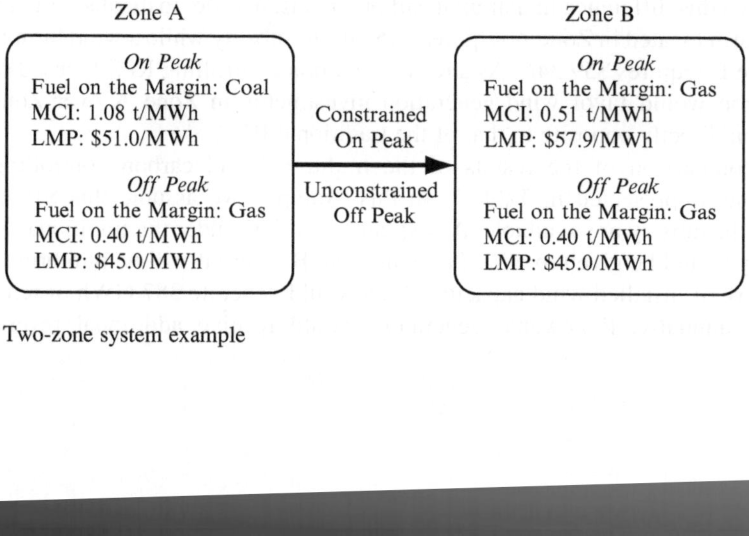

Fig. 6 Two-zone system example

Table 6 Two-zone system characteristics

Zone	A			B		
	On-peak	Off-peak	Annual	On-peak	Off-peak	Annual
Hours	4,160	4,600	8,760	4,160	4,600	8,760
Fuel on the margin	Coal	Gas	n/a	Gas	Gas	n/a
Market heat rate (Btu/kWh)	12,000	7,000	n/a	9,000	7,000	n/a
Fuel price ($/MMBTU)	2	5	n/a	5	5	n/a
CO_2 cost or tax ($/t)	25	25	25	25	25	25
Wind capacity factor	20%	40%	31%	20%	40%	31%
Marg. Renewable Inc. (MWh/MWh)	1	1	1	1	1	1
MCI (t/MWh)	1.08	0.40	0.74	0.51	0.40	0.46
Marg. CO_2 offsets (t/MWh)	1.08	0.40	0.61	0.51	0.40	0.44
LMP ($/MWh)	51.00	45.00	48.74	57.86	45.00	52.06
Wind energy revenue ($/MW)	42,430	82,800	125,230	48,140	82,800	130,940
Wind capital requirements ($/MW)			200,000			200,000
Wind subsidy needed ($/MW)			74,770			69,060

regulatory policy with an underlying price of $25/t of CO_2, natural gas and coal prices at $5/MMBtu and $2/MMBtu, respectively and an annual carrying cost of building a wind farm of $200,000/MW of installed capacity. Based on these assumptions, in order to be financially viable, a wind farm in Zone A requires a $27.98/MWh in subsidy while a similar wind farm in Zone B requires subsidy of $25.85/MWh only. A traditional RPS program focused on the total renewable energy would favor wind generation investments in Zone B over Zone A.

However, as shown in Table 6, marginal carbon offsets generated by renewable energy differ significantly between Zones. During on-peak hours, 1 MWh of renewable energy generated in Zone A displaces 1.08 t of CO_2, compared to 0.51 t for 1 MWh of renewable energy generated in Zone B. Due to transmission congestion during on-peak hours, renewable energy in Zone A displaces coal-fired generation with CO_2 emission rate of 1.08 t/MWh. At the same time, in Zone B marginal technology is gas-fired with a much lower emission rate of 0.51 t/MWh. In off-peak hours, in absence of congestion, renewable energy generated and Zone A or Zone B have identical carbon offsets of 0.4 t/MWh. On average over a year, marginal carbon offset of renewable generation in Zone A is 0.61 t/MWh in Zone A and 0.44 t/MWh in Zone B.

Given this difference in marginal carbon offsets, in order to displace 1 t of CO_2, a wind farm located in Zone A requires $45.74/t in subsidy while a wind farm located in Zone B requires $59.34/t. As a result, a carbon controlling RPS focused on CO_2 reduction would favor wind generation investments in Zone A over Zone B, a decision directly opposite to that of the traditional RPS.

A comparison of the results of the traditional and carbon controlling RPS programs is presented in Table 7. In this example, we assume that $10 million dollars in subsidy are available. As explained earlier, under the traditional RPS all subsidy would be given to wind farms in Zone B where it will help to bring online 145 MW of installed wind capacity which would generate 387 GWh of renewable energy annually. Renewable generators would receive additional revenues of

Table 7 Results for traditional and carbon-controlling RPS programs

RPS type	Traditional			Carbon-controlling		
Budget constraint	$10,000,000			$10,000,000		
Zone	A	B	Total	A	B	Total
Required subsidy ($/MWh renewable energy)	27.98	25.85	25.85	27.98	25.85	27.98
Required subsidy ($/t carbon offset)	45.74	59.34	59.34	45.74	59.34	45.74
Subsidized wind capacity (MW)	–	145	145	134	–	134
Subsidized wind energy (GWh)	–	387	387	357	–	357
CO_2 emissions offset (1,000 t)	–	169	169	219	–	219
RPS price ($/MWh)			25.85			27.98
RPS price ($/t CO_2)			59.34			45.74

$25.85/MWh for energy they inject into the system. In other words, the RPS price for this program is $25.85/MWh. The resulting annual carbon offset under this program would be 169,000 t of CO_2. Given the marginal carbon offset of this investment of 0.51 t/MWh, the corresponding marginal cost of CO_2 reduction in the form of the RPS subsidy is $59.34.

In contrast, under the carbon-controlling RPS all the subsidy would be diverted to support wind farms in Zone A. Given that more subsidy is needed there per unit of installed capacity, only 134 MW of wind power will be brought online, compared to 145 MW under the traditional RPS. Only 357 GWh of renewable energy would be produced, compared to 387 GWh under the traditional RPS. However, the carbon-controlling RPS will displace 219,000 t of CO_2 which exceeds the carbon offset of the traditional RPS of 169,000 t by 30%. In order to administer the carbon controlling RPS, renewable generators would be paid the RPS price of $45.74/t of CO_2 assessed on the basis of the negative hourly carbon footprint of the wind farm. The marginal cost of subsidy per MWh of renewable energy in this case is $27.98/MWh which is higher than under the traditional RPS.

These results are exactly what should be expected: when the program is designed to maximize the renewable energy, it results in the least expensive way of doing just that and yields a higher level of renewable energy produced and lower marginal cost of renewable energy than the alternative. When the program is designed to maximize the amount of carbon displaced (by means of minimizing the total level of carbon emitted), it finds the least expensive way of achieving this goal and yields the higher level of displaced mass of CO_2 emissions and the lower marginal cost of CO_2 abatement than the traditional RPS program.

5.5 RPS Implementation Issues

According to the US Department of Energy [16], presently there are 29 states plus the District of Columbia that have RPS policies in place. Together these states account for more than half of the electricity sales in the United States. Five other states, North Dakota, South Dakota, Utah, Virginia, and Vermont, have nonbinding goals for adoption of renewable energy instead of an RPS. All known RPS programs are designed in a manner which requires electricity providers to obtain a minimum percentage of their power from renewable energy resources by a certain date. In other words, all existing RPS programs in the US simply target the total amount of renewable energy generated regardless of the efficiency of CO_2 abatement ultimately achievable by these programs. The American Clean Energy and Security Act [17], passed by the US House of Representatives and presently under consideration by the US Senate, sets the national Renewable Energy Standard also specified in terms of the minimum percentage of total electricity production which has to be provided by renewable resources.

Based on these design goals, as discussed in the previous section, rational developers of renewable generation would prefer renewable projects at locations that are most financially attractive and would seek transmission interconnection for these projects to areas with highest electricity prices regardless of the level of CO_2 emissions these projects would be able to displace. While there is nothing wrong with this strategy on the part of the developers, the current design of RPS programs is highly questionable. If the proclaimed goal of these programs is to promote renewable generation in order to displace CO_2 emissions, the design is not consistent with this goal and will most likely lead to suboptimal levels of CO_2 emission reductions.

Existing RPS programs goals are set in physical terms (e.g. annual level of renewable generation). Trading of renewable credits set RPS prices. Similarly, the goal of a carbon controlling RPS could be set in physical terms, i.e. in terms of the total carbon footprint of renewable resources and the RPS prices could be determined through trading between buyers and sellers. This approach would provide market participants with a rule leading to an optimal strategy for investment in renewable resources. Alternatively, the program could be designed by directly setting the RPS price and providing the RPS subsidy to renewable generators based on (45).

In order to implement a carbon-controlling RPS program, the following need to happen: (1) electricity market operators should calculate and publish locational marginal carbon intensities for all nodes in the system which should be used to compute marginal carbon offsets for renewable generators; (2) the RPS goal should be set either in physical terms or by specifying the RPS price of carbon; (3) the subsidy should be paid to renewable generators according to (45); and (4) an RPS cost recovery mechanism should be established in order to equitably allocate RPS costs among electricity consumers. Addressing these issues goes beyond the scope of this paper.

6 Conclusions

A widespread interest in understanding carbon footprint associated with economic activity of individuals and organizations requires means for a precise measurement of that footprint. As an industry, power generation is a single largest source of carbon emissions. For example, power generation accounts for over 40% of carbon emissions in the United States [18]. This paper presents a rigorous methodology for determining the carbon footprint of electricity consumption based on the economics and engineering rules of the power industry. Understanding and measuring the temporal and locational nature of the carbon footprint associated with electricity consumption is important for the design of all carbon abatement policies affecting the location and timing of the use of electric energy. The range of potential application of this methodology spreads from carbon footprint reporting to very specific policy implications for power industry stakeholders.

There are a number of public policy and individual economic decisions that are directly affected by the discussion of the temporal and locational nature of CO_2. Understanding of this concept provides analytical means for analyzing the efficiency of the design of Renewable Portfolio Standards, among other things. However, the proposed RPS analysis is only one among other potential applications of this concept.

The RPS approach presently implemented in the United States and in other countries should not be considered efficient as long as ultimate goal of the RPS policy is the reduction of greenhouse gases. An alternative approach based on the subsidy rule compensating renewable generation for the negative carbon footprint they provide achieves the optimal strategy in targeting carbon emissions through RPS. This alternative design would require a precise calculation of locational marginal carbon intensity which could be provided by system operators responsible for the operations of regional electrical grids and a design of renewable cost allocation rules by regulatory agencies.

The alternative RPS program described in this paper can be combined with other carbon regulations, such as cap and trade programs or carbon taxes, to offer an effective two-prong approach to CO_2 emissions reduction. The carbon regulation program would determine, either implicitly or explicitly, the price of carbon emissions, in effect penalizing carbon emitters. The alternative RPS program would provide direct incentives for the installation of renewable generation to complement the penalties imposed by the other programs and to ensure that the electricity generation resource mix transitions into a low emissions one.

Appendix

This appendix provides a proof of (4), (8) and (9) by computing sensitivities of the optimal generation dispatch $\mathbf{G}(t)$ with respect to changes in parameters that affect the dispatch. An example of a varied parameter is the load level at a given node. First, the dispatch model is introduced and then the sensitivities are obtained

Dispatch Model

Assume that the linearized lossless DC assumptions hold, so that the flows $\mathbf{F}$ on transmission elements and flow gates are modeled by

$$\mathbf{F} = \mathbf{F}^o + \Psi(\mathbf{G} - \mathbf{G}^o - (\mathbf{L} - \mathbf{L}^o)). \tag{47}$$

where the superscript ^{o} indicates a base case power flow quantity, and $\mathbf{G}$ and $\mathbf{L}$ are the vectors of nodal power generation and loads, respectively. The *transmission sensitivity matrix* Ψ [19] gives the variations in flows due to changes in the nodal injections, with the reference bus assumed to ensure the real power balance. For a given point in time, the system operator dispatches the committed units to minimize the total costs of operations. Assume that the generation costs are piecewise linear, and denote the vector of nodal generation variable costs in \$/MWh by $\mathbf{c}$. These costs account for fuel costs $\mathbf{c}_\mathrm{f}$, variable operation and maintenance costs $\mathbf{c}_o$, and emissions costs,

$$\mathbf{c} = \mathbf{c}_f + \mathbf{c}_o + \mathbf{R}_e\mathbf{c}_e. \tag{48}$$

The matrix $\mathbf{R}_e$ gives the emissions rates of different pollutants for each generator, and the vector $\mathbf{c}_e$ gives the cost of emitting each pollutant per unit of emission (e.g., in the form of an emission allowance price under a cap-and-trade program, or a carbon tax).

The economic dispatch solved is a lossless linearized DC OPF that takes into account environmental costs, along the lines of the ones discussed in [20, 21],

$$\min_{\mathbf{p}} \mathbf{c}'\mathbf{G} \tag{49}$$

$$\text{s.t } \mathbf{1}'\mathbf{G} = \mathbf{1}'\mathbf{L} \leftrightarrow \lambda \tag{50}$$

$$\Psi(\mathbf{G} - \mathbf{L}) \le \overline{\mathbf{F}} \leftrightarrow \overline{\mu} \tag{51}$$

$$\underline{\mathbf{G}} \le \mathbf{G} \le \overline{\mathbf{G}} \leftrightarrow \underline{\gamma}, \overline{\gamma}. \tag{52}$$

Constraint (50) ensures the load-generation balance, (51) enforces the flow limits on transmission elements, and (52) models the lower and upper generation limits. Note that the base power flow quantities from (47) are included in the limits in (51), for notational simplicity. The nodal prices are given by

$$\pi = \lambda\mathbf{1} - \Psi'\overline{\mu}. \tag{53}$$

The *marginal units* are those that may change their output at an optimal redispatch in response to a small variation of a system parameter. They consist of the units that are dispatched between their generating limits, and of those that are at the limit but with the corresponding dual variables $\overline{\gamma}$ and $\underline{\gamma}$ equal to 0. These units have their nodal prices equal to their costs. We denote the cost vector for the marginal units by $\widetilde{\mathrm{c}}$ and the generation vector for marginal units by $\widetilde{\mathbf{G}}$.

Differential Load Perturbation

Assume that the solution of (49)–(52) exists and is unique,[11] that the solution of the dual problem to (49)–(52) exists and is unique as well,[12] and that the optimal values of the dual variables for all binding inequality constraints are non-zero.[13] Consider a differential change ds in a parameter of (49)–(52). To accommodate this differential change, the system operator uses the solution of (49)–(52) and solves a differential redispatch to correct for the parameter change. Under the assumptions stated above, the solutions of the primal and dual problems are continuous functions of the parameter in a neighborhood of its base case value. Hence, since the parameter change is infinitely small and the dual variables for the binding inequality constraints are non-zero, the inequality constraints that do not bind in (49)–(52) would not bind in the redispatch, and the inequality constraints that bind in (49)–(52) would bind in the redispatch. In particular, the parameter change results in dispatch changes for the marginal generators only, i.e., $d\mathbf{G}_n$ is nonzero only if generator n is marginal.

Consider a differential increment dL_k in the load at a specified node k.[14] As discussed in the previous paragraph, a differential change in load brings a differential change in the dispatch of the marginal generators only. Moreover, since the line limits and the binding constraints do not change, the flows on the binding lines also do not change. Let us denote the change in the dispatch of marginal generators by $d\widetilde{\mathrm{G}}$ and construct the reduced sensitivity column vector ψ_k and reduced sensitivity matrix $\widetilde{\Psi}$ consisting of the elements of Ψ which contain the impact of injections at node k and at the marginal generator buses, respectively on binding constraints.

Using these reduced matrices, we can formulate the optimization problem that is solved for the redispatch as

$$\min_{d\widetilde{\mathbf{G}}} \widetilde{\mathbf{c}}' \, d\widetilde{\mathbf{G}} \tag{54}$$

[11] Uniqueness rules out cases such as the existence of multiple marginal generators at the same bus.

[12] Existence and uniqueness of the dual problem solution implies a non-empty feasible set for the differential redispatch problems discussed next (problems (54)–(56) and (60)–(63)). Alternatively, one could assume regularity of the primal problem solution, i.e., the gradients of the binding constraints are linearly independent.

[13] This strict complementary slackness assumption ensures that differential perturbations will not change the set of binding inequality constraints.

[14] Or, equivalently, a differential increment $-dL_k$ in the node k generation.

$$\text{s.t} \qquad 1' d\widetilde{\mathbf{G}} = dL_k \leftrightarrow \widetilde{\lambda} \tag{55}$$

$$\left[\psi_k, \widetilde{\Psi}\right]\begin{bmatrix} -dL_k \\ d\widetilde{\mathbf{G}} \end{bmatrix} = \mathbf{0} \leftrightarrow \widetilde{\boldsymbol{\mu}}. \tag{56}$$

The differential redispatch problem (54)–(56) is feasible, since the optimal value of the dual variables for (49)–(52) exists. Moreover, since the solution of (49)–(52) is unique, and the optimal values of the dual variables for all binding inequality constraints are non-zero and are continuous functions of the varied parameter it is easy to demonstrate that the solution of (54)–(56) is also unique.

The Lagrangian for (54)–(56) is

$$\mathscr{L} = \widetilde{\mathbf{c}}' d\widetilde{\mathbf{G}} + \widetilde{\lambda}\left(1' d\widetilde{\mathbf{G}} - dL_k\right) + \widetilde{\mu}'\left[\psi_k, \widetilde{\Psi}\right]\left[-dL_k, d\widetilde{\mathbf{G}}'\right]'. \tag{57}$$

The optimal solution for (54)–(56) satisfies

$$\frac{\partial \widetilde{\mathscr{L}}}{\partial d\widetilde{\mathbf{G}}} = \widetilde{\mathbf{c}}' + \widetilde{\lambda}\mathbf{1}' + \widetilde{\boldsymbol{\mu}}'\widetilde{\Psi} = 0. \tag{58}$$

From (58), we obtain that the nodal prices at the marginal generators' buses do not change with the differential change in load. The necessary condition for optimality is also sufficient in this case, since the objective function and the constraints are linear, and therefore convex [22].

Due to the strict complementary slackness assumption, (59) is satisfied for every differential perturbation $d\widetilde{\mathbf{G}}$. Thus, given the uniqueness of the solution of (54)–(56) we obtain that the feasible set has only one element, and that is

$$\beta_k = \frac{d\widetilde{\mathbf{G}}}{dL_k} = \begin{bmatrix} \mathbf{1}' \\ \widetilde{\Psi} \end{bmatrix}^{-1} \begin{bmatrix} 1 \\ \psi_k \end{bmatrix}, \tag{59}$$

where (59) was obtained by a rearrangement of (55)–(56). This proves (4). Existence of the optimal solution necessarily guarantees that the matrix in the left hand side in (59) can be inverted.

Differential Change in Line Rating

Now consider a differential increment dF_ℓ in the line rating of a specified binding line l. As before, the differential change in line rating brings a differential change in the dispatch of the marginal generators only, and the flows on the binding lines do not change, except for the flow on line ℓ, which increases exactly by dF_ℓ.

Proceeding as in the previous subsection, denote the change in the dispatch of the marginal units by $d\widetilde{\text{G}}$, and construct the reduced transmission sensitivity row vector $\widetilde{\psi}_l$ and matrix $\widetilde{\psi}_{-l}$ consisting of flow sensitivities for the line l and for the rest of the flow elements with binding limits, respectively, with respect to injections at the marginal generator buses. The optimization problem that is solved for the redispatch is

$$\min_{d\widetilde{\mathbf{G}}} \widetilde{\mathbf{c}}' \, d\widetilde{\mathbf{G}} \tag{60}$$

$$\text{s.t.} \qquad \mathbf{1}' \, d\widetilde{\mathbf{G}} = 0 \leftrightarrow \widetilde{\lambda} \tag{61}$$

$$\tilde{\psi}_\ell d\tilde{\mathbf{G}} = dF_\ell \leftrightarrow \tilde{\mu}_\ell \tag{62}$$

$$\tilde{\Psi}_{-\ell} d\tilde{\mathbf{G}} = \mathbf{0} \leftrightarrow \tilde{\boldsymbol{\mu}}_{-\ell}. \tag{63}$$

Using the same argument as in the previous section, one concludes that the feasible set of (60)–(63) has only one element, and that is

$$\gamma_l = \frac{d\tilde{\mathbf{G}}}{df_\ell} = \begin{bmatrix} \mathbf{1}' \\ \tilde{\psi} l \\ \tilde{\Psi} - \ell \end{bmatrix}^{-1} \begin{bmatrix} 0 \\ 1 \\ \mathbf{0} \end{bmatrix}, \tag{64}$$

where (64) was obtained by a rearrangement of (61)–(63). This proves (8).

Relations Between Nodal and Transmission Shadow Carbon Intensities

Let bus 0 be the reference bus, and $\text{MC}I_0(t)$ be the marginal carbon intensity at the reference bus. By the definition of the transmission sensitivity matrix, $\psi_0 = 0$.

Hence,

$$MCI_0(t) = \tilde{\sigma}' \begin{bmatrix} 1' \\ \tilde{\Psi} \end{bmatrix}^{-1} \begin{bmatrix} 1 \\ \mathbf{0} \end{bmatrix}, \tag{65}$$

where $\tilde{\sigma}$ is the vector of emission rates of marginal units.

Consider now the marginal carbon intensity $MCI_k(t)$ node k,

$$MCI_k(t) = \tilde{\boldsymbol{\sigma}}' \begin{bmatrix} \mathbf{1}' \\ \tilde{\Psi} \end{bmatrix}^{-1} \begin{bmatrix} 1 \\ \psi_k \end{bmatrix} = \tilde{\boldsymbol{\sigma}}' \begin{bmatrix} \mathbf{1}' \\ \tilde{\Psi} \end{bmatrix}^{-1} \begin{bmatrix} 1 \\ 0 \end{bmatrix} + \sum_{\ell, \overline{\mu}_\ell > 0} \psi_{lk} \tilde{\boldsymbol{\sigma}}' \begin{bmatrix} \mathbf{1} \\ \tilde{\psi}_\ell \\ \tilde{\Psi}_{-\ell} \end{bmatrix}^{-1} \begin{bmatrix} 0 \\ 1 \\ \mathbf{0} \end{bmatrix}$$
$$= MCI_0(t) - \sum_l \psi_{lk} SCI_l(t). \tag{66}$$

References

1. The Regional Greenhouse Gas Initiative Inc (2005) Regional greenhouse gas initiative memorandum of understanding, New York. http://www.rggi.org
2. Schwarzenegger A (2008) Exec. Order S-14-08, Office of the Governor of the State of California, Sacramento [Online]. http://gov.ca.gov/executive-order/11072/
3. Midwest ISO (2012) Multi value projects portfolio results and analyses, Indiana. http://www.midwestiso.org
4. Shavel I (2009) Testimony before the Federal Energy Regulatory Commission on behalf of ITC Green Power Express, LLC. FERC, Washington, Docket ER09-681-000
5. Spiegel RJ, Greenberg DL, Kern EC, House DE (2000) Emissions reduction data for grid-connected photovoltaic power systems. Solar Energy 68(5):475–485
6. Bettle R, Pout C, Hitchin E (2006) Interactions between electricity-saving measures and carbon emissions from power generation in England and Wales. Energy Policy 34(18): 3434–3446
7. Grubb M, Neuhoff K (2006) Allocation and competitiveness in the EU emissions trading scheme: policy overview. Clim Policy 6(1):7–30
8. Chen Y, Sijm J, Hobbs BF, Lise W (2008) Implications of CO [Trial mode] emissions trading for short-run electricity market outcomes in northwest Europe. J Regul Econ 34(3):251–281
9. Ruiz P, Rudkevich A (2010) Analysis of marginal carbon intensities in constrained power networks. In: Proceedings of the 43rd Hawaii international conference systems science, Koloa
10. Rudkevich A (2010) Locational carbon footprint and renewable portfolio standards. In: Proceedings of the 7th conference economics energy markets, Toulouse [Online]. http://www.idei.fr///doc/conf/eem/rudkevich.pdf
11. Holland SP, Mansur ET (2006) The short-run effects of time-varying prices in competitive electricity markets. Energy J 27(4):127–155
12. The Carbon Trust (2007) Carbon footprinting: an introduction for organizations. The Carbon Trust, Witney, Publication CTV033 [Online]. http://www.carbontrust.co.uk
13. Scientific Applications International Corporation (SAIC) (2006) Life cycle assessment: principles and practices. U.S. Environmental Protection Agency, Washington, EPA 600/R06/060
14. Clean Development Mechanism (2009) Tool to calculate the emission factor for an electricity system, UNFCCC, EB 50 Report, Annex 14. http://cdm.unfccc.int/EB/050/eb50_repan14.pdf
15. Schweppe FC, Caramanis MC, Tabors RD, Bohn RE (1987) Spot pricing of electricity. Kluwer, Norwell
16. U.S. Department of Energy (2009) States with renewable portfolio standards, Washington. http://apps1.eere.energy.gov/states/maps/renewable_portfolio_states.cfm

17. United States Congress (2009) American Clean Energy and Security Act. H. R. 2454, 111th United States Congress, Washington
18. Lindstrom P (2008) Emissions of greenhouse gases report. Energy Information Administration, Washington, Report DOE/EIA-0573, Chap. 2 [Online]. http://www.eia.doe.gov/environment
19. Cheng X, Overbye TJ (2006) An energy reference bus independent LMP decomposition algorithm. IEEE Trans Power Syst 21(3):1041–1049
20. Delson JK (1974) Controlled emission dispatch. IEEE Trans Power App Syst PAS-93: 1359–1366
21. Talaq JH, El-Hawary F, El-Hawary ME (1994) A summary of environmental/economic dispatch algorithms. IEEE Trans Power Syst 9(3):1508–1516
22. Bertsekas DP (1999) Nonlinear programming, 2nd edn. Athena Scientific, Belmont

Optimal Operational Strategies for CO_2 Emission Reduction in Sustainable Energy Systems

Pierluigi Mancarella and Gianfranco Chicco

Abstract In the energy sector, the new millennium is bringing unprecedented challenges related to global warming, which are representing a major driver for change in energy system planning, design and operation. In this context, and on the basis of the state-of-the-art literature in the field, the aim of this chapter is to illustrate the fundamental aspects relevant to energy system operation in the presence of CO_2 emission-related issues and constraints, with an indicative time horizon from 1 day to 1 week. Within this time frame, the role of CO_2 emissions in optimal operational strategies to exploit the energy system equipment is opportunely identified, illustrated and modelled. In particular, this chapter discusses a number of optimisation problem formulations set up in different frameworks, with the aim of highlighting their characteristics and at the same time synthesising the main relevant points of the studies.

Keywords CO_2 • Operational planning • Optimisation • Emission reduction • Energy efficiency • Environmental impact • Sustainable energy

1 Introduction

The environmental challenges of the new millennium are leading to substantial changes in the way energy systems are operated. In particular, current debates put the focus on CO_2 emissions as the major greenhouse gas (GHG) contributing to

P. Mancarella (✉)
School of Electrical and Electronic Engineering, The University of Manchester, Ferranti Building, M13 9PL, Manchester, UK
e-mail: p.mancarella@manchester.ac.uk

G. Chicco
Dipartimento di Ingegneria Elettrica, Politecnico di Torino, corso Duca degli Abruzzi 24, 10129 Torino, Italy
e-mail: gianfranco.chicco@polito.it

Q.P. Zheng et al. (eds.), *Handbook of CO_2 in Power Systems*, Energy Systems,
DOI 10.1007/978-3-642-27431-2_9, © Springer-Verlag Berlin Heidelberg 2012

global warming of anthropogenic origin. The key relevance of carbon dioxide as GHG depends on its massive production in thermally-driven energy supply systems based on burning fossil fuels. The emitted CO_2 could remain in the environment for several decades, therefore adequate long-term as well as short-term strategies need to be timely devised in order to ensure *sustainability* of the energy system and prevent dramatic consequences for the future generations. Extensive policy actions are being undertaken with the aim of reducing the volume of CO_2 in the atmosphere, including taxation schemes, incentives for low-carbon generation, market framework developments, and so forth, and challenging emission reduction targets have been set in a number of jurisdictions worldwide. This will eventually lead (and is already leading) to unprecedented changes in the classical operation and development paradigm of power and energy systems. At the same time, an increasing number of relevant studies are being performed in order to formulate advanced models and solutions to run power and energy systems economically while enhancing their overall environmental efficiency and meeting classical and new constraints, in case imposed by emerging regulations.

The aim of this chapter is to illustrate issues and models relevant to power and energy system *operation* that take into account CO_2 emissions. The general discussion is conducted across four basic dimensions:

1. *Time*: Effective measures for decarbonising the power system can be identified and developed along different time horizons, from long-term (planning horizon, years) to short-term (operation horizon, days). More specifically, on the operational side, positive contributions may come from exploiting suitable control strategies for the equipment used for energy production (Section 2.4).
2. *Space*: CO_2 emission reduction can be targeted in both the large scale (regional, national or worldwide, because of the global GHG impact) and relatively small scales (up to local plants, to address the applications of specific technologies of different size). Both centralised energy systems and distributed energy systems are considered here, with particular focus on multi-generation solutions due to their characteristics in improving energy efficiency.
3. *Fuel*: A third dimension that could commonly appear in studies of CO_2 emission effects is represented by considering different numbers of input–output interactions within the energy system in terms of energy vectors. This translates into taking into account different types of fuel used to supply the energy system (also including possibilities of and solutions with multiple fuel switching capability) and different types of energy vectors that are generated, with relevant characterisation and modeling of the system environmental performance. In this respect, significant benefits on CO_2 emission reduction refer to the adoption of renewable energy sources (RES) and their integration in energy systems.
4. *Economics*: Overall, the economics dimension is driving the development of acceptable technical solutions, also regarding their operation. Environment and economics need to be addressed at the same time, otherwise the system might be ‘green’ but completely uneconomical and then unacceptable, or the system

might be economical but not 'green' (as in the business-as-usual case in many jurisdictions).

This chapter is organised as follows. Section 2 introduces the main concepts concerning time scales, space dimensions and fuel variables in energy system analyses. The relations between CO_2 emissions and other aspects of energy system studies, such as energy efficiency, cost reductions, emissions of local pollutants, emission trading schemes and carbon capture systems are summarised. These relations serve as basic components for the analyses illustrated in Section 3, where CO_2 emissions are incorporated into a comprehensive optimisation framework, describing different approaches including energy system operation. Section 4 contains the concluding remarks. The list of references mainly points to relevant publications in books and journal papers, especially with reference to contributions issued in recent years.

2 CO_2 Emissions in Energy System Analysis

2.1 *Time Scales and Space Dimensions*

Different *time scales* are usually involved in energy system analysis, spanning a broad range of durations from seconds (real-time electrical dynamics) to hours (operation for economic transactions, for instance) up to dozens of years (power plant and network strategic investment planning).

Among the major categories of analysis that can be identified according to the time dimension, it is possible to mention for the purpose of environmental studies:

- *Real-time* problems, involving both electrical and thermal phenomena from seconds to hours, which may entail optimal operational strategies;
- *Operational planning* problems, where simulations of the energy system operation are carried out with a certain time resolution (for instance, hourly) with the purpose of addressing the planning problem in deeper detail;
- *Expansion planning* or *design* problems, addressing the need for new investment in generation systems or networks to pursue specific objectives such as meeting load growth;
- *Strategic* problems, in which the problem of new (relatively long-term) investment is dealt with, not necessarily needing detailed simulation of the energy system but where all the major factors affecting energy system economics and environmental outputs (such as system siting, fuel types, and so forth) are taken into account.

The operational *time* scale addressed in this chapter indicatively ranges from 1 min to one or a few days. Electrical dynamics, typically much faster than thermal dynamics, are not represented in detail for the purpose of studies concerning CO_2 emissions. Time-dependent quantities are expressed by considering their average values in a given time interval of defined duration, rather than by their instantaneous values, so that the energy referring to the time interval duration also defines the average power.

From a *space* dimension standpoint, the studies can be typically applied to different resolutions, ranging from the individual piece of equipment or local energy system (for instance, to address energy efficiency aspects) to a city (to assess pre-specified environmental targets) up to countries (to study the overall power system for combined generation/network investment planning strategies and assess the compliance to objectives of international environmental commitments).

Various levels of aggregation of different sites can be taken into account, for instance when emission limits are set for the generation portfolio of a specific generation owner, network operator or energy trader. In this case, the spatial distribution of the energy system involved is less relevant, as only the *global* aspect of CO_2 emissions becomes significant for this type of studies.

Considering time and space, the energy system analysis and optimisation including CO_2 emissions addressed here is formulated with reference to operation problems and mixed problems with interactions between planning and operation (operational planning).

2.2 Emission Factors, Energy Efficiency and Fuel Switching

Energy generation from fossil fuels causes emissions of carbon dioxide from oxidation of carbon with oxygen in the burning reaction. Considering CO_2 emissions from fossil fuels, a widespread approach to characterise the emissions from a generic combustion device is represented by the *emission factor* model [1–3]. Let us consider the mass $m^X_{CO_2}$ (typically expressed in [g]) of CO_2 emitted to produce the useful energy output X, for instance electrical energy W [kWh$_e$] or heat Q [kWh$_t$], and the *emission factor* $\mu^X_{CO_2}$ or *specific mass emissions* of CO_2 per unity of X, expressed in [g/kWh]. The mass $m^X_{CO_2}$ can be represented by using the following model:

$$m^X_{CO_2} = \mu^X_{CO_2} X \tag{1}$$

As a special case, it is possible to substitute X with the primary energy F released when burning a given fuel, which yields the definition of the emission factor $\mu^F_{CO_2}$ related to the fuel thermal energy content [2, 3].

The mass of CO_2 emitted when burning a given type of fuel can be assessed with good approximation according to the nominal characteristics of the relevant chemical reactions, and depends in particular on the carbon content and the lower heating value (*LHV*) of the specific fuel [2]. At first approximation, the fuel-related emission factor $\mu^F_{CO_2}$ is a function of the fuel type only, and can be considered constant under various operating conditions [4]. For instance, typical emission factor values for natural gas are in the order of 200 g/kWh$_t$ based on the *LHV*. Relevant figures for other fuels can be found for example in [5]. On these bases, a generic energy *output-related* emission factor model can be estimated by taking into account that the fuel input F and the generic useful energy output X are linked through the relevant generation efficiency η_x. For instance, the electrical efficiency η_e takes into account the generation of useful electricity W in a power plant from the fuel input F. Thus, the relation among output-related and fuel-related emission factors becomes

$$\mu_{CO_2}^X = \frac{\mu_{CO_2}^F}{\eta_x} \tag{2}$$

The expression (2) is the key formula to understand the role of both *energy efficiency* and *fuel switching* for fossil-based energy systems. In fact, with roughly constant $\mu_{CO_2}^F$ for a given fuel, the inverse proportionality of energy efficiency and CO_2 emissions provides evidence on the direct link between CO_2 emission reduction and reduction in the fuel consumption. Technologies able to supply the same useful energy outputs by using less fuel and bringing energy efficiency improvements are then particularly useful to limit CO_2 emissions. On the other hand, given a certain value of conversion efficiency η_x, the specific emissions per unit of energy output decrease if a fuel with lower carbon value is used, which represents the main rationale for fuel switching from coal to natural gas in a number of countries.

While CO_2 is the main GHG of interest in energy system applications involving fossil fuels, also other emissions could be of concern in terms of global warming [6], such as methane emissions, which are for instance the large majority of the unburned hydrocarbons in natural gas-fuelled units. In this regard, the models illustrated in this work, focused on CO_2 emissions, can be readily generalised to other GHG by exploiting the fact that the effects of a generic GHG can be measured against the effects of CO_2 through the relevant Global Warming Potential (*GWP*) [7], normalised with respect to the CO_2 impact. Hence, $GWP_{CO_2} = 1$, and the term GWP_{GHG} represents the mass of CO_2 equivalent to the emission of a unity of mass of the generic GHG under consideration. By introducing the emission factor μ_{GHG}^X related to the generic GHG rather than CO_2, the CO_2 equivalent emission factor for a generic GHG pollutant can thus be expressed as

$$\mu_{CO_2\,eq,GHG}^X = GWP_{GHG}\mu_{GHG}^X \tag{3}$$

As far as RES are concerned, there is no fuel burning involved in their operation, and as such the emission factor model cannot be applied. However, within the above general emission factor framework, moving from fossil fuels to RES can be seen as an extreme case of fuel switching to a zero-carbon 'fuel'. Indeed, the major changes in energy system and particularly power system evolution in the last decade have been driven by the uptake of a number of RES such as wind, photovoltaics, solar thermal power, and so on, following the concept that these technologies are virtually CO_2-free [8, 9]. On the other hand, the extension of the study boundaries beyond the operational stage of such plants, with consideration of all the cradle-to-grave emissions, calls for the application of full life cycle assessment (LCA) concepts and techniques [10, 11]. However, these details are beyond the operational aspects considered in this chapter.

2.3 Environmental Efficiency and Polygeneration

An effective way to increase the overall efficiency is, when possible, to generate additional energy vectors besides electricity, leading to the concept of cogeneration or Combined Heat and Power (CHP), trigeneration or Combined Cooling Heat and Power (CCHP), and in general multi-generation or polygeneration [12–16]. For thermal power plants, decreasing the input quantity while providing the same useful output leads to energy efficiency increase and CO_2 emissions reduction.

The assessment of environmental efficiency in polygeneration systems is not straightforward, and adequate models are needed. For instance, emission reduction from cogeneration and seasonal trigeneration is discussed in [1], and further studies are reported in [3, 4, 17]. Generalisation to multi-generation is addressed in [18], by formulating a synthetic approach to assess environmental efficiency of multi-generation systems in terms of CO_2 emission reduction with respect to conventional separate production (SP) technologies for the generation of the same energy vectors [15]. In this approach, the energy system is seen as a black-box characterised by equivalent performance indices on the basis of the only input–output energy flows, without going into the detailed representation of the internal components [19]. An application to environmental modelling of integrated CHP-electric heat pumps is provided in [16].

Let us consider the set **X** containing the energy vectors considered for polygeneration. For SP, two further terms are introduced, with $m_{CO_2}^{F,SP}$ representing the CO_2 mass [g] that would be emitted if the energy vectors were produced in SP, estimated through the *reference* emission factors in separate production for the energy vectors $X \in \mathbf{X}$. For instance, in a cogeneration plant the relevant emission factors would be $\mu_{CO_2}^{F}$ $[g/kWh_t]$ for the fuel input to the CHP system, $\mu_{CO_2}^{W,SP}$ $[g/kWh_e]$ for the conventional separate production of electricity, and $\mu_{CO_2}^{Q,SP}$ $[g/kWh_t]$ for the conventional separate production of heat [20].

According to the general model presented in [18], the Polygeneration CO_2 Emission Reduction (*PCO₂ER*), or Polygeneration Carbon Dioxide Emission Reduction (*PCDER*) indicator is defined as

$$PCDER = \frac{m_{CO_2}^{F,SP} - m_{CO_2}^{F}}{m_{CO_2}^{F,SP}} = 1 - \frac{F\mu_{CO_2}^{F}}{\sum_{X \in \mathbf{X}} X\mu_{CO_2}^{X,SP}} \tag{4}$$

Positive values of the indicator (4) correspond to positive CO_2 emission reduction brought by the polygeneration system relative to the conventional SP references.

The *PCDER* indicator can be used in different time scales, including the ones of interest for operation purposes. In [18] the *PCDER* indicator is derived under a unified approach to energy efficiency and emission reduction assessment. The energy efficiency side is represented by the Polygeneration Primary Energy Saving (*PPES*) indicator, that generalises the classical Primary Energy Saving (*PES*) indicator used for cogeneration [21] and the Trigeneration Primary Energy Saving (*TPES*) indicator introduced for trigeneration in [22]. Considering the fuel energy (primary energy) that would be needed in SP to generate the same energy vectors X through conventional SP benchmark models, and the conventional SP efficiency of the generic type of energy x associated to the energy vector X produced, the *PPES* is defined as

$$PPES = \frac{F^{SP} - F}{F^{SP}} = 1 - \frac{F}{\sum_{X \in \mathbf{X}} \frac{X}{\eta_x^{SP}}} \tag{5}$$

Furthermore, it is possible to introduce the CO_2 emission equivalent efficiency, defined as [18]

$$\eta_{CO_2,x}^{F,SP} = \frac{\mu_{CO_2}^{F}}{\mu_{CO_2}^{X,SP}} \tag{6}$$

Thus, the *PCDER* indicator can be rewritten in formal analogy with the *PPES* indicator:

$$PCDER = 1 - \frac{F}{\sum_{X \in \mathbf{X}} \frac{X}{\eta_{CO_2,x}^{F,SP}}} \tag{7}$$

In this way, the *PCDER* and the *PPES* indicators have essentially the same structure, the only difference being the efficiency 'weights'. These 'weights' incorporate the conceptual difference between energy and emissions. When assessing energy transformations starting from given outputs, what actually matters are only the efficiencies and not the type of fuel used in the various systems. A 'blind' black-box model can be formulated by introducing the efficiencies relevant to the specific technologies. Yet, in CO_2 emission assessment not only the technologies (that is, efficiencies), but also the fuel inputs to the systems or equivalent black-boxes are relevant [4], as different fuels exhibit different carbon contents and therefore generate different CO_2 emissions. The fuel information is then introduced by the CO_2 emission equivalent efficiency (6).

Considering different types of fuels and equipment characteristics, with different CO_2 emission profiles, the numerical value of (6) changes depending on the polygeneration and SP fuel inputs, making environmental assessment more complicated than energy efficiency assessment. For instance, for electricity generation $\mu_{CO_2}^{W,SP}$ could refer to an average or marginal value modelling the bulk power system emissions, or could refer to a specific power plant type (e.g., natural gas-fuelled combined cycle gas turbines). Likewise, the relevant CO_2 emission equivalent electrical efficiency $\eta_{CO_2,e}^{F,SP}$, calculated on the basis of (6), would depend on the specific fuel input (for instance, diesel or bio-masses) through $\mu_{CO_2}^{F}$.

In the assessment methodologies that make reference to conventional models (of SP, in this specific case), the equivalent characteristics of the SP benchmarks play a key role. In particular, the conventional values assigned to the SP entries involved in the specific analysis may depend on the purpose of the analysis, and may represent matter of policy and regulatory discussion (as for the *PES* indicator at the European level [21]).

2.4 Marginal Operation and Control Strategies

At a system level, another major difference between conventional fossil-based technologies and RES also lies in the fact that RES cannot be dispatched as conventional generation, in that they depend on availability and uncertainty of the underlying natural source. This brings about the issue of what technologies and plants are actually to be considered as alternatives in energy system studies. In fact, while average emission factors for overall energy generation brings important information on the carbon intensity of a given energy system, more detailed studies may need an assessment based on high time resolution (for instance, hourly emissions), identifying the types of generation technologies involved in the operation, principally as *marginal* generators. These aspects are particularly relevant when assessing, for instance, different generation paradigm such as large conventional generation against small distributed generation, or energy systems involving the production of multiple energy vectors.

Utilisation of average (and for instance yearly-based) emission factors rather than (hourly) marginal emission factors may lead to substantially different results. In fact, in some contexts marginal operation is usually related to dispatchable and fossil-based power plants, whose emissions can be quite higher than for base-load power plants (nuclear and hydro sources, combined cycles gas turbines, and so forth) and RES that make up the overall generation portfolio [15, 20]. This indirectly means that the environmental assessment of alternative scenarios may contain subjective components that may be dependent on the purpose of the study, as well as on political views and strategies. From this outlook, overlooking the marginal operation of energy systems could generate flawed environmental results or send biased market signals [23–25]. However, this type of studies could be quite complicated, and competitive energy markets even exacerbate the difficulty of running such analyses. In fact, fast-evolving market scenarios could prompt the deployment of different sources in spite of being environmentally unfriendly and claimed to be phased out soon. A typical example is represented by coal power plants that, although low-efficient and high-polluting, can be widely utilised in the presence of high gas prices, displacing less polluting but eventually more costly gas plants. Therefore, it is clear how adequate measurement and information frameworks should be set up to boost optimal environmental energy system operation, and policy action in this respect should be carefully planned.

Directly related to the system dispatchability of fossil generators is the adoption of predefined *control strategies*. For large-scale centralised power plants, the control strategies are driven by market operation and bilateral contracts for both energy and ancillary service provision. However, for smaller (distributed) generation the operational strategies to be adopted may depend on different factors. In particular, the following strategies may be typically found:

- *On-off*: particularly for small plants (below few MW_e), in many cases the operators prefer to run their plants without applying on-line control strategies, deciding to maintain the plant *on* for a certain period of time (for instance, during daytime) and switching it *off* when it is not convenient to run it (for instance, during the night). This happens in particular under regulated tariffs for buying/selling electricity, switching the plant *off* in the periods in which the cost of buying electricity from the electrical distribution systems is relatively low and, correspondingly, the price for selling the electricity produced would be low as well.
- *Load tracking* (or *load-following*): in this case, the plant is operated by following the (local) load. A particularly significant case concerning this type of operation refers to combined generation plants producing both electricity and heat, for which the plant operation can in case and most often be driven by the thermal needs rather than the electrical needs.
- *Spark-spread*: in this case, the operation of the plant is modulated according to the relative value/price of electricity output (or in case electricity input, if the plant is switched off and electricity is bought from the grid to supply the local load) and fuel input. This would typically occur in a deregulated market, where ideally both fuel (typically gas) and electricity prices would change for instance hourly, so that the operator can evaluate based on the market spot prices what the best strategy would be. This type of strategy could also be adopted for large power plants, with the most stringent constraints being represented by issues such as minimum up/down times, maximum ramping rates, maximum number of switch on/off operations, and so on (see also Section 3.1.3).

2.5 Taxes, Emission Trading Schemes and Market Interactions

It is apparent that the promotion of high-efficiency and low-emission energy solutions is strongly related to the associated economics, even when technical and energy performance can be outstanding. In this light, several regulatory actions, incentives and schemes have been introduced worldwide with the aim of promoting efficient technologies and a more rational energy use for developing more sustainable energy networks [26, 27]. In particular, specific *energy-related* markets [28], such as markets for emission allowances and renewable energy certificates [29], have been developed to deal with different energy vectors such as heat and above all electricity. These kinds of markets, although not directly trading energy commodities, have a notable influence on how the electricity prices are formed, with *cross-market* interactions that, in spite of emerging only in the latest years, are likely to become much more significant in the future.

The origin of these markets lies in the need for internalising external effects, above all from an environmental outlook, and for boosting energy efficiency and renewable production while decreasing CO_2 emissions [30]. In fact, as externalities are a form of market failure, government interventions are justified in order to minimise their distorting market influence and impact on the community. Hence, reduction of CO_2 emissions can be pursued through the regulatory lever by introducing specific taxes on more polluting technologies [31]. The effects of these taxes should be to promote investments for CO_2 sequestration [32] or for switching from one *fuel* to another (including renewable options), provided that the related costs are lower than the tax amount. Under a market approach, including externalities and other distorting influences into the design of competitive energy markets is a logical evolution of governments' responsibility, so that a number of market options or regulatory incentives frameworks have been or can be established. Internalisation can indeed take place through market mechanisms (trading) or alternative regulatory mechanisms (feed-in tariffs, financial discounts, emission limits, etc.). In this respect, several types of market-related schemes are already shaped up in different countries. *Green markets* are for instance an area of importance, referring to various market-based mechanisms established in order to promote cost-effective expansion of renewable electricity. They are typically implemented in the form of tradable certificates (*green certificates* or *renewable quotas*) which can be used to meet regulatory requirements. *White certificates* represent a similar form of incentives, in which quotas earned upon proof of and proportionally to specific energy saving actions can be traded to increase energy efficiency (*efficiency market*). *Cap-and-trade* schemes represent a further typical form of energy-related markets, such as the European Emission Trading System (ETS) [33] for CO_2 emission allowances. For instance, an interesting application is shown in [34] using emission allowances as a constraint and calculating specific economic indicators representing the social welfare in function of the CO_2 emissions. Likewise, *pollution permits* might be set out as well to limit negative effects of local air pollution, and this should be taken into account within a framework aiming at minimising also CO_2 emissions, due to the possible conflicting objectives concerning the reduction of different types of emissions.

2.6 *Operational Flexibility of CO_2 Capture Systems*

CO_2 capture is an essential practice for continuing the use of coal-based plants, due to the prominent role of these plants in determining CO_2 emissions. The CO_2 capture system becomes an overstructure to be integrated in the local energy system, needs power to be supplied and energy to be exploited, and results in additional costs for system operation [35].

CO_2 capture is investigated in most cases by taking into account the overall energy effect of the capture systems, considering a specified level of consumption in regards to the plant capacity. Recent studies [36] have been dedicated to the evaluation of the CO_2 capture effects during the energy system *operation*, accounting for the daily variation of the electricity demand and prices. The CO_2 capture systems are switched on and off depending on the loading level, in particular switching the CO_2 capture system off at peak loading in order to avoid limiting the amount of power available to serve the load in the most demanding conditions for the system. In this way, the costs to build additional capacity to supply the CO_2 capture system at peak load is limited as well. Flexibility of turning on/off the CO_2 capture system is not generally feasible [37], but can be used in some cases, such as in post-combustion CO_2 absorption.

A comparison among different operation strategies for exploiting the CO_2 capture system for coal-fired power plants is carried out in [36]. The options for the CO_2 capture system include always-on operation, system switch-off at daily peak hours each day, system switch-off for the entire day in high loading days, and system switch-off only for some days in high loading seasonal periods. Conversely, CO_2 capture for natural gas-fired plants would require to treat a volume of flue gas greater than in coal-fired plants to capture a given amount of CO_2, because of the lower CO_2 emission factor of natural gas. This would lead to greater energy requirement and cost per amount of CO_2 avoided [32].

Possible use of additional power for CO_2 capture in high loading periods would also impact on the reduction of the reserve margins in such periods. In any case, the CO_2 capture system has to be switched off when the reserve margin becomes too limited. In the operational scenarios used in [36] this is reinterpreted as the definition of a load limit beyond which the CO_2 capture system is switched off. In alternative to a load limit, the CO_2 capture systems can be switched off when electricity prices are high. In this case, these systems are used as means to modify the price elasticity of demand [38].

3 Role of CO_2 Emissions in Optimisation Problems

The concepts discussed in the previous sections form the background to interpret the incorporation of CO_2 emission terms in various optimisation problems. Optimisations can be formulated with different perspectives and may refer to different time scales. In order to keep the focus of this chapter on energy system operation, in the sequel only formulations in which the optimisation scope involves operational aspects taking into account CO_2 emissions are considered. Indeed, in a carefully designed plant the operation strategies should have been already considered at the design stage [39]. The chosen strategies also impact on the size and characteristics of the energy supply system. Changing an operation strategy into another one not considered during system design is generally inappropriate and could even provide technically inapplicable solutions (for instance, with undersized components or links).

Extended literature on *economic dispatch* of power plants has been developed with reference to production costs, in some cases including emission reduction as additional objective. However, the emissions considered in this case are generally from *local* pollutants (e.g., NO_2, SO_x) with no direct reference to the CO_2 impact [40, 41]. Emission reduction of CO_2 and local pollutants could sometimes represent *conflicting* objectives, and this aspect needs to be specifically taken into account in the formulation of the optimisation problems.

3.1 Optimisation Problem Formulations

3.1.1 Types of Optimisation Problems

The global nature of CO_2 emissions implies that the relevant issue is the overall effect of these emissions. At first glance, it could seem that the location and the moment in time in which the CO_2 is emitted has no direct impact on the definition of the operational strategies. On this standpoint, the design problem could be solved without specifying the CO_2 emissions in operational terms, just resorting to the determination of the cumulative quantity of these emissions. However, in this section it will be shown how, under specific conditions, both location and time aspects in reference to CO_2 emissions can play a key specific role in energy system analysis.

The definition of the operational conditions requires a sufficiently detailed specification of the time scale (for instance, with hourly time intervals). It is then possible to consider problem formulations in which:

1. *Coupling-in-time* of the operational states is relevant: coupling-in-time occurs in the presence of dynamic constraints (with electrical, mechanical or thermal transients, ramp-rate limits, start-up and shut-down conditions, minimum switch-on/switch-off duration, etc.), or considering the effects of the interaction with storage systems of different type (such as hydro, heat or cooling storage, batteries, plug-in hybrid electric vehicles, and so forth) [42–44]. In these formulations, the operational state of the equipment at a given time moment depends on the past history, and the effects of the decisions taken at that time moment may affect the operating states that can be reached in the future. In these

cases, the solutions are obtained either through multiple time-domain simulations, or by adopting suitable numerical techniques (such as dynamic programming, multi-interval linear programming or heuristics) with which it is possible to handle time coupling effects.

2. Coupling-in-time is *not* relevant: the solution can be obtained by analysing each hour independently of the other hours. In addition, it is possible to aggregate groups of hours carrying consistent characteristics. These groups of hours could be sequential in time, or even non-sequential in time. The latter case corresponds to the use of duration curves [45] based on discretisation of the relevant entry into multiple levels, regardless of the hour to which the entry is referred. If different entries are used, it has to be reminded that in the approach based on duration curves the information on the possible simultaneity among the various entries is lost.

The *location* issue becomes relevant in the presence of regional constraints to CO_2 emissions, such as those introduced by the application of emission trading schemes. Likewise, the moment in time could become relevant owing to the possibility of using a certain energy source to substitute for another source or energy mix with higher CO_2 emissions in specific time periods. Indeed, CO_2 emissions depend on the fuel used, in turn depending on the composition of the energy mix deployed to produce electricity, that changes during the day. For instance, in a context in which CO_2 emissions from the energy mix are higher at some hours of the day, exploiting cogeneration systems during those hours could bring better advantages than exploiting the same systems at different hours of the day. However, the analysis of the benefits requires to consider an overall problem taking into account aspects concerning energy efficiency, emission saving (for CO_2 and other global and local pollutants), and economic benefits. All these aspects lead to a variety of formulations for the optimisation problems involving CO_2 emissions. In this respect, the following distinction is introduced here to identify different types of problems:

- *Type 0*: CO_2 emissions are *not* included in the objective function(s), but are explicitly considered as a *constraint* in the formulation of the optimisation problem;
- *Type 1*: CO_2 emissions are included in the *single* objective function formulated in the optimisation problem;
- *Type 2*: CO_2 emissions are included among the objective functions of a *multi-objective* problem.

The details of the three types of problems are addressed in the sequel, with an overview of a sample of recent literature references.

Other salient aspects impacting on the problem formulations, that could also heavily condition the choice of the solution technique to be adopted, depend on the need for taking into account:

(a) The *uncertainty* of the variables involved in the study, requiring to define random variables (RVs) with suitable probability distributions, to handle the correlations among the RVs, and to identify suitable indicators, for instance based on exceeding probability or risk analysis concepts;

(b) The level of *interaction* with energy or energy-related *markets* (including specific certificates and emission allowances), in order to manage the relevant economics by considering different conditions for the participating entities as price makers, price takers or competing players;

(c) The nature of the *domain* of definition of the relevant variables used in the analysis, with possible non-connected domains due to the existence of the switch-off condition and of a range of operational conditions from the technical minimum to the capacity limit, as well as possible couplings among different energy vectors (such as in the case of cogeneration), and finally the possible presence of non-convex domains [46];

(d) The interactions with different *energy networks*, that introduce further limits to the energy vectors that can be delivered.

3.1.2 Individual Optimisation Objectives

The individual objectives can be classified, according to their nature, as:

- *Technical*, mainly related to energy saving that can result from the adoption of different technologies to serve given energy loads. Generally energy saving assessment is carried out by comparing the effects of local provision of energy with respect to the separate production from external sources or networks. These objectives are often expressed as energy efficiency indicators, such as the *PPES* (Section 2.3).
- *Economic*, including various types of costs and benefits, depending on the time scale of analysis. The *cost* entries include investment costs, operation and maintenance costs (also with start-up and stop costs, especially for large generation units, modelled as functions of exponential [47] or piecewise linear [48] type), costs of fuel depending on markets, tariff systems and contract provisions, costs associated to reliability (i.e., in reference to power and energy not supplied), general costs (e.g., for administration), taxes and penalties of various kinds [49]. The *benefit* entries take into account the revenues obtained from selling energy or energy-related services in market-orientated frameworks, and the effects of possible incentives. Considering operational scenarios, in which the type and size of the technologies are fixed, the economic indicators can be formulated in terms of gross operational margin, obtained by calculating the difference between the operational costs of the solution under study with respect to a predefined reference solution. Other indicators like net present value, payback period and internal rate of return are suitable for long-term planning. In some cases, the operational strategies are directly associated to the alternatives to be compared in the planning study for technologies of various type and size.

- *Environmental*, taking into account global emissions of CO_2 and more generally of greenhouse gases [4], or local emissions of various hazardous pollutants in areas of defined size. CO_2 emission saving may be an alternative objective function with respect to primary energy input from non-renewable sources [50].
- *Social*, dealing with the effects of hazardous pollutants on the human health and with the consumers' preferences to satisfy their energy needs by exploiting local energy production from the energy vectors seen as most suitable.

3.1.3 Constraints

In the formulation of the optimisation problems, the *equality* constraints are generally given by power or energy balances for all the energy vectors involved, at every plant location and for every individual equipment in a multi-generation system. Furthermore, the *inequality* constraints can be of various types, for instance:

- *Network* constraints can be imposed in multi-site energy systems interconnected through various types of energy networks. In addition, in a multi-generation system the links among interconnected units can be described by using independent dispatch factors, constrained by the topology of the system, and further constraints may be included to represent irreversibility of the energy flows in input to or output from each unit [15, 19].
- *Capacity* constraints include maximum and minimum output of generation units and storage systems. For storage systems, also the maximum input and output rates are significant.
- *Operational* constraints, that may depend on the type and size of the generation unit. For instance, the number of switch-on (and switch-off) operations in a given time interval (for instance, one day) can be a critical constraint particularly for large- and medium-size units, heavily impacting on equipment aging and auxiliary circuits maintenance, while it becomes less relevant for micro-generation units. Including the number of switch-on (-off) operations creates a coupling-in-time extended to the entire time period of analysis. Coupling-in-time is also introduced by ramp-rate constraints, depending on the maximum allowable rate of change of power during transients and thus limiting the maximum or minimum output available at a certain time moment, and other constraints such as minimum up and down times; these constraints are usually modelled with non-linear functions, but can be converted into linear constraints [48]. For combined production of different energy vectors, such as electricity and heat, further constraints refer to the feasibility regions defined in terms of electrical and thermal power simultaneously available from various types of technologies [51].
- *Reserve* constraints, imposing that the maximum capacity of the available generation exceeds the loading level by at least an amount given by the minimum level of reserves that must be available in the system in a given time period [52]. Effective assessment of reserve availability needs to take into account the

energy source availability, uncertainty and possibility of finding adequate ways for delivering power in case of need through non-constrained energy corridors. Accurate definition of the ramp-rate constraints is also required in order to describe the possibility of effectively deploying the reserves when needed.
- *Environmental* constraints, that can be imposed on a global or a local scale. *Global* constraints are given by caps imposed by specific emission trading schemes [24, 51, 53, 54] and their definition typically involves multiple jurisdictions. On the *local* scale, the typical constraints refer to the cumulative emissions of specified hazardous pollutants imposed by legislations at international, national or local levels, also taking into account the background effect due to other causes of emission of these pollutants, such as road traffic or other activities.

3.1.4 Solution Methods

The various types of problems call for the exploitation of suitable *solution techniques*. The computational method has to be chosen according to the nature of the problem to be solved. Classical computational models (e.g., based on gradient techniques) could be unable to handle problems with multiple local optima, non-connected domain of definition of the variables, or multiple objectives.

In general, it is possible to make a distinction among the computational techniques used to address:

- A *single objective* function, formed by one objective only, or by the weighted sum of a set of partial objectives;
- *Multi-objective* problems, requiring specifically developed solution algorithms.

For a single objective function, available solution techniques can be based on:

- *Linear programming* (LP), used when it is possible to formulate a linear problem, including the cases in which non-linear aspects are handled by constructing piecewise linear curves or piecewise planar surfaces;
- *Mixed integer-linear programming* (MILP), used when it is not possible to handle all variables through linearisation of the problem, for instance due to the presence of a non-connected domain with individual points; this happens for the domain of operation of a given equipment with an integer switch-off condition and linearised portions of the domain from the minimum technical limit to the maximum capacity;
- *Non-linear programming* (NLP), used when the problem formulation contains non-linear expressions that cannot be approximated in a linear or piecewise linear way, such as in the presence of products of decision variables [19];
- *Mixed integer-nonlinear programming* (MINLP), used when the problem contains non-linear expressions and the domain of definition of one or more variables is non-connected;
- *Meta-heuristics* handling variables defined on *discrete domains*; discretisation of the domain is always possible, provided that the discretisation step is small

enough to avoid losing the characteristics of the domain itself; the choice of the discretisation step clearly impacts on the computation time and on the accuracy of the solution; isolated points and non-connected domains are easily represented within discrete domains.

In the *multi-objective* case, different situations may occur depending on the nature of the individual objectives. In the case of individual objectives *conflicting* with each other, some particularly significant notions are:

- *Non-dominated solution*: a solution is defined as *non-dominated* when no other solution exists with better values for *all* the individual objective functions.
- *Pareto front*: is the set of all the non-dominated points that can be defined for the multi-objective optimisation problem. The solutions located on the Pareto front contain both the optimal points corresponding to the application of the individual optimisation criteria, as well as a number of *compromise solutions* in which none of the individual optima is achieved. The compromise solutions are relevant because they provide viable solution alternatives. Establishing the importance and ranking these alternatives is up to the decision maker. For this purpose, the points on the Pareto front can be ranked by using suitable numerical techniques [55, 56].

Among the solution techniques specifically developed for *multi-objective* problems, the cases reported here for optimisations involving CO_2 emissions refer to the construction of the Pareto front:

- By calculating the *weighted sum* of the outcomes from the individual objectives;
- By introducing part of the objectives as constraints and successively *relaxing* the corresponding constraints;
- With *direct* construction of the Pareto front by using specific techniques to find a subset of the non-dominated solutions denoted as the *best-known* Pareto front; a review of these techniques is illustrated in [57].

Table 1 shows how the types of problems concerning CO_2 emissions and the corresponding techniques of analysis have been addressed in a sample set of recent literature papers.

3.1.5 Integrated Software Packages

Various aspects of energy system planning and operation can be carried out by using specific integrated software packages dealing with multiple types of analysis and optimisation. Some packages are orientated to large-scale analyses and do not model the energy system operation details. In other cases, data can be represented at hourly (or even shorter) rates, thus enabling introducing the system operation characteristics. A non-exhaustive list of packages addressing the explicit representation of CO_2 emissions and system operation in the energy optimisation models include BALMOREL (model for analyses of the electricity and CHP markets in the

Table 1 Types of problems and solution techniques from recent references

Problem type	Type 0	Type 1				Type 2	
Solution technique/ objective(s)	Annual or total costs	Net profit	Cost (with internalised CO_2 emission costs) or environomic	Weighted sum of energy, CO_2 emissions and economic savings	Individual objectives (CO_2 emissions, energy saving, costs)	CO_2 emissions and costs	Costs, CO_2 emissions and emissions of other pollutants
(Not indicated)				[79] 2009, [80] 2009	[24] 2007		
Linear programming	[34] 2006	[51] 2005	[81] 2005		[85] 2005	[94] 2003, [34] 2006	
Mixed integer-linear programming tools			[52] 2009, [86] 2010				
Non-linear programming			[91] 2002		[19] 2009		
Genetic algorithms	[76] 2006		[89] 2000, [90] 2000	[78] 2010	[83] 2003		
Goal programming			[92] 2007				
Minimum least square error							[83] 2003
Inspection on normalised objectives					[82] 2009		
Pareto front calculation							
Convex sum						[77] 2006	
ε-constrained method						[34] 2006, [93] 2008	[92] 2007
Clustering evolutionary algorithms						[95] 2003, [96] 2004, [97] 2005, [98] 2006	
Direct construction with evolutionary algorithms						[102] 2006	[103] 2010

The year of publication is indicated after the reference number

Baltic Sea Region) [58, 59], DEECO (Dynamic Energy Emission and Cost Optimization) [42, 60, 61], DER-CAM (Distributed Energy Resource Customer Adoption Models) [62, 63], EFOM (Energy Flow Optimization Model) [64], EHTO (Electricity and Heat Trade Optimization) [46], EnergyPLAN [65–67], HOGA (Hybrid Optimization by Genetic Algorithms) [68], HOMER (Hybrid Optimization Model for Electric Renewables) [69], MODEST (Model for Optimisation of Dynamic Energy Systems with Time Dependent Components and Boundary Conditions) [70, 71], OPTIGEN (Optimal Generation Mix Model) [72], PROMIX (Production Mix) [39, 73], and TIMES (The Integrated MARKAL-EFOM System) [74].

3.2 Type 0 Problems: CO_2 Emissions as Optimisation Constraint

Classical optimisations including operational aspects are formulated by making use of economic objectives, for instance to minimise the annual costs incurred by the consumers. In this formulation, CO_2 emissions can be considered as a constraint, imposing a cap on emission allowances, as in [34]. By varying the CO_2 emission constraint it is possible to study the variations in the economic impact on consumers, and to investigate and compare the effects of stricter CO_2 emission limits on the different operation strategies of the equipment. The results obtained also depend in a significant way on the energy production context, that is, on the emission factor values for electricity production in the reference system (Section 2.2).

In a wider sense, CO_2 emission reduction can be achieved by acting not only on the production side, but also on the demand side, by reducing energy consumption, especially in the time intervals in which CO_2 emissions are higher. In this respect, generally consumers tend to reduce their own costs, and their responsiveness to participation in CO_2 reduction objectives can be rather limited. Participation can be enhanced by setting up proper incentives, for instance within demand response programmes [75] with voluntary reductions of consumption associated to an economic worth. Moreover, another degree of freedom for consumer participation to the operational side refers to changing the type of thermal energy consumption, for instance exploiting dual-fuel equipment upon availability of suitable technology (whose choice depends on the solution of a long-term planning problem, of which the operational aspects can be one of the contents). In such a way, the minimum cost optimisation problem can be reformulated in order to identify suitable operational strategies of the production units also involving the consumer side. This aspect impacts on the price elasticity of demand [38], upon proper deployment of load management strategies capable to respond to economic signals or to be flexibly integrated into operational schedules set up on the basis of the optimisation results.

Among the references introducing CO_2 emissions as constraints or parameters, reference [76] addresses the combined design and operation of a distributed multi-generation system [15]. Optimisation of the net present value is considered, and specific scenarios regarding the introduction of taxes for CO_2 and NO_x are taken into account. System operation assessment includes modelling the partial load operation of the equipment. Furthermore, an operational optimisation in which an energy broker is able to get low emission levels while minimising the costs is carried out in [53]. No explicit reference is made to any type of emission, however the scheme can be adapted to deal with CO_2 emissions in an emission trading context, solving the operational planning problem by using linear programming techniques. The explicit energy system operation is not described in a time-sequential way, but different loading levels are identified and aggregated on the basis of the load duration curve. In the solution process, a first stage uses minimum emission dispatch for determining the maximum emissions that can be traded by each entity participating in emission trading. Then, the second stage deals with the broker, maximising the total cost savings subject to meeting emission targets for each entity and satisfy the limits imposed to the input and output quantities.

3.3 Type 1 Problems: Single Objective Function

According to the concepts discussed in Section 2.2 to describe the connections between energy efficiency and CO_2 emissions, optimisation problems related to energy efficiency aspects in thermally-driven applications can be assimilated or restated as optimisation problems aimed at minimising CO_2 emissions. In other cases, a single objective function can be defined by summing up different objective functions in a weighted way, by introducing weights representing the willingness of the decision maker to give more importance to some terms compared with the others. Other ways to obtain a single objective function starting from multiple objective functions are to resort to a goal programming approach or to apply an *environomic* approach in which all entries are translated into costs and the objective is total cost minimisation. Finally, the individual objectives can be compared to each other through inspection of normalised objectives. Some aspects referring to these techniques are illustrated below.

3.3.1 General Formulation of the Optimisation with Single Objective Function

Given a vector **x** containing the decision variables, the optimisation problem with a single objective is generally formulated considering an objective function f(**x**) to be minimised, a vector function **h**(**x**) containing the equality constraints, and a vector function $\psi(\mathbf{x})$ containing the inequality constraints, as

$$\min_{\mathbf{x}} \{f(\mathbf{x})\} \tag{8}$$

subject to

$$\mathbf{h}(\mathbf{x}) = \mathbf{0} \tag{9}$$

$$\psi(\mathbf{x}) \leq \mathbf{0} \tag{10}$$

The contribution of CO_2 emissions to the objective function can be expressed according to different approaches. Energy system analysis is often aimed at comparing the results obtained in different scenarios of variation of operational schedules for different equipment of given type and size. Let us represent with the variable $\xi = 1,\ldots, \Xi$ the scenarios to be analysed.

A first approach considers a number of generation points $n = 1,\ldots, N$, and a partitioning of the time interval of analysis (for instance, 1 year) into $s = 1,\ldots, S$ time periods (e.g., seasons or other significant partitioning of the days of the year, also considering heating periods for buildings, or explicitly separate holiday periods) [77]. Let us consider an elementary time interval of given duration τ (e.g., 1 h), and let us assume that the time period s contains $i = 1,\ldots, I(s)$ time intervals of duration Δt_{si} each (for instance, associated to day and night, to specific electricity tariff bands, or to predefined load duration curve levels). The duration of each time interval is assumed to be a multiple of the elementary time interval τ. The rationale for using different time intervals lies in the possibility of assigning different emission factors to the different components of the energy mix which is expected to be exploited in various loading conditions. For this purpose, for the time interval $i = 1,\ldots, I(s)$ related to the time period $s = 1,\ldots, S$, let us consider the emission factor $\mu_{CO_2,si}^{X_{nj}^{(\xi)}}$ with reference to the energy input $X_{nj}^{(\xi)}$ used to supply the generation point $n = 1,\ldots, N$ with the energy vector $j = 1,\ldots, J(n)$ in the scenario $\xi = 1,\ldots, \Xi$, and let us denote as $X_{njsi}^{(\xi)}$ the *hourly* energy input in that time interval.

The overall CO_2 emissions in regards to the ξth scenario in the period of analysis are

$$E_{CO_2}^{(\xi)} = \sum_{n=1}^{N} \sum_{j=1}^{J(n)} \sum_{s=1}^{S} \sum_{i=1}^{I(s)} X_{njsi}^{(\xi)} \mu_{CO_2,si}^{X_{nj}^{(\xi)}} \frac{\Delta t_{si}}{\tau} \tag{11}$$

For instance, considering a generation point with electricity and natural gas inputs ($J = 2$), the above formulation has $X_{n1} = W$ for electricity and $X_{n2} = F$ for natural gas, and the emission factors involved are $\mu_{CO_2,si}^{W}$ and $\mu_{CO_2,si}^{F}$, to be adapted to the various scenarios considered.

The formulation (11) can be also written by assuming the time intervals of duration Δt_{si} to be composed of an integer number of non-successive elementary time intervals. In this case, the set of hours belonging to time interval $i = 1,\ldots, I(s)$ is denoted as $\boldsymbol{H}_{si}$, and Δt_{si} = dimension $\{\boldsymbol{H}_{si}\}$ hours. It is then possible to represent the hourly-based energy inputs $X_{njsih}^{(\xi)}$ and, if needed, to specify the hourly emission factors $\mu_{CO_2,sih}^{X_{nj}^{(\xi)}}$, expressing the overall CO_2 emissions related to the ξth scenario as

$$E_{CO_2}^{(\xi)} = \sum_{n=1}^{N} \sum_{j=1}^{J(n)} \sum_{s=1}^{S} \sum_{i=1}^{I(s)} \sum_{h \in H_{si}} X_{njsih}^{(\xi)} \mu_{CO_2,sih}^{X_{nj}^{(\xi)}} \qquad (12)$$

The CO_2 emissions calculated according to the above expressions can be used individually or in combination with other objectives to generate single objective functions with multiple components. Some examples are listed below.

In the approach used in [78], the objective function is constructed as the weighted sum of three types of savings (energy, CO_2 emissions, and economic). Each saving is expressed as the relative variation occurring by adopting one technology (with specific control strategy) with respect to the reference case of separate production. For CO_2 emissions, the indicator used recalls the *PCDER* described in Section 2.3. The optimisation problem is solved through a genetic algorithm.

In [79], optimisation with primary energy, operational costs and CO_2 emission objectives is applied to a specified CHP system. Each objective has to be minimised and is represented by an indicator given by the ratio of the objective functions in the case under study and in the conventional case of separate production. The operational rule is that the CHP unit is run if the corresponding indicator is lower than unity. An overall indicator is also formulated as the sum of the three indicators. The same approach is used in [80] to address a trigeneration system.

Hourly trigeneration is addressed in [81] with the objective of minimising simultaneously an economic function whose entries are the production and purchase costs of the energy components (electricity, heat and cooling) and the CO_2 emission costs. The structure of the problem concerning the three energy components is analysed in detail in order to set up an efficient linear programming model called Tri-Commodity Simplex.

In [82], optimisation is performed with reference to a primary energy saving (*PES*) objective, in order to achieve energy savings higher than those obtainable from cost-based operational strategies. The specific contribution is to analyse an operational strategy for trigeneration systems aimed at minimising CO_2 emissions, taking into account the characteristics of the building with heating, ventilation and air conditioning systems, and assuming the size of the combined production system as parameter. The energy saving (or emission reduction) indicator is calculated as the ratio between the building primary energy (or emission of pollutants, determined by taking into account the appropriate emission factors) in separate production and in combined production (of cooling and heating). According to these definitions, the combined production system can be conveniently operated when the indicators assume values higher than unity, otherwise the system should be supplied from separate production. The limit conditions for convenience are by definition consistent with the breakeven conditions $PPES = 0$ and $PCDER = 0$ indicated in Section 2.3. The results show that the operational strategy optimised for CO_2 emission reduction corresponds to an optimal size of the equipment different from the ones obtained by optimising primary energy consumption and operation costs. The differences are evaluated by inspection on the normalised objectives.

If the individual objectives are expressed in function of the same parameter, this technique may be useful to recognise to what extent the individual objectives are conflicting with each other, by comparing the optimal conditions reached by considering the various objectives.

In [83], costs and multi-emissions in the operation of a cogeneration system are minimised under a set of constraints on fuel mix, operation and emissions. In this case, primary energy saving is not explicitly included among the objectives, considering CO_2 emission reduction to be not conflicting with energy efficiency improvement. The individual optimisations are solved by minimising the corresponding objective functions through a genetic algorithm-based approach.

Optimisation of the short-term operation of a trigeneration system is addressed in [84]. The trigeneration system is supplied by a single type of fuel, so that minimising CO_2 emissions is equivalent to minimising fuel consumption. A further objective considered is minimisation of the energy cost by taking into account the possibility of buying electricity from (or selling it to) the electrical distribution system. The trigeneration plant is represented by using a black-box model and is modelled through an overall efficiency matrix. Separate hourly optimisations are solved for each objective function through a sequential quadratic programming approach, considering three representative days for the winter, intermediate and summer seasons. The optimal operational strategies found show additional benefits compared to the conventional electricity tracking and heat tracking control strategies, especially in the intermediate and summer seasons, owing to the better flexibility of exploiting the interactions among the equipment to satisfy the cooling demand. The optimal strategies for energy cost minimisation and energy saving maximisation in general lead to different solutions, and additional information can be obtained by reformulating the problem under a multi-objective approach.

An LP model is formulated in [85] to minimise GHG emissions for a biomass-based energy system. The optimisation model is based on environmental considerations, minimising the net balance between CO_2-equivalent emissions (using *GWP*s to represent the overall effect of the GHGs involved in the study) of the integrated cogeneration system and the actual configuration of the energy supply. The objective function entries are the GHG sources in the cogeneration system and the avoided emissions due to substitution for current fossil-fuel boilers with district heating connections.

Minimisation of the total cost of energy supply from distributed generation to serve the specified energy demand is carried out in [86]. The total cost considered is composed of the sum of electricity and fuel costs, annual costs of capital, operation and maintenance, carbon tax costs, and costs associated with start-up and shut-down of the equipment. The revenues due to the electricity sold are subtracted. Time-of-use tariffs are considered for electricity and fuel. The decision variables are setup in integer and continuous forms. The integer variables include the number of generation units of different types, the on/off operation status, and the presence of energy storage devices. The continuous variables include the input and output energy flows referring to the system components. The solution is obtained through a

MILP approach, determining the units to install and their operational schedules for a test year.

3.3.2 Operational Aspects in Energy Markets and Emission Trading Frameworks

The operational aspects related to energy production and emission trading issues can be included in the formulation of optimisation problems that take into account the interactions with electricity or energy markets and energy-related markets [28] trading for instance emission allowances. As such, the above formulation can be extended to represent the interactions with further aspects related to CO_2 emissions such as participation in emission trading schemes. Coordinated optimisation of operation of a cogeneration system and CO_2 emissions trading is addressed in [51] by formulating a multi-period stochastic optimisation problem. System operation is represented as an hourly production process. This process interacts with the emission trading process only through the variable indicating the CO_2 emissions. The production and trading processes are run with independent objectives and solution techniques. Thus, production optimisation becomes a nested operational process in the overall emission trading scheme. This process is set up to satisfy the heat demand and is run to maximise the net production profit, that is, the difference between the production profit and the CO_2 emission costs. Linear programming techniques are used to obtain the optimal solution.

The operational problem with emission trading refers to a formulation synthesised in the following way. By recalling the notation used in (11), each emission trading time interval within the time period $s = 1,\ldots, S$ is represented as the time interval $i = 1,\ldots,(s)$. Each emission trading time interval is assumed to be composed of an integer number of hours. At each generation point $n = 1,\ldots, N$ there is a set of useful energy outputs $Y_n^{(\xi)}$ dependent on the scenario $\xi = 1,\ldots, \Xi$ considered, such that at hour $h \in \boldsymbol{H}_{si}$ the output $y_{sih} \in Y_n^{(\xi)}$ is sold to the relevant consumer or to the grid at the price ρ_{sih}^{y}.

At the beginning of the trading period $i*$ belonging to a time period not earlier than $s*$, the net production profit $\Pi_{s*i*}^{(\xi)}$ (to be maximised, in this case) for scenario $\xi = 1,\ldots, \Xi$ is calculated depending on the emission allowance prices $\chi_{si}^{(\xi,i*)}$ established by the current instances of the external cycle dedicated to emission trading optimisation, for $s = s*,\ldots, S$ and $i = i*,\ldots, (s)$, as:

$$\Pi_{s*i*}^{(\xi)} = \sum_{n=1}^{N}\sum_{s=s*}^{S}\sum_{i=i*}^{I(s)}\sum_{h\in\boldsymbol{H}_{si}}\left(\sum_{y\in Y_n^{\xi}}\left(\rho_{sih}^{y}\,y_{sih}\right) - \sum_{j=1}^{J(n)} X_{njsih}^{(\xi)}\left(c_{jsih}^{X_{nj}^{(\xi)}} + \mu_{CO_2,si}^{X_{nj}^{(\xi)}}\chi_{si}^{(\xi,i*)}\right)\right) \tag{13}$$

In particular, the term $\mu_{CO_2,si}^{X_{nj}^{(\xi)}}\chi_{si}^{(\xi,i*)}$ represents a penalty added to the fuel price $c_{jsih}^{X_{nj}^{(\xi)}}$ to incorporate the effect of the CO_2 emissions due to the fuels used to generate the energy input $X_{njsih}^{(\xi)}$. The profit obtained by selling the useful energy outputs is then reduced by the amount of this penalty. In more detail, the entry $X_{njsih}^{(\xi)}$ also depends on the partial-load operation of the generation units and, in case of combined production such as cogeneration systems, it depends on the characteristics of the feasible operating point region. Highlights on these aspects are provided in [51].

Further entries for the objective function (13) are introduced in [52], such as the start-up costs explicitly represented for large generation plants. The optimisation method set up and solved in [52] minimises the sum of generation costs, start-up costs and costs of selling or buying emission allowances in the period of analysis. The generation schedules obtained satisfy, in addition to the power balance and to the equipment limits, several time-dependent constraints such as ramp-rate constraints and minimum duration of switch-on and switch-off operations. The influence of generation offers in the electricity market and costs of emission trading on the decision process of establishing the market clearing quantities and prices at different times is investigated. Although emission trading is developed in a separate market with respect to the electricity market, the decisions undertaken in the emission trading framework may affect the electricity market outcomes, markedly by making the use of generators with higher CO_2 emissions less profitable. Under this perspective, emission trading may have a significant impact on the strategies of definition of convenient generation unit portfolios, as well as on the strategies of definition of the offers in the electricity markets for provision of energy and reserves.

3.3.3 Operational Aspects with Distributed Energy Resources and Microgrids

Potential benefits can result from coordinated operation of distributed energy resources (DER) [24, 87]. The operating costs and the CO_2 emissions are used as individual objectives. Possible participation of DER in CO_2 emission trading markets is also studied. The emissions are considered as average emissions from the generation mix, as well as emissions from marginal units. The main hypothesis is that the DER penetration is low, thus affecting the hourly production but not the choice of the units in operation. The issue of knowing the emission data at different rate is addressed by considering cases in which information is known at an annual, monthly and hourly rate. Operation aspects are more conveniently specified in the latter case. The analysis is developed in the framework of electricity markets, assuming the marginal generation unit (determining the market clearing price) and the corresponding CO_2 emissions are known at each hour. The application is focused on a microgrid, whose centralised supervisor is also acting as the local market operator. Finally, renewable energy sources (RES) are considered to be dispatched with priority with respect to the other DER.

The effectiveness of using DER depends on the type of DER used and to the characteristics of the marginal production unit displaced by the introduction of cheaper DER. In this respect, as also pointed out in [88], the objective of maximising the use of RES in the microgrid would produce almost the same results of minimising CO_2 emissions.

Effective assessment of the benefits of DER introduction is linked to the possibility of correctly estimating the expected hourly marginal units in a future period. This assessment depends on the hourly schedules determined by clearing the day-ahead energy markets, and has to be solved through simulation of that market on the basis of various components, including the nature and availability of the generation mix, as well as load and electricity price forecasting. In addition, DER participation in an emission trading market can be addressed by considering the further objective of profit maximisation from such participation, linked to emission reductions and variations in the operational schedules. In [24] the marginal costs of DER are modified by subtracting from the operational costs the cost due to the avoided CO_2 emissions due to the DER production instead of grid supply. Different scenarios based on annual, monthly or hourly emissions are taken into account to establish the remuneration criteria for the avoided CO_2 emissions. The results obtained show that setting up the objective of CO_2 emission minimisation may reduce the revenues from DER with respect to minimum cost operation, but participation in CO_2 emission trading may conveniently limit this reduction. Furthermore, it is pointed out that errors in the estimation of the pollutants could result in wrong estimation of the DER benefits with respect to the displaced generation, preventing the optimisation procedures from reaching the actual optimal solution.

3.3.4 Environomic Optimisation

One of the approaches used to obtain a single objective function synthesising multiple objectives is based on transforming all objectives into entries related to *economics*, thus defining the objective function in terms of overall costs and benefits. This leads to the so-called environomic approach, in which costs and benefits referring to CO_2 and other environmental pollutants are internalised and included in the general objective function formulation.

By summarising some literature contents [89–91], the objective function in environomic optimisation can be written by separating two components, including aspects referred to as:

- The *long-term* (planning) time scale, such as general costs (e.g., for administration), capital costs in regards to the equipment, fixed component of the maintenance costs for scheduled maintenance, costs of abatement systems for reducing pollutants other than CO_2, costs of the resources for producing the abatement systems used for reducing pollutants other than CO_2, costs for CO_2 sequestration, and costs of the resources needed for CO_2 sequestration;
- The *short-term* (operation) time scale, such as fuel costs, operational costs referring to pollutants other than CO_2, component of the maintenance costs depending on the system operation, penalties and taxes related to pollutants other than CO_2, taxes referring to CO_2, and benefits from production revenues.

Among these terms, the fuel costs (referring to the type of fuel) and the taxes in regards to CO_2 emissions depend on the CO_2 emissions in the operation time scale.

The environomic model is applied in [89, 90] for optimisation of a district heating network including centralised and decentralised heat pumps. The comprehensive approach used involves optimal synthesis, design and operation of the network. The problem formulated is highly non-linear and the domain of definition of the decision variables is non-contiguous. The solution is found through a genetic algorithm. Furthermore, [91] reports the results of a study to determine the break-even point for paying a carbon tax with respect to adopt CO_2 sequestration (and abatement of other pollutants) for three different types of plants. The CO_2 sequestration factor is assumed to vary from 0 to 90%. The results are shown with a parametric analysis based on plant capital costs.

3.3.5 Goal Programming Approach

Goal programming can be used to convert a multi-objective problem to a single objective one. Since the beginning, the decision maker defines the goals for each objective function. Then, a set of deviational variables are defined. These variables are positive if the goal is over-achieved (representing the useful margin with respect to the goal) and negative if the goal is under-achieved (representing the missing amount to reach the goal, indicated as a positive value). The goal programming objective function is then the minimisation of the sum of under-achievements. Let us consider $i = 1,\ldots, G$ goals, each of which is associated to the goal value g_i and to the weighting factor ρ_i set by the decision maker to represent the importance given to that goal, and for each of which the deviational variables are denoted as Δf_i^+ and Δf_i^-. The optimisation problem referring to the decision variables included in the vector **x** is stated as:

$$\min\left\{\sum_{i=1}^{G} \rho_i \Delta f_i^-\right\} \tag{14}$$

subject to, for $i = 1,\ldots, G$:

$$f_i(\mathbf{x}) + \Delta f_i^+ - \Delta f_i^- - g_i = 0 \tag{15}$$

In the multi-objective optimisation problem presented in [92], dealing with CO_2 together with other pollutants, the objective is to find the mixture of best available technologies (BATs) leading to the best solution concerning the net present value of the projects and the reduction in the emission of CO_2 and other pollutants. The operational aspects appear in the problem because of the definition of BAT given in the Directive 96/61/EC, according to which a BAT is the most effective and advanced stage in the development of activities *and their methods of operation* which indicate the practical suitability of particular techniques for providing in principle the basis for emission limit values. Synthesising the goal programming approach formulated in [92], the goals include different extents of CO_2 emission reductions. The emission reduction obtained in N installations is calculated with respect to the baseline case (existing technology, with no action to replace it with a BAT) for different types of emissions (the case of CO_2 emissions is considered here). A number M_n of BATs is assumed to be available for each installation $n = 1, \ldots, N$. The binary decision variables contained in the set $\boldsymbol{B} = \{b_{nj}\}$ indicate whether the BAT $j = 1, \ldots, M_n$ is being considered ($b_{nj} = 1$) in the installation $n = 1, \ldots, N$, thus leading to the emission reduction $r_{nj}^{(co_2)}$ with respect to the CO_2 emissions $e^{(co_2)}$ in the baseline case, otherwise $b_{nj} = 0$. Considering the set containing the binary decision variables, the optimisation problem concerning CO_2 emissions is formulated as

$$\max_{b \in B} \left\{ \sum_{n=1}^{N} \sum_{j=1}^{M_n} \frac{r_{nj}^{(\mathrm{CO_2})}}{e^{(\mathrm{CO_2})}} b_{nj} \right\} \tag{16}$$

A similar objective function is defined for the other pollutants considered. A minimum emission reduction constraint is imposed to consider the solution acceptable. Furthermore, only one BAT can be adopted for each installation and there can be no installation without a BAT, resulting in the constraints $\sum_{j=1}^{M_n} b_{nj} = 1$ for $n = 1, \ldots, N$.

3.3.6 Least Square Error Minimisation

Starting from the results of the optimisation of cost and multi-emissions as individual objectives, a multi-objective approach can be set up and solved by transforming the different objectives into a single objective function. Considering the decision variable vector $\mathbf{x}$, for each objective $i = 1, \ldots, N$ the minimum value $f_i^{\min}$ is the one obtained by minimising the corresponding objective function. An upper bound $f_i^{\max}$ is also computed as the maximum of the values obtained by optimising the other objectives. Then, a new objective function is formulated by considering the weighted sum of a function of the emissions defined as

$$T(\mathbf{x}) = \sqrt{\sum_{i=1}^{N} w_i \left(\frac{f_i(\mathbf{x}) - f_i^{\min}}{f_i^{\max} - f_i^{\min}} \right)^2} \tag{17}$$

In addition to the technical equality and inequality constraints, further constraints are $f_i^{\min} \le f_i(\mathbf{x}) \le f_i^{\max}$ for $i = 1, \ldots, N$, and the weights sum up to unity, $\sum_{i=1}^{N} w_i = 1$. The objective function $T(\mathbf{x})$ is solved through a minimum least square error approach, updating the weights until the values assumed by a set of satisfaction factors are considered acceptable. The satisfaction factors σ_i, for

$i = 1,\ldots, N$, are defined by using the values of the decision variables $\mathbf{x}^*$ resulting from $T(\mathbf{x})$ minimisation, as

$$\sigma_i(\mathbf{x}*) = \frac{f_i^{\max} - f_i(\mathbf{x}*)}{f_i^{\max} - f_i^{\min}} \tag{18}$$

This approach is applied in [83] to cogeneration system operation. Clearly, in a multi-objective formulation with conflicting constraints it is not possible to obtain the best values for all the objectives, and the extent to which σ_i ($\mathbf{x}^*$) approaches unity for $i = 1, \ldots, N$ depends on every objective. Hence, the decision on whether to accept the solution or proceed to further iterations by changing the weights is up to the decision maker.

3.4 Type 2 Problems: Multi-Objective Optimisation

3.4.1 General Formulation of the Multi-Objective Optimisation

The interactions among the various objectives can be studied under a comprehensive multi-objective optimisation framework. Given a vector $\mathbf{x}$ containing the decision variables of the problem, and considering a column vector of objective functions $\mathbf{f}(\mathbf{x}) = [f_1(\mathbf{x}), f_2(\mathbf{x}), \ldots, f_K(\mathbf{x})]^{\mathrm{T}}$, the multi-objective optimisation problem (in the hypothesis of minimisation) is generally formulated as

$$\min_{\mathbf{x}} \{\mathbf{f}(\mathbf{x})\} \tag{19}$$

subject to constraints like the ones indicated in (9) and (10).

Considering overall minimisation, the individual objective functions to be minimised are introduced in $\mathbf{f}(\mathbf{x})$ with positive sign, while the sign of the objective functions to be maximised is changed before being introduced in $\mathbf{f}(\mathbf{x})$.

3.4.2 Weighted Sum-based Construction of the Pareto Front

The rationale used for constructing the Pareto front in a *convex* case is based on constructing the weighted sum of the individual conflicting objective functions f_k, for $k = 1, \ldots, K$:

$$f = \sum_{k=1}^{K} \lambda_k f_k \tag{20}$$

where each weight $\lambda_k \in [0, 1]$. In addition, the sum of the weights is constrained to be equal to unity:

$$\sum_{k=1}^{K} \lambda_k = 1 \tag{21}$$

The Pareto front is obtained by minimising (20) for different combinations of the weights that respect the constraint (21). For small numbers of individual objective functions, the Pareto front can be visualised by plotting in 2D or 3D the points corresponding to the individual objective functions. This technique is applied in [77] to deal with optimal operation of fuel cells in residential applications. The overall objective function, containing annual energy costs C (including the cost of electricity and natural gas supplied from external sources) and CO_2 emissions E_{CO_2}, is expressed in the form:

$$f = \lambda \frac{C}{C^{\max}} + (1 - \lambda) \frac{E_{CO_2}}{E_{CO_2}^{\max}} \tag{22}$$

where the terms $C^{\max}$ and $E_{CO_2}^{\max}$, used as normalising factors to make the two objectives comparable, are determined as the maximum values of annual energy costs and CO_2 emissions occurring by applying different operational strategies to cope with predefined seasonal demand variations.

3.4.3 Construction of the Pareto Front by Constraint Relaxation

One of the techniques for determining the Pareto front is called ε-constrained method. This method considers one of the objective functions as the target to be optimised, and introduces all the other objectives as constraints, setting for each of these objectives a limit defined on the basis of a threshold ε, whose value is chosen in such a way that some acceptable solutions can be found. The non-dominated solutions forming the Pareto front are then found by comparing the solutions obtained by imposing successive threshold variations. This method is also suitable for problems with non-convex Pareto front. The ε-constrained method is used in [93] to address the multi-objective problem of minimising costs and CO_2 emissions.

In the application developed in [94] for an urban energy system, the objective function is initially set to the consumers' costs, that are minimised by imposing a cumulative CO_2 emission constraint. Various outputs such as consumers' costs and CO_2 emissions are calculated using linear programming models by assuming different energy pricing scenarios. The solutions with CO_2 emission constraint respected are then chosen and used to evaluate the non-dominated solutions in the plane containing consumers' costs versus CO_2 emissions. The procedure is repeated various times, by relaxing each time the CO_2 emission constraints, updating the composition of the Pareto front on the basis of the new points chosen.

The same application is further exploited in [34] to compare three kinds of energy service system in urban areas, with the overall objectives of minimising the CO_2 emissions and the cost to consumers.

Furthermore, Pareto optimal solutions are generated by using the ε-constrained method in [92], with reference to the problem of determining the optimal BAT mixture mentioned in Section 3.3.5, adopting an interactive filtering procedure to select the most preferred Pareto optimal solution.

3.4.4 Pareto Front Construction with Clustering Evolutionary Algorithms

A series of papers report on the development and application of a dedicated procedure for determining the Pareto front. The paper [95] addresses simultaneous optimisation of design, integration and operation of an energy system with electricity, heat and cooling generation. The objectives are the annual total generation costs and the annual CO_2 emissions rate. The approach used provides the global Pareto-optimal frontier by using a dedicated procedure based on evolutionary algorithms. In addition, suitable clustering algorithms are applied to find and retain many local Pareto optimal frontiers, corresponding for instance to consistent technological choices or other parameter settings. A similar approach is used in [96, 97] for optimisation of integrated energy systems with economic (investment and operational costs) and CO_2 emission reduction objectives, and in [98] to simultaneously optimise CO_2 emission rate and cost of electricity without considering emission cost internalisation. A two-level approach is presented in [99] to deal with design and operation procedures for decentralised energy systems. The design procedure is formulated as multi-objective optimisation considering minimum annual costs and CO_2 emissions, and is solved with an evolutionary programming approach to find Pareto-optimal solutions. The operation procedure minimises the operational costs through a linear programming approach. The CO_2 emissions are calculated after solving the operational optimisation. No dedicated constraint is directly imposed on CO_2 emissions in the linear programming solution.

3.4.5 Direct Construction of the Pareto Front

For combinatorial multi-objective optimisation problems, the identification of the entire Pareto front is practically infeasible [101]. As such, dedicated procedures are developed to *directly* obtain a subset of the Pareto front, called best-known Pareto front, containing a number of non-dominated solutions. These techniques are typically based on evolutionary programming algorithms [57, 100, 101]. Starting from an initially empty Pareto front, multiple non-dominated solution are obtained in a single execution of the algorithm, with successive iterations in which multiple solutions are generated and the solution set is then reduced to maintain only the non-dominated solutions.

In [102] a multi-objective design of isolated hybrid systems is developed for electricity generation from a hybrid photovoltaic-wind-diesel system. The system operation is represented with hourly values, taking into account different control strategies of the plant. The objective functions considered are the total net present cost of the system (including the investment costs and the discounted present values of future costs in the operational life of the plants) and the annual CO_2 emissions. The best-known Pareto front is directly generated by using the Strength Pareto evolutionary algorithm [57].

Multi-objective optimisation is formulated in [103] for a trigeneration system by considering an economic objective (net present value), an energy saving objective (primary energy saving) and an emission reduction objective (with reference to an array of pollutants such as CO_2 and others). The decision variables include two planning entries (sizes of CHP and chiller) and three operational variables (winter and summer operational strategies, and tariff). Optimisation is solved through the NSGA II algorithm [57, 100], with optimal contours drawn in the planning variable space.

4 Conclusions

This chapter has presented a general framework regarding the analysis of optimal operational strategies of energy systems while taking into account CO_2 emission minimisation. Making reference to the recent scientific literature, the operational optimisation problems have been classified within three originally developed categories. In these categories a distinction is made by considering CO_2 emissions either as a constraint in optimisation problems with different objectives, or as a direct component of an individual objective function or a part of a multi-objective function that allows construction of a Pareto front.

The analysis has been backed by the introduction of suitable models and indicators to quantify the relevant environmental benefits. On these bases, it has been highlighted how operational flexibility can be enhanced by switching among energy vectors and performing different operational strategies. Polygeneration of multiple energy vectors has been seen as a significant potential contributor to increasing generation efficiency and therefore reducing GHG emissions.

The categorisation indicated in this paper can serve as a basis for identifying the common underlying principles and synthesising the characteristics of the different methods used to formulate and solve the relevant optimisation problems. This will facilitate the development and implementation of upgraded versions of methods and solution techniques for operational optimisation taking into account CO_2 emissions.

References

1. Meunier F (2002) Co- and tri-generation contribution to climate change control. Appl Therm Eng 22(6):703–718
2. Cârdu M, Baica M (2002) Regarding the greenhouse gas emissions of thermopower plants. Energy Convers Manage 43(16):2135–2144
3. Minciuc E, Le Corre O, Athanasovici V, Tazerout M (2003) Fuel savings and CO_2 emissions for trigeneration systems. Appl Therm Eng 23:1333–1346
4. Chicco G, Mancarella P (2008) Assessment of the greenhouse gas emissions from cogeneration and trigeneration systems. Part I: models and indicators. Energy 33(3):410–417
5. United States Environmental Protection Agency (2008) Catalogue of CHP technologies (Online). http://www.epa.gov/
6. Dincer I (1999) Environmental impacts of energy. Energy Policy 27(14):845–854
7. Kreider JF (2001) Handbook of heating, ventilation and air conditioning. CRC Press, Boca Raton
8. Krauter S, Rüther R (2004) Considerations for the calculation of greenhouse gas reduction by photovoltaic solar energy. Renew Energy 29:345–355
9. Luickx PJ, Delarue ED, D'haeseleer WD (2009) Effect of the generation mix on wind power introduction. IET Renew Power Gen 3:267–278
10. Pehnt M (2006) Dynamic life cycle assessment (LCA) of renewable energy technologies. Renew Energy 31(1):55–71
11. Finnveden G, Hauschild MZ, Ekvall T, Guinée J, Heijungs R, Hellweg S, Koehler A, Pennington D, Suh S (2009) Recent developments in life cycle assessment. J Environ Manage 91:1–21
12. Horlock JH (1997) Cogeneration-combined heat and power (CHP). Krieger, Malabar
13. Wu DW, Wang RZ (2006) Combined cooling, heating and power: a review. Prog Energy Combust Sci 32:459–495
14. Chicco G, Mancarella P (2009) Distributed multi-generation: a comprehensive view. Renew Sustain Energy Rev 13:535–551
15. Mancarella P, Chicco G (2009) Distributed multi-generation systems: energy models and analyses. Nova, New York
16. Mancarella P (2009) Cogeneration systems with electric heat pumps: energy-shifting properties and equivalent plant modelling. Energy Convers Manage 50:1991–1999
17. Mancarella P, Chicco G (2008) Assessment of the greenhouse gas emissions from cogeneration and trigeneration systems. Part II: analysis techniques and application cases. Energy 33 (3):418–430
18. Chicco G, Mancarella P (2008) A unified model for energy and environmental performance assessment of natural gas-fueled poly-generation systems. Energy Convers Manage 49 (8):2069–2077
19. Chicco G, Mancarella P (2009) Matrix modelling of small-scale trigeneration systems and application to operational optimization. Energy 34(3):261–273
20. Mancarella P, Chicco G (2010) Distributed cogeneration: modeling of environmental benefits and impact. In: Gaonkar DN (ed) Distributed generation. In-Teh, Vukovar, Chapter 1
21. Cardona E, Piacentino A (2005) Cogeneration: a regulatory framework toward growth. Energy Policy 33:2100–2111
22. Chicco G, Mancarella P (2007) Trigeneration primary energy saving evaluation for energy planning and policy development. Energy Policy 35(12):6132–6144
23. Voorspools KR, D'haeseleer WD (2003) The impact of the implementation of cogeneration in a given energetic context. IEEE Trans Energy Convers 18(1):135–141
24. Tsikalakis AG, Hatziargyriou ND (2007) Environmental benefits of distributed generation with and without emissions trading. Energy Policy 35(6):3395–3409
25. Hawkes AD (2010) Estimating marginal CO_2 emissions rates for national electricity systems. Energy Policy 38:5977–5987

26. Marbe Å, Harvey S (2005) The role of policy instruments for promoting combined heat and power production with low CO_2 emissions in district heating systems. Int J Energy Res 29 (6):511–537
27. Karlsson M, Gebremedhin A, Klugman S, Henning D, Moshfegh B (2009) Regional energy system optimization – potential for a regional heat market. Appl Energy 86(4):441–451
28. Chicco G, Mancarella P (2007) Exploiting small-scale cogeneration in energy-related markets. In: Proceedings of the IEEE PowerTech 2007, Lausanne, 1–5 July 2007. The IEEE, Piscataway, paper 319
29. Conde Lázaro E, Ramos Millán A, Reina Peral P (2006) Analysis of cogeneration in the present energy framework. Fuel Process Technol 87:163–168
30. Gulli F (2006) Social choice, uncertainty about external costs and trade-off between intergenerational environmental impacts: the emblematic case of gas-based energy supply decentralization. Ecol Econ 57(2):282–305
31. Siddiqui AS, Marnay C, Edwards JL, Firestone R, Ghosh S, Stadler M (2005) Effects of carbon tax on microgrid combined heat and power adoption. J Energy Eng 131(3):2–25
32. Rubin ES, Chen C, Rao AB (2007) Cost and performance of fossil fuel power plants with CO_2 capture and storage. Energy Policy 35(9):4444–4454
33. European Commission (2003) Emission trading system (Online). http://ec.europa.eu/environment/climat/emission/index_en.htm
34. Aki H, Oyama T, Tsuji K (2006) Analysis of energy service systems in urban areas and their CO_2 mitigations and economic impacts. Appl Energy 83(10):1076–1088
35. Praetorius B, Schumacher K (2009) Greenhouse gas mitigation in a carbon constrained world: the role of carbon capture and storage. Energy Policy 37:5081–5093
36. Cohen SM, Rochelle GT, Webber ME (2008) Turning CO_2 capture on and off in response to electric grid demand: a baseline analysis of emissions and economics. In: Proceedings of energy sustainability 2008, Jacksonville, 10–14 Aug 2008, paper 54296
37. Chalmers H, Chen C, Lucquiaud M, Gibbins J, Strbac G (2006) Initial evaluation of carbon capture plant flexibility. In: Proceedings of the 8th international conference on greenhouse gas control technologies. Elsevier, Trondheim
38. Newell RG, Pizer WA (2008) Carbon mitigation costs for the commercial building sector: discrete-continuous choice analysis of multifuel energy demand. Resour Energy Econ 30 (4):527–539
39. Voorspools KR, D'haeseleer WD (2000) The influence of the instantaneous fuel mix for electricity generation on the corresponding emissions. Energy 25(11):1119–1138
40. Abido MA (2003) Environmental/economic power dispatch using multiobjective evolutionary algorithms. IEEE Trans Power Syst 18(4):1529–1537
41. Palanichamy C, Sundar Babu N (2008) Analytical solution for combined economic and emissions dispatch. Electr Power Syst Res 78:1129–1137
42. Lindenberger D, Bruckner T, Morrison R, Groscurth HM, Kümmel R (2004) Modernization of local energy systems. Energy 29:245–256
43. Willett K, Tomic J (2005) Vehicle to grid fundamentals: calculating capacity and net revenue. J Power Sources 144(1):268–279
44. Haeseldonckx D, Peeters L, Helsen L, D'haeseleer W (2007) The impact of thermal storage on the operational behaviour of residential CHP facilities and the overall CO_2 emissions. Renew Sustain Energy Rev 11(6):1227–1243
45. Cormio C, Dicorato M, Minoia A, Trovato M (2003) A regional energy planning methodology including renewable energy sources and environmental constraint. Renew Sustain Energy Rev 7(2):99–130
46. Makkonen S, Lahdelma R (2006) Non-convex power plant modelling in energy optimisation. Eur J Oper Res 171:1113–1126
47. Gjengedal T (1996) Emission constrained unit commitment (ECUC). IEEE Trans Energy Convers 11(1):132–138

48. Arroyo JM, Conejo AJ (2000) Optimal response of a thermal unit to an electricity spot market. IEEE Trans Power Syst 15(3):1098–1104
49. Voorspools K, Peersman I, D'haeseleer W (2005) A comparative analysis of energy and CO_2 taxes on the primary energy mix for electricity generation. Int J Energy Res 29:879–890
50. Lindenberger D, Bruckner T, Groscurth HM, Kümmel R (2000) Optimization of solar district heating systems: seasonal storage, heat pumps and cogeneration. Energy 25:591–608
51. Rong A, Lahdelma R (2007) CO_2 emissions trading planning in combined heat and power production via multi-period stochastic optimization. Eur J Oper Res 176(3):1874–1895
52. Kockar I, Conejo AJ, McDonald JR (2009) Influence of the emissions trading scheme on generation scheduling. Electr Power Energy Syst 31:465–473
53. Chattopadhyay D (1995) An energy brokerage system with emission trading and allocation of cost savings. IEEE Trans Power Syst 10(4):1939–1945
54. Bertoldi P, Huld T (2006) Tradable certificates for renewable electricity and energy savings. Energy Policy 34(2):212–222
55. Belton V, Stewart TJ (2002) Multiple criteria decision analysis. An integrated approach. Kluwer Academic Publishers, Boston
56. Hwang CL, Yoon K (1981) Multiple attribute decision making. Methods and applications: a state-of-the-art survey, Springer-Verlag, Berlin, Germany
57. Shukla PK, Deb K (2007) On finding multiple pareto-optimal solutions using classical and evolutionary generating methods. Eur J Oper Res 181:1630–1652
58. Hindsberger M, Ravn HF (2001) Multiresolution modelling of hydro-thermal systems. In: Proceedings of the power industry computer applications (PICA 2001), Sydney
59. Karlsson K, Meibom P (2008) Optimal investment paths for future renewable based energy systems-using the optimisation model Balmorel. Int J Hydrogen Energy 33(7):1777–1787
60. Bruckner T, Groscurth HM, Kummel R (1997) Competition and synergy between energy technologies in municipal energy systems. Energy 22(10):1005–1014
61. Bruckner T, Morrison R, Handley C, Patterson M (2003) High-resolution modeling of energy-services supply systems using DEECO: overview and application to policy development. Ann Oper Res 121(1–4):151–180
62. Siddiqui AS, Firestone R, Ghosh S, Stadler M, Edwards JL, Marnay C (2003) Distributed energy resources customer adoption modeling with combined heat and power applications, Report LBNL-52718
63. Zhou N, Gao W, Firestone R, Stadler M, Marnay C, Nishida M (2006) An analysis of the DER adoption climate in Japan using optimization results for prototype buildings with U.S. comparisons. Energy Build 38(12):1423–1433
64. Grohnheit PE (1991) Economic interpretation of the EFOM model. Energy Econ 13 (2):143–152
65. Lund H, Münster E (2006) Integrated energy systems and local energy markets. Energy Policy 34(10):1152–1160
66. Lund H (2007) Renewable energy strategies for sustainable development. Energy 32 (6):912–919
67. Lund H (2009) EnergyPLAN: advanced energy system analysis computer model. http://energy.plan.aau.dk/
68. Dufo-López R, Bernal-Agustín JL (2005) Design and control strategies of PV-diesel systems using genetic algorithms. Sol Energy 79(1):33–46
69. National Renewable Energy Laboratory (NREL) (2006) HOMER: the optimization model for distributed power. http://www.nrel.gov/homer/
70. Henning D (1997) MODEST: an energy-system optimisation model applicable to local utilities and countries. Energy 22(12):1135–1150
71. Henning D, Amiri S, Holmgren K (2006) Modelling and optimisation of electricity, steam and district heating production for a local Swedish utility. Eur J Oper Res 175:1224–1247
72. Nagata Y (2005) Quantitative analysis of CO_2 emissions reductions through introduction of stationary-type PEM-FC systems in Japan. Energy 30:2636–2653

73. Haeseldonckx D, D'haeseleer W (2008) The environmental impact of decentralised generation in an overall system context. Renew Sustain Energy Rev 12:437–454
74. Remme U, Goldstein GA, Schellmann U, Schlenzig C (2001) MESAP/TIMES-advanced decision support for energy and environmental planning. In: Chamoni P, Leisten R, Martin A, Minnemann J, Stadtler H (eds) Operations research proceedings 2001 – Selected papers of the international conference on operations research (OR 2001), 3–5 Sep 2001. Springer, Duisburg, pp 59–66. www.uni-duisburg.de/or2001
75. Yusta JM, Khodr HM, Urdaneta AJ (2007) Optimal pricing of default customers in electrical distribution systems: effect behavior performance of demand response models. Electr Power Syst Res 77(5–6):548–558
76. Li H, Nalim R, Haldi PA (2006) Thermal-economic optimization of a distributed multi-generation energy system – a case study of Beijing. Appl Therm Eng 26(7):709–719
77. Aki H, Yamamoto S, Ishikawa Y, Kondoh J, Maeda T, Yamaguchi H, Murata A, Ishii I (2006) Operational strategies of networked fuel cells in residential homes. IEEE Trans Power Syst 21(3):1405–1414
78. Wang JJ, Jing YY, Zhang CF (2010) Optimization of capacity and operation for CCHP system by genetic algorithm. Appl Energy 87:1325–1335
79. Mago PJ, Chamra LM, Hueffed A (2009) A review on energy, economical, and environmental benefits of the use of CHP systems for small commercial buildings for the North American climate. Int J Energy Res 33(14):1252–1265
80. Mago PJ, Chamra LM (2009) Analysis and optimization of CCHP systems based on energy, economical, and environmental considerations. Energy Build 41:1099–1106
81. Rong A, Lahdelma R (2005) An efficient linear programming model and optimization algorithm for trigeneration. Appl Energy 82:40–63
82. Fumo N, Mago PJ, Chamra LM (2009) Emission operational strategy for combined cooling, heating, and power systems. Appl Energy 86:2344–2350
83. Tsay MT (2003) Applying the multi-objective approach for operation strategy of cogeneration systems under environmental constraints. Electr Power Energy Syst 25:219–226
84. Chicco G, Mancarella P (2008) Optimal operational strategies for multi-generation systems. In: Proceedings of the power systems computational conference, Glasgow, 14–18 July 2008, paper 528
85. Chinese D, Meneghetti A (2005) Optimisation models for decision support in the development of biomass-based industrial district-heating networks in Italy. Appl Energy 82(3): 228–254
86. Ren H, Gao W (2010) A MILP model for integrated plan and evaluation of distributed energy systems. Appl Energy 87:1001–1014
87. Schwaegerl C, Tao L, Mancarella P, Strbac G (2010) A multi-objective optimization approach for assessment of technical, commercial and environmental performance of microgrids. Eur Trans Electr Power. doi:10.1002/etep. 472
88. Spitzley DV, Keoleian GA, Baron SG (2007) Life cycle energy and environmental analysis of a microgrid power pavilion. Int J Energy Res 31:1–13
89. Curti V, von Spakovsky MR, Favrat D (2000) An environomic approach for the modeling and optimization of a district heating network based on centralized and decentralized heat pumps, cogeneration and/or gas furnace. Part I: methodology. Int J Therm Sci 39:721–730
90. Curti V, Favrat D, von Spakovsky MR (2000) An environomic approach for the modeling and optimization of a district heating network based on centralized and decentralized heat pumps, cogeneration and/or gas furnace. Part II: application. Int J Therm Sci 39:731–741
91. Borchiellini R, Massardo AF, Santarelli M (2002) Carbon tax vs CO_2 sequestration effects on environomic analysis of existing power plants. Energy Convers Manage 43:1425–1443
92. Mavrotas G, Georgopoulou E, Mirasgedis S, Sarafidis Y, Lalas D, Hontou V, Gakis N (2007) An integrated approach for the selection of best available techniques (BAT) for the industries in the greater Athens area using multi-objective combinatorial optimization. Energy Econ 29:953–973

93. Wang C, Larsson M, Ryman C, Grip CE, Wikström JO, Johnsson A, Engdahl J (2008) A model on CO_2 emission reduction in integrated steelmaking by optimization methods. Int J Energy Res 32(12):1092–1106
94. Aki H, Oyama T, Tsuji K (2003) Analysis of energy pricing in urban energy service systems considering a multiobjective problem of environmental and economic impact. IEEE Trans Power Syst 18(4):1275–1282
95. Burer M, Tanaka K, Favrat D, Yamada K (2003) Multi-criteria optimization of a district cogeneration plant integrating a solid oxide fuel cell-gas turbine combined cycle, heat pumps and chillers. Energy 28(6):497–518
96. Xia H, Koyama M, Leyland G, Kraines S (2004) A modularized framework for solving an economic-environmental power generation mix problem. Int J Energy Res 28(9):769–784
97. Pelet X, Favrat D, Leyland G (2005) Multiobjective optimisation of integrated energy systems for remote communities considering economics and CO_2 emissions. Int J Therm Sci 44(12):1180–1189
98. Li H, Maréchal F, Burer M, Favrat D (2006) Multi-objective optimization of an advanced combined cycle power plant including CO_2 separation options. Energy 31(15):3117–3134
99. Weber C, Maréchal F, Favrat D, Kraines S (2006) Optimization of an SOFC-based decentralized polygeneration system for providing energy services in an office-building in Tokyo. Appl Therm Eng 26(13):1409–1419
100. Deb K, Pratap A, Agarwal S, Meyarivan T (2002) A fast and elitist multiobjective genetic algorithm: NSGA-II. IEEE Trans Evol Comput 6(2):182–197
101. Konak A, Coir DW, Smith AE (2006) Multi-objective optimization using genetic algorithms: a tutorial. Reliab Eng Syst Safe 91:992–1007
102. Bernal-Agustín JL, Dufo-López R, Rivas-Ascaso DM (2006) Design of isolated hybrid systems minimizing costs and pollutant emissions. Renew Energy 31:2227–2244
103. Kavvadias KC, Maroulis ZB (2010) Multi-objective optimization of a trigeneration plant. Energy Policy 38:945–954

Impact of GHG Emission Reduction on Power Generation Expansion Planning

F. Careri, C. Genesi, P. Marannino, M. Montagna, S. Rossi, and I. Siviero

Abstract In this work the impact of greenhouse gas (GHG) emission reduction on Power Generation Expansion Planning (PGEP) is investigated. An overview of several PGEP models, which also consider environmental constraints and GHG emission limits, is presented. After a short introduction on regulations about GHG emission Cap and Trade System in Europe and in the United States and a survey on state of the art PGEP, a new approach to assess the effects of fuel and electric energy price volatility on long term generation planning by a GENCO (GENeration COmpany) is proposed. The objective function, to be maximized, consists of the total revenue obtained by the GENCO over a certain time horizon into the future. A general model is developed to find both present and future generation mixes of a given GENCO and the Lagrangian Relaxation method is used to solve the large scale mixed integer problem. However, its results will be not really suitable to define a generation planning strategy, unless the uncertainties of costs, prices and construction times are considered. For this reason, a Monte Carlo simulation procedure is implemented including an expansion planning computation at each step. In the model renewable resources, like off-shore and on-shore wind, biomass, mini hydro, geothermal, solar thermodynamic and photovoltaic power plants, are also taken into account. The results are presented with reference to an hypothetical GENCO, by setting all the scenario variables on the basis of available historical data.

Keywords Emission Reduction • Power Generation Expansion Planning • Generation Company • Risk Management • Lagrangian Relaxation • Monte Carlo Simulation

---

F. Careri • C. Genesi (✉) • P. Marannino • M. Montagna • S. Rossi • I. Siviero
Department of Electrical Engineering, University of Pavia, via Ferrata 1, 27100 Pavia, Italy
e-mail: camillo.genesi@unipv.it

Q.P. Zheng et al. (eds.), *Handbook of CO_2 in Power Systems*, Energy Systems,
DOI 10.1007/978-3-642-27431-2_10, © Springer-Verlag Berlin Heidelberg 2012

1 GHG Emission Reduction: Cap and Trade System in Europe and in the United States

A cap-and-trade bill would limit greenhouse gas emissions from specified sources and enforce the limit by requiring firms to surrender allowances to cover their regulated emissions. The government would issue a declining number of allowances each year according to the limits specified by law (the "cap" part of cap-and-trade). Firms could trade allowances so that those with lower abatement costs could reduce emissions and sell excess allowances to firms with higher abatement costs (the trade part of cap-and-trade). Thus the program allows market forces to steer abatement to the most cost effective approaches, lowering the overall cost of achieving the environmental goal.

To meet its obligations to reduce greenhouse gas concentrations under the Kyoto Protocol, the European Union (EU) established the first cap-and-trade system for carbon dioxide emissions in the world starting in 2005. Proposed in October 2001, the EU's Emissions Trading System (EU-ETS) was up and running just over 3 years later. The first 3-year trading period (2005–2007)—a trial period before Kyoto obligations began—is now complete and has been heavily scrutinized.

Recognizing their lack of experience with cap-and-trade and the need to build knowledge and program architecture, EU leaders began by covering only one gas (carbon dioxide) and a limited number of sectors. Once the infrastructure was in place, other GHGs and sectors could be included in subsequent phases of the program when more significant emission reductions were needed.

The development of the EU-ETS has not, however, proceeded without its challenges due to some controversies:

- Due to a lack of accurate data in advance of the program, allowances to emitters were overallocated;
- Concerns about program volatility emerged when initially high allowances prices (driven largely by high global energy costs) dropped precipitously in April 2006 upon the release of more accurate, verified emissions data. Late in the trial phase, there was another sharp decline in allowance price because there were no provisions for banking emissions reductions for use in the second phase of the program;
- Windfall profits [1] by electric power generators that passed along costs (based on market value) of their freely issued allowances resulted in improved understanding of how member country electricity sector regulations affect the market.

In the United States of America the Congress continues to debate a potential cap-and-trade program for the control of greenhouse gas (GHG) emissions. The first comprehensive greenhouse gas reduction bill ever to be reported out of a committee emerged from the Senate Environment and Public Works Committee in December 2007. As debate continues on this landmark legislation, the House of Representatives has signaled its intention to design its own emissions trading program. The emissions reduction targets were first proposed by Representatives

Waxman and Markey in their April 2009 discussion draft (DD) bill before the Energy and Commerce Committee of the House of Representatives.

To build the U.S. cap-and-trade scheme, it is necessary to consider that EU-ETS is a classic cap-and-trade system. However, it also contains some significant design differences from those reflected in cap-and-trade systems for other emissions (SO_2, NO_x) that have been implemented in the U.S. The common features are that:

- An absolute quantity limit (or cap) on CO_2 emissions has been placed on some 12,000 emitting facilities located in the European Union;
- Tradable allowances have been distributed to these facilities (typically for free) in an amount equal to the cap;
- These facilities must measure and report their CO_2 emissions and subsequently surrender an allowance for every ton of CO_2 they emit during annual compliance periods.

The primary differences from U.S. experience with cap-and-trade mechanisms relate to how the cap is set, the process for allocating emission allowances, banking and borrowing provisions, the monitoring, reporting and verification procedures, and the linking or off-system provisions [2].

The time taken to develop and implement the relatively complex EU-ETS was much less than the case for the simpler U.S. SO_2 cap-and-trade program. The initial Bush Administration proposal that became the basis of the eventual U.S. legislation was put forward in April 1989, almost 7 years before the intended start in 1996. Subsequent changes in Congress moved the start date to 1995, but even so the legislation was voted out of both Houses of Congress and signed by the first President Bush in November 1990, a little more than 4 years before the programs start. More importantly, the aggregate SO_2 emission cap and the initial allocations had been determined during the legislative process so that attention could focus thereafter on implementing regulations and the development of the registry, all of which were in place when the program started on January 1, 1995.

In the last years, there were studies in the U.S. aimed at understand the European ETS scheme and its applicability to North America energy system [2]. These studies analyze what happened in the European ETS trial period, in order to identify the problems that occurred, like those listed above, and to propose some corrective actions including:

- More accurate emission data and a centralized cap-setting and reporting process;
- The emission cap should be sufficiently binding;
- Improved data quality and provisions for unrestricted banking between compliance periods will help to moderate price fluctuations in the future;
- Allowance auctioning to avoid windfall generators' profits.

Warwick et al. [3] examine the relationships between emission reductions and cost across different pathways to similar long run objectives. Especially McKibbin et al. explore the relationship between short run, long run and cumulative emissions targets and the economic effects of the policy over time. Paltsev et al. [4] underline how important the details of a cap-and-trade program can be in influencing its environmental and economic performance.

There is an increased attention both in Europe, where the ETS has been already implemented and tested, and in the United States of America, where the House of Representatives and many researchers are working to identify the best framework of cap-and-trade system.

From the authors' point of view, the goal of emission reduction in electric power generation industry could be achieved by encouraging renewable energy resources and low-carbon technologies. In this way, the ETS could be an indirect incentive to the spreading of "green power" generators and a deterrent for those power plants with high CO_2 emissions.

In the past, researchers have designed lots of generation planning (GP) models, by exploiting different mathematical techniques and considering different objective functions and constraints. Nowadays, the impact of ETS (thus the CO_2 price allowances) on generators profits and costs has to be included and analyzed in the GP scheme as well.

2 State of the Art in Generation Expansion Planning (GEP) Methodologies

Generation expansion planning is an important planning activity for utility companies. Since it takes a long time for the development of power resources from planning to operation, it is necessary for the company to obtain a best scheme aiming for the well-balanced generation planning in an overall view. In recent years, along with the development of deregulation in electric power industry, competition is being introduced in power generation, power transmission, and power supply. Further the environmental impact on the portfolio and merit order ranking must be taken into account by generation utility to meet the environmental challenges under various regulatory regimes such as EU-ETS and other emissions cap-and-trade systems.

In the following subsection we present a comprehensive survey on different GEP models taking into account environmental constraints and the various ways to implement emission limits.

2.1 Traditional Deterministic Models

When the problem of emission limit appeared, technicians began to implement new constraints in the traditional deterministic models already available. The papers [5] and [6] propose an approach to obtain the optimal generation expansion planning including IPPs (Independent Power Providers), within the CO_2 emission limits.

The optimal mix problem is formulated for minimizing the total cost including investment and operational cost at the target year, considering nuclear, coal, oil and gas fed power plants for utility generation planning and three types of IPPs classified on the basis of power generation characteristic (Base-type, Middle-type and Peak-type). The mentioned model, formulated as an integer problem and solved by means of the Dynamic Programming (DP) method, enforces the respect of target supply reliability indices like loss of load probability (LOLP) and expected energy not served (EENS). The environmental constraint consists of an upper bound on the sum of generation technologies emissions (proportional to generated energy by means of an emission coefficient).

A combined relationship among reliability, generation and transmission coordination is illustrated through the exponential analytic model of a multi-area system in the paper [7]. Sun et al. use exponential approximation for system risk to implement relationships between multi-area reliability and installed generation and transmission capacity. The price response of the consumers is taken into account by means of an affine inverse demand function while the authors suppose that pollution emission of generation units cause health damage cost locally; thus each area has its emission limit to respect. Two models are built up. Both maximize the difference between consumer benefit and the costs for generators and scheduled lines. The first model takes into account only the impact of generation capacity changes (due to generation planning) on reliability risk damage, while the second one also considers the flowgate increased capacity impact. Other constraints of the models, solved by means of a mixed integer programming technique, are flowgates and reliability damage risk limits.

Many papers consider environmental impact on generation expansion planning, by means of an upper bound on total generator emissions. Genoese et al. [8] show some possibilities to analyze the impact of different emission allowances allocation methods (in the emission trading system) by optimizing energy system models. The starting point for the model-based analysis is the PERSEUS (Program package for Emission Reduction Strategies in Energy Use and Supply) model, an optimizing inter-regional long-term model of the European electricity sector. In particular the general objective of energy system model PERSEUS-NAP (PERSEUS-National Allocation Plans) is to provide an analysis tool for qualification of economic and technological impacts that a CO_2 emission trading system and the design of emission allocation plans may have on electricity prices, technology choices, CO_2 allowance prices and interregional power exchanges. PERSEUS-NAP is an energy and material flow model applying a multi-periodic linear programming approach. The target function demands a minimization of all decision-relevant expenditure within the entire energy supply system. This basically comprises fuel supply and support costs, transmission fees, fixed and variable costs of the physical assets (operation, maintenance, load variation costs, etc.) and investment costs for new plants. The relevant techno-economic characteristics of the real supply system have been considered by implementing further equations covering technical, ecological and political restrictions. The most important technical restrictions are physical energy and material balances, capacity restrictions and plant operations.

The PERSEUS-NAP model consists of 42 regions covering all Western, Central and Eastern European countries, in some case even splitting one country into several regions. National allocation plans of the emission trading system are integrated into this linear programming model by a set of additional equations. Within the model, it is assumed that market prices of homogenous goods, in a open and fully competitive market, are set by marginal costs; then the authors use these indicators of the mathematical model to indicate both the resulting electricity prices as well as CO_2 allowances prices.

2.2 Portfolio Theory Models

Taking into account the CO_2 allowances allocation, paper [9] discusses a methodology for the electricity generators for optimizing their operations in the wake of Emission Trading Scheme (ETS) in Europe and another model for long term generation portfolio optimization using the portfolio theory in finance. To implement this last theory, first the expected portfolio cost and the expected portfolio risk (conditional expected value) are calculated and are plotted on a graph for various available portfolio alternatives. Thus a frontier of the portfolios with the lowest expected cost for a particular level of risk undertaken is obtained. The generator can select any of the portfolios falling on the efficient frontier making a risk-cost trade-off based on their risk aversion. In defining the expected generation portfolio cost, carbon abatement cost, capital expenditure, operating expenditure etc. have been considered. The cost component carbon abatement cost is a limiting factor from a cost minimization perspective; abatement cost is compensated through Emission Trading Schemes and allocation mechanisms like grandfathering. The expected portfolio risk is defined by means of standard deviation of holding period return (HPR) of technologies' annual costs and their correlations. The primary objective of the model is to predict the probability that the HPR for a particular technology will not exceed a certain value for the current year (if the holding period is 1 year). The model can be configured to set the holding period at less than a year (quarter, month etc.) or more than a year. By convention, the holding period is 1 year. The sequential mathematical procedures are simple. A Monte Carlo technique is used to generate simulated "cost" stream (also called the forward path/future path) for each cost type within a particular technology. From each simulated forward path for a particular cost type, a forward path of HPRs of that cost type is obtained. The same process is iterated to all the other cost types within the same technology. Finally, by aggregating the forward paths of HPRs of all the cost types within a particular technology, a forward path of HPR of that technology itself is determined. HPR of a technology is calculated in the model schema. The same set of iterative procedures is repeated for all the other technologies in the portfolio. Finally, the HPR at Risk for each technology in the portfolio is calculated from its respective HPR forward paths. Simulated "cost" streams take into account the fuel availability and renewable resources generation quantity.

2.3 Carbon Dioxide Mitigation Methods

To investigate carbon dioxide mitigation ways, Hashim et al. [10] built up an integrated model that combines the objectives of both the economic and environmental operating modes through the use of an external pollution index as a conversion factor from pollution to cost. The problem, formulated as a mixed integer linear problem, considers two carbon dioxide mitigation options: fuel balancing and fuel switching. Fuel balancing is the optimal adjustment of the operation of generating plants to reduce CO_2 emissions without making structural changes to the generation park. In contrast, fuel switching, which is switching from carbon-intensive fuels (i.e. coal) to less carbon-intensive fuels (i.e. natural gas), involves structural changes to the generation park. A sensitivity analysis study is carried out to investigate the effect of parameter uncertainties such as uncertainties in natural gas price, coal price and retrofit costs (associated with switching the fuel from coal to natural gas for the fossil-fuel stations) on the optimal solution.

Paper [11] exploits a model similar to the previous one, but it investigates a further CO_2 mitigation strategies: enhanced use of alternative energy and advanced technologies.

To reduce greenhouse gases emissions and to increase production from less attractive renewable resources (like wind, biomass, solar etc.), Yamgar et al. [12] built up a model named Integrated Resource Planning and Analysis (IRPA) developed at the Asian Institute of Technology (AIT). IRPA is a least cost generation expansion-planning model, which determines optimum schedule for new capacity addition to meet the given demand for a given planning period. The IRPA model not only determines the optimum expansion plan but also calculates the total cost and emission levels of different pollutants. The energy tax (used to indirectly stimulate production from less attractive RES) is a duty levied on the fuel used in proportion to their heat content. The tax is levied on electricity generation based on fossil fuels, nuclear and hydro sources. Renewable energy sources (except hydro) are exempted from the energy tax. In [12] energy tax is applied as an increment to the fuel cost for fossil fuel fired power plants and as an increment to the operating cost for hydro and nuclear power plants.

2.4 Handling Uncertainties

The utility deals with uncertainties in all aspects of generation expansion planning, from economic factors such as interest and inflation rates to technical considerations such as the availability and costs of alternative generation, fuels, and new emission reduction technologies. There are also uncertainties in projections of demand and energy requirements, and in the expected impacts of such load alteration measures as demand side management and conservation. Addressing these uncertainties increases the number of conditions that must be considered when evaluating alternative decisions. Paper [13] presents a technique for modeling and studying the effects of uncertain environmental factors in the electric utility planning process. This takes into account the traditional utility planning process and shows how it can be facilitated with the use of advanced decision analysis tools. The concept of an influence diagram is introduced in the utility planning arena. Using the interdependent data analysis (IDA) technique, a method for providing the probability estimates necessary for the influence diagram, is shown. The IDA technique allows the use of expert opinions and intuitive judgments to develop the necessary probability estimates. Emissions caps and emissions trading have added a new dimension to the problem. Estimates on the market price of emission allowances vary, as well as experts' opinions on the extent to which government legislation and regulations would impact on utility decisions. An influence diagram consists of decision nodes, chance nodes and value nodes whose relationships are shown using arrows that represent either conditioning arcs (into chance and value nodes) or informational arcs (into decision nodes). Especially for one specific decision node, an array of emission abatement options (technologies options) is considered, depending on the resource mix information available from the first. These technologies have their corresponding costs and emission reduction efficiencies which can be treated as chance variables. After having estimated (by means IDA) the probability distributions of the chance events (based on the variety of expert opinions which are then represented as discrete probabilities for the decision analysis), the influence diagram is then solved using decision analysis software like the Decision Programming Language to determine the least cost alternative considering capital and operating costs, the cost of emissions and the likelihoods of the different scenarios.

Another technique, named agent based model (ABM), is used to consider the relationship between technical systems and social networks that make up the today's electricity infrastructure. In this way, Chappin et al. [14] built up an ABM to elucidate the impact of CO_2 Emission-Trading on the European electricity production system-of-systems. The model is defined by the sequence of potential operations and strategic actions of agents (agent definitions range from a mere subroutine to a conscious entity), exogenous trends modeled in scenarios, the definitions of agents' style and behavior and by the portfolio of available technologies and installed facilities. A major characteristic of an agent-based model is that agents autonomously decide whether or not to undertake specific actions. Thus, production facilities decide autonomously on investment or disinvestment and on the bids they put out to sell the electricity they produce. They arrive at these decisions using available information and applying hard criteria, such as the profitability of their options. Also softer factors, such as nuclear fear and environmental friendliness may enter the equation. Each agent may apply a unique weighing of the information and apply these criteria in a different manner. The result is an explicit model of the disaggregate decision-making that shapes the complex interrelated system of electricity production systems. Important parts of the market economy like fossil fuel markets are modeled as a set of exogenous variable that evolve in specific scenarios. One of the agents is the Emission Trading

Scheme that influences the overall response of the system of electricity production systems. This effect is not straightforward because CO_2 emissions are the result of disaggregate investment decisions by individual electric power producers. In order to obtain a robust image of Carbon Emission Trading (CET)-impact, parameter sweeps over the entire scenario and parameter space were completed by running some of 450 simulations. The results obtained were statistically analyzed by using the SPSS software. In the analyses no values of single run were used, but average values over the scenario and parameter space.

2.5 Modeling Emission Limits

About modeling emission limits, paper [15] divides emission limits into three categories: periodical emission limits, overall emission limits and tradable emission limits. Correspondingly, three different mathematical models are presented to integrate emission limits in product planning. This last problem could be transformed into a generation planning problem considering energy as a product. All the previous papers impose emission limits as periodical emission limits and overall emission limits. In the first case, the typical rule or regulation to limit emission is that the plants emission, in a fixed period, cannot exceed an upper limit, while in the second case, it is the sum of plants emissions from multiple periods that must be lower than a suitable upper bound. Along with the tradable emission markets developing, emissions can be exchanged between generators and a steady pricing mechanism is generated. To take into account this mechanism, it is necessary to add in the objective function another term considering the emission allowances acquisition cost and to modify the emission constraints increasing emission limits of the allowances bought at the market.

2.6 Multiobjective Models

Previous works consider only one goal: the revenues maximization and thus the costs minimizations under certain constraints like flowgates limit, environmental safeguard etc. This formulation is absolutely correct in the cases where the electricity market is completely liberalized and the GENeration COmpanys (GENCOs) pursue the profit maximization considering emission costs. In the presence of vertical integrated framework, the monopolistic GENCO must include in its objective function, the minimization of the systems cost, the maximization of the systems reliability, and the minimization of environmental degradation. It is also possible that several of previous aims could be conflicting.

In this way a long-term multiobjective model for the power generation expansion planning of electric systems is described and evaluated in paper [16]. The model simultaneously optimizes multiple objectives (i.e., minimizes costs, environmental impact, imported fuel and fuel price risks) and decides the location of the planned generation units in a multiperiod planning horizon. Among the attributes considered in the model are the investment and operation cost of the units, the environmental impact, the amount of imported fuel, and the portfolio investment risk. What concerns the environment impact, the goal is the minimization of the total generator emissions during the planning horizon. The constraints of the multiobjective (MO) problem are the same used in the previous works plus maximum investment limit for each unit type during each sub period (due to natural reasons like space, lack of resources etc.) and fuel balance (consumed for the generation and those domestic products or imported). The emissions limit is not considered because the environment damage control is performed by means of objective function. The approach to solve this problem is based on multiobjective linear programming and the analytical hierarchy process.

In the same way, Heggeler Antunes et al. [17], present a multiple objective mixed integer linear programming (MOMILP) model for power generation expansion planning that allows the consideration of modular expansion capacity values of supply-side options. The integrated resource planning (IRP) feature is to consider a demand-side option in the planning process (demand-side management DSM is most useful way for utilities peak clipping). This requires that the electricity market in the target region or country is organized in such a way that an important part of the load is supplied in a franchise environment and generation capacity expansion is mostly centrally planned. This still corresponds to a very large number of cases around the world, either because such electricity markets have not been liberalized or because the transformations kept some fundamental characteristics of the traditional market organization. The model takes into account the full cost of implementation and operation of the used DSM option, including lost revenues as an operational cost. In fact, DSM has been modeled as an equivalent DSM generating unit, which is used from the point of view of satisfying the forecasted demand, replacing peaking units and contributing to the generation system reliability when available, even if not in effective action. DSM actions focus on changing the pattern of demand for electricity, including generic load shape objectives such as peak clipping, valley filling, load shifting, flexible load shape, strategic load growth and strategic conservation. The same type of parameters used to characterize the supply-side generating units is used to model the DSM unit: implementation costs (installation and removal costs, initial contacts with the customers and incentives paid at the beginning of the DSM program), operation costs (equipment operation and maintenance charges, incentives paid on an annual basis and lost revenues) and environmental impact. About the last parameter, one of the aims of the model is to find how many DSM units are needed to reduce pollutant emissions.

The model presented in [17] pursues the minimization of three objective functions: the first is the sum of generation costs (already mentioned), the second concerns the environment impact associated with the generating capacity to be installed throughout the planning period and the third envisages the monetized environment externalities associated with the energy output from each type of power plant. The environmental impact associated with the generating capacity is obtained by means of a coefficient computed by using a weighted average of the impact evaluation criteria associated with the installation of the generating facilities: land use, large accidents and effects on ecosystems. The monetization of environment externalities associated with the energy output is calculated by using the coefficients obtained by means a weighted average of impact evaluation criteria associated with energy output: emissions/public health, water use and occupational accidents.

The model under examination [17] does not take into account the multi area framework while it includes in the constraints list, the upper bounds for sulphur-dioxide (SO_2), nitrogen-oxide (NO_2) and carbon-dioxide (CO_2) emissions for taking into account regulations on pollutant emissions resulting from electricity production.

2.7 *Methods Based on Artificial Intelligence*

Recently, new algorithms based on an Artificial Intelligence (AI) have been developed, such as Genetic Algorithm (GA) and expert systems. The results reported were promising and encouraging for further research in this direction. In the last several years, Tabu Search (TS) appeared as another promising heuristic algorithm for handling combinatorial optimization problems. Standard TS has been applied to various power system optimization problems with impressive success. However, some disadvantages of standard TS is that tabu list cannot adjusted properly in each iteration, and the current neighborhood may contain a large number of non-effective moves. Zhang et al. [18] have developed an Improved Tabu Search (ITS), which can overcome the aforementioned problem of standard TS. The proposed ITS incorporating a new feature of self-reforming candidate list is used to solve a least-cost GEP problem under constrained environmental impact consideration by introducing biomass energy sources for CO_2 reduction countermeasure and by including carbon tax charge to the objective function.

Papers [19] and [20] focus on the solution, with different heuristic techniques, of the same mixed-integer non-linear optimization problem. The model minimizes the investment and operation cost of existing and expanding plants, the energy purchase expenditure from independent power providers and the CO_2 tax. The constraints concern power balance, capacity limits, reliability level (expressed by means suitable limits of LOLP and EENS) and bargain condition on energy purchase price from IPP. The constraints are set on total CO_2 emission limit and on each plant emission limit. Specifically, the CO_2 emission model is assumed to be a combination of polynomial and exponential terms and the CO_2 tax is tripled when the plant emission limit is not respected.

In paper [19] the model is solved by means of an Immune Algorithm (IA) based on the operation of the human immune system. The immune system is a basic bio-defense system against viruses and disease-causing organisms. This complex defense mechanism combines genes to deal with the in breaking antigens. IA takes advantage of the conventional GA and Tabu Search, but the choice of a proper probability for crossover and mutation is still a dilemma.

Chen et al. [19] propose Refined Immune Algorithm (RIA) which further improves IA. In order to avoid prematurity, the crossover and mutation mechanism is refined by a competitive and auto-adjustable mechanism.

Shayeghi et al. [20] use as a solving method Particle Swarm Optimization (PSO) algorithms based on population stochastic search methods, derived on bird flocks and fish schools social behavior analysis. Each particle represents a possible solution and it is associated with position and velocity vectors: at each iteration a particle, which knows its and group's previous best positions and velocity, moves towards a optimum solution assessed by calculating the values of an appropriate fitness function. Search movement is based on a balance of three components: inertia term (depends by previous search direction) cognitive term (related to individual exploration) and social term (connected to information sharing).

Similarly in [21], a GEP model, slightly less detailed than the previous one (especially on emission models), is solved through Hybrid Coded Genetic Algorithm (HCGA) because random variables are a combinations of integer and real parameters. Integer variables are the numbers of the new units while real ones are the energy purchase prices from IPPs. In this method crossover can take place only at the boundary of two random variables. The mutation operator selects one of the existing random variables and changes its value randomly. The reproduction operator, similar to standard form, produces each chromosome proportional to the value of its objective function. Therefore, the chromosomes which have better objective functions will be selected more probably than other chromosomes for the next population (Elitist strategy).

Paper [22] proposes an innovative technique called "GA-Benders' Decomposition (GA-BD) method" (a combination of GA-heuristic-based method and Benders' decomposition method) for solving the GEP problem (written as a mixed-integer non-linear problem).The proposed GA-BD heuristic applies the decomposition structure of Benders to the GEP problem. Similar to integer programming (IP), the master problem (MP) deals with the integer variables representing the selection of power generating plants but also includes the Benders' cuts as its additional constraints. Given each investment plan prescribed by MP, a linear programming (LP) subproblem is solved for an optimal operational plan. For the proposed GA-BD, instead of solving the MP to optimality, the GA is proposed for solving those MP. In the GA part of the proposed GA-BD, only the best chromosomes are solved for LP optimality. This can shorten the computational runtimes, whereas the pure GA solves all chromosomes for LP optimum taking longer runtimes. The primary objective function is to determine an investment schedule to meet the customer power demand and operating reserve levels at the minimal cost (the sum of the present worth of investment and operation cost) and, at the same time, to meet all security constraints imposed on the electricity system for reliability purposes as well as the emission constraints for the environmental control purposes. The model incorporated reliability constraints, DSM programs, environmental requirements, and location constraints to determine an optimal electricity generation plan that minimizes the expected sum of discounted investment costs and variable costs that comprise fuel costs, operating costs, and environmental costs of each new generating unit of power generation. Emission constraints are written to consider the maximum allowance concentrations and background concentrations of pollutants (Ambient Air Quality Standards) at various receptors for each year.

2.8 Models Based on Fuzzy Logic

Fuzzy logic, is a multi-valued logic that allows a continuum of intermediate values to be defined between conventional Boolean valuations, in order to apply a more human-like way of thinking in the computer programming: appropriate membership functions can represent the degree of membership using values from 0 to 1. The fuzzy approach can be useful in power system multi-objective optimization problems because it allows consideration of more realistic models, taking into account several variations due to uncertainties, randomness or inaccuracies in assessment of various system parameters.

In [23] a methodology is exposed to solve a long-term multi-stages best generation mix (BGM) problem using a fuzzy linear programming (LP) approach, considering renewable energy sources (RES) as wind turbine generators (WTG) and solar cell generators (SCG) as well as conventional generation technologies.

The economic criteria is to satisfy the total cost minimization (sum of the construction and operating cost) under several constraints: energy demand and production balance, different pollutants emission upper bound (SO_x, NO_x and CO_2). Since the BGM assessment is conditioned by classic formulation inaccuracy, the fuzzy set theory approach may be a useful and flexible methodology. Budget and reliability criterion in the target year are expressed in terms of fuzziness, so the problem is formulated in order to maximize the satisfaction level, under the same constraints mentioned above.

Through this approach, the BGM for the Korean power system (time period 2008–2030) was determined: base year parameters, annual loads, generation types, yearly capacity factors (CFs), economic (as the apparent escalation rates CERs), consumption and environmental data are known or hypothesized.

Two types of sensitivity analyses were conducted:

- By varying CFs for RES in their usual ranges; as these CFs raise, the total cost decreases, because the operating cost reduction is greater than other technologies' construction cost increase; furthermore, high CFs increase satisfaction level;
- By varying the apparent escalation rates of construction material for RES generation type (CERsRES). As technological development will take place, it is expected that construction cost for wind turbines and especially for solar cells will decrease.

Through these analyses, appropriate values for CFs and CERsRES can be found in order to allow RES to win the conventional generation.

3 Risk Management in Long Term Generation Planning

In this section, a procedure [24, 25] is presented to evaluate the impact of energy and fuel price volatility on the long term generation planning of a generation company. The proposed model maximizes the Net Present Value (NPV) that can be obtained by a given generation expansion plan, taking into account the risk related to energy market and fuel supplying. A nonsequential Monte Carlo simulation is used to evaluate the risk by creating a large number of different scenarios obtained by suitably generating random values for energy prices, fuel prices, CO_2 emission and green certificates prices (i.e. the government incentives for renewable energies). All random variables are generated according to a specific probability distribution associated to each price (a lognormal distribution).

For each randomly generated scenario, the generation mix maximizing the total NPV is obtained by solving a large scale mixed integer linear programming problem, to which the Lagrangian Relaxation (LR) method is applied. The constraints include the energy balance, the minimum generation from renewable sources and CO_2 emission limits. To complement the LR solution, another optimization problem is set up to find the (integer) number of new plants to be installed to maximize NPV while enforcing a limit on the amount of money the GENCO is willing to invest in expansion planning. This problem is solved at the end of each LR solution, by an integer linear programming routine.

By simulating a sufficiently large number of different scenarios, a probability distribution for the expected NPV and for the optimum number of plants pertaining to different generation technologies is obtained. The procedure is implemented in MATLAB and applied to a real case study to determine a generation plan that takes the price volatility risk into account.

The main symbols, used to define most frequently used indices, parameters and variables, are listed below to ease understanding of the model.

3.1 Nomenclature

3.1.1 Parameters

r: discount rate used to find the present value of future cash flows;

N_y: number of years of the planning horizon considered;

N_s: number of sub-intervals in 1 year;

ρ^e_{ik}: expected energy market price [euro/MWh] for the (kth) sub-interval of the (ith) year of the planning horizon;

$\rho^{CO_2}_{ik}$: expected price of the emission credit [euro/t];

η_i: minimum thermal production percentage (2% in 2004 with an increase of 0.35% per year), to be covered yearly by renewable resources generation or by acquisition of a proper amount of green certificates (according to Italian decree 79/99);

N^{ex}_{tot}: set of all the existing generation plants;

N_{tot}^{new}: set of all the candidate generation plants which could be planned for building;
N_{hydro}: set of all the (existing or new) hydro power plants;
N_{eol}: set of all the (existing or new) wind power plants;
N_{geot}: set of all the (existing or new) geo-thermal power plants;
N_{biom}: set of all the (existing or new) power plants exploiting biomass or waste;
N_{therm}: set of all the (existing or new) conventional thermo-electric power plants;
N_{nuc}: set of all the (existing or new) nuclear power plants;
N_{ren} $= N_{hydro} \cup N_{eol} \cup N_{geot} \cup N_{biom}$: set of all the (available or planned) power plants exploiting renewable energy;
I_{jik}: investment cost of the (*j*th) new power plant incurred in the specified sub-interval;
ψ_j: linear coefficient relating electric energy generated by wind power plant *j* to the cube of wind speed;
v_{jik}: wind speed for wind power unit *j* relative to a specified subinterval;
$\varepsilon_j^{CO_2}$: amount of CO_2 released to the atmosphere per MWh of energy produced.

3.1.2 Indices

i: year index, $i = 1, 2, \ldots, N_y$;
k: sub-interval index, $k = 1, 2, \ldots, N_s$;
j: generation unit index.

3.1.3 Variables

E_{ik}^{sold}: total amount of energy sold at the market in the specified sub-interval;
G_{ik}^{sold}: green certificate sold at the market corresponding to an equal amount of energy produced by renewable resources;
G_{ik}^{bou}: green certificate bought at the market corresponding to an equal amount of energy produced by renewable resources;
Q_{ik}^{sold}: CO_2 emission right sold at the market;
Q_{ik}^{bou}: CO_2 emission right bought at the market;
$C_j(\ldots)$: cost function of unit *j* (generally expressed as a function of the energy produced);
E_{jik}: electric energy produced by existing plant *j* during sub-interval *k* of the year *i*;
E_{jik}^{new}: electric energy produced by new plant *j* during sub-interval *k* of the year *i*;
E_{jik}^{max}: upper bound on the electric energy produced by existing plant *j* during sub-interval *k* of the year *i*;
$E_{jik}^{new,max}$: upper bound on the electric energy produced by new plant *j* during sub-interval *k* of the year *i*;
w_{jik}: binary variable associated with candidate new power plant *j*: $w_{jik} = 1$ if plant *j* is installed and operating in subinterval *i*, *k*; $w_{jik} = 0$ if plant *j* is not installed in the specified *i*, *k* subinterval.

3.2 Formulation of Single GENCO Model

The objective function, to be maximized, consists of the total revenue obtained by the GEneration COmpany (GENCO) over a time horizon of N_y years into the future.

$$max \sum_{i=1}^{N_y} \sum_{k=1}^{N_s} \frac{1}{(1+r)^{(i-1)}} \Bigg[\rho_{ik}^{e} E_{ik}^{sold} + \rho_{ik}^{gc} (G_{ik}^{sold} - G_{ik}^{bou}) + \rho_{ik}^{CO_2} (Q_{ik}^{sold} - Q_{ik}^{bou}) + \\ - \sum_{j \in N_{tot}^{new}} w_{jik} (C_j(E_{ik}^{new}) + I_{jik}) - \sum_{j \in N_{tot}^{ex}} C_j(E_{ik}) \Bigg] \tag{1}$$

The incomes of the GENCO derive from the sales of energy E_{ik}^{sold} and green certificates G_{ik}^{sold} at the market; the expenses derive from the purchase of green certificates G_{ik}^{bou} as well as from the costs of generating energy by the power plants already in operation and by the new ones. With reference to the power units planned for installation, the investment cost I_{jik} is to be taken into account.

The maximization of the objective function (1) is constrained by a number of limitations depending on both physical and regulatory obligations. In particular constraint:

$$\sum_{j \in N_{tot}^{ex}} E_{jik} + \sum_{j \in N_{tot}^{new}} w_{jik} E_{jik}^{new} = E_{ik}^{sold} i = 1, ..., N_y \;\; k = 1, ..., N_s \tag{2}$$

relates the amount of energy sold at the market to the production of existing and new power units. Equation 2 must be enforced for each sub-interval of the whole planning horizon.

The fulfillment of the compulsory renewable energy quota, according to the Italian decree 79/99, results in the following constraint:

$$\sum_{k=1}^{N_s} \Bigg[\sum_{j \in N_{ren}} (E_{jik} + w_{jik} E_{jik}^{new}) + G_{ik}^{bou} \Bigg] = \sum_{k=1}^{N_s} \Bigg[\sum_{j \in N_{therm}} \eta_i (E_{jik} + w_{jik} E_{jik}^{new}) + G_{ik}^{sold} \Bigg] \\ i = 1, ..., N_y \tag{3}$$

that needs to be satisfied for every year of the planning horizon. Constraint (3) is supplemented by the following non-negativity condition on the amount of traded green certificates:

$$G_{ik}^{bou} \geq 0; \;\; G_{ik}^{sold} \geq 0 \;\; i = 1, ..., N_y \;\; k = 1, ..., N_s \tag{4}$$

The combined effect of constraints (3) and (4) implies that, for each year, the considered GENCO will be either selling or buying green certificates, but not both.

$$\sum_{k=1}^{N_s}\left[\sum_{j\in N_{ren}} \varepsilon_j^{CO_2}(E_{jik} + w_{jik}E_{jik}^{new}) + (Q_{ik}^{sold} - Q_{ik}^{bou})\right] = A_i^{CO_2} \quad i = 1,...,N_y \qquad (5)$$

$$Q_{ik}^{bou} \geq 0; \quad Q_{ik}^{sold} \geq 0 \quad i = 1,...,N_y \quad k = 1,...,N_s \qquad (6)$$

Constraints (5) and (6) are analogous to (3) and (4) respectively; they take into account CO_2 emission limitations with respect to the total yearly allowance $A_i^{CO_2}$.

The remaining constraints to be accounted for are those pertaining to the different generation technologies available to the GENCO as either existing or candidate generation units.

All generation plants are characterized by simple bound constraints of the following type:

$$0 \leq E_{jik} \leq E_{jik}^{max} \quad i = 1,...,N_y \quad k = 1,...,N_s \quad j = 1,...N_{tot}^{ex} \qquad (7)$$

$$0 \leq w_{jik}E_{jik}^{new} \leq E_{jik}^{new,max} \quad i = 1,...,N_y \quad k = 1,...,N_s \quad j = 1,...,N_{tot}^{new} \qquad (8)$$

which refer to existing units, (7) or candidate new plants (8); the value of the upper bound E_{jik}^{max} or $E_{jik}^{new,max}$ must be evaluated according to the amount of producible energy computed with reasonable accuracy on the basis of historical data analysis, for existing units, or predicted for new plants by comparison with similar projects.

Instead of using producible energy data, the values of E_{jik}^{max} or $E_{jik}^{new,max}$ can also be determined by making use of a more detailed model of the plant in case all the relevant data are available.

The availability of more detailed data about the wind frequency distribution of a given site allows a better estimate of the energy that can be obtained from an existing or planned wind power plant according to the following formulas:

$$0 \leq E_{jik} \leq \psi_j V_{jik}^3 \quad i = 1,...,N_y \quad k = 1,...,N_s \quad j = 1,...N_{eol}^{ex} \qquad (9)$$

$$0 \leq w_{jik}E_{jik} \leq w_{jik}(\psi_j V_{jik}^3) \quad i = 1,...,N_y \quad k = 1,...,N_s \quad j = 1,...,N_{eol}^{new} \qquad (10)$$

in which V_{jik}^3 is the average value of probability distribution of the third power of wind speed. This computation requires some assumptions regarding the probability distribution of wind speed and the characteristics of the wind power generator installed.

3.2.1 Comment on SGM

SGM is a large scale problem with mixed (real and integer) variables. It is worth noting that the majority of the constraints, namely those labeled (7, 8, 9, 10), involve only one generation plant at a time. In contrast, constraints (2, 3, 4, 5, 6) involve the energy generation of many power plants considered at the same sub-interval of the planning horizon. For ease of reference, we define "decoupled" all the constraints that, like (7, 8, 9, 10), refer to a single power plant. Constraints (2, 3, 4, 5, 6), are called "coupling" since they have the effect of binding the energy productions of possibly all power plants.

To tackle the mathematical complexity of SGM, it would be desirable to relax the coupling constraints in such a way that the resulting relaxed problem could be more easily solved for one generation unit a time; of course some iterative re-adjustment of solutions will be required to enforce the coupling constraints.

In the case of SGM, constraints (2) can be eliminated from the model in an exact way by simply substituting the amount of energy sold E_{jik}^{sold} into the objective function; after substitution the objective function takes the following form:

$$max \sum_{i=1}^{N_y} \sum_{k=1}^{N_s} \frac{1}{(1+r)^{(i-1)}} \Big[\rho_{ik}^{gc} (G_{ik}^{sold} - G_{ik}^{bou}) + \rho_{ik}^{CO_2} (Q_{ik}^{sold} - Q_{ik}^{bou}) + \\ + \sum_{j \in N_{tot}^{new}} w_{jik} (\rho_{ik}^{e} E_{jik}^{new} - C_j(E_{ik}^{new}) - I_{jik}) + \sum_{j \in N_{tot}^{ex}} \rho_{ik}^{e} E_{jik} - C_j(E_{ik}) \Big] \tag{11}$$

Unfortunately it is not possible to eliminate constraints (3, 4, 5, 6) by the same simple substitution used for constraints (2). This is because constraints (3) and (5) are not relative to a single sub-interval, but they encompass a whole year of production by all the thermal and renewable energy power plants.

The mathematical technique used to deal with these coupling constraints is often addressed to as Lagrangian Relaxation. In principle, the idea of the Lagrangian Relaxation method is quite simple; each of the N_y constraints (3) is rewritten as follows:

$$\sum_{k=1}^{N_s} \left[(G_{ik}^{bou} - G_{ik}^{sold}) + \sum_{j \in N_{ren}} (E_{jik} + w_{jik} E_{jik}^{new}) - \sum_{j \in N_{therm}} \eta_i (E_{jik} + w_{jik} E_{jik}^{new}) \right] = 0$$

and, similarly each of the N_y constraints (5) is rewritten in the form:

$$A_i^{CO_2} - \sum_{k=1}^{N_s} \left[\sum_{j \in N_{therm}} \varepsilon_j^{CO_2} (E_{jik} + w_{jik} E_{jik}^{new}) + (Q_{ik}^{sold} - Q_{ik}^{bou}) \right] = 0$$

Then both constraint expressions are added to the objective function after multiplication by the proper Lagrangian multiplier (λ_i for the green certificate limit and μ_i for the emission constraints). The resulting modified objective function takes the following form:

$$\sum_{i=1}^{N_y}\sum_{k=1}^{N_s}\Bigg[\left(\frac{\rho_{ik}^{gc}}{(1+r)^{(i-1)}}-\lambda_i\right)(G_{ik}^{sold}-G_{ik}^{bou})+$$
$$\left(\frac{\rho_{ik}^{CO_2}}{(1+r)^{(i-1)}}-\mu_i\right)(Q_{ik}^{sold}-Q_{ik}^{bou})+$$
$$+\sum_{j\in N_{tot}^{new}}\frac{w_{ik}^{gc}}{(1+r)^{(i-1)}}\left[\rho_{ik}^{e}E_{jik}^{new}-C_j(E_{jik}^{new})-I_{jik}\right]+$$
$$+\sum_{j\in N_{tot}^{ex}}\frac{1}{(1+r)^{(i-1)}}\left[\rho_{ik}^{e}E_{jik}-C_j(E_{jik})\right]+\lambda_i\sum_{j\in N_{ren}}(E_{jik}+w_{jik}E_{jik}^{new})+$$
$$-\lambda_i\sum_{j\in N_{therm}}\eta_i(E_{jik}+w_{jik}E_{jik}^{new})-\mu_i\sum_{j\in N_{therm}}\varepsilon_j^{CO_2}(E_{jik}+w_{jik}E_{jik}^{new})\Bigg] \quad (12)$$

The new objective function is maximized taking into account the decoupled constraints (7, 8, 9, 10) only, in the course of the so-called "dual iteration" of Lagrangian Relaxation; in the computation, the multipliers λ_i and μ_i are assumed known either by approximation or as the result of previous iterations. The solution of the dual problem results to be particularly simple thanks to the form of the decoupled constraints. In fact the behaviour of each generation unit along the planning horizon can be considered independently of the behaviour of all the remaining generators. In addition to the physical generator units, a fictitious generator, corresponding to the traded green certificates and emission credit, will need to be considered.

The dual iteration is followed by a "primal iteration" in which the relaxed constraints (3, 4, 5, 6) are enforced while keeping the binary variables w_{jik} fixed at the values obtained during the dual iteration. By comparing the values of the primal and dual objective functions, it is possible to check the convergence of the overall Lagrangian Relaxation algorithm. The procedure is stopped after a sequence of alternate primal and dual iterations when the difference between the primal and dual objective functions (the so called "duality gap") falls under a pre-specified tolerance; in general the convergence of the Lagrangian Relaxation method becomes quite fast after the values of the binary variables w_{jik} stabilize to their optimal values.

If convergence is not achieved yet, a further iteration cycle is initiated after properly updating the values of the multipliers λ_i and μ_i so as to enforce constraints (3, 4, 5, 6). The Lagrangian Relaxation algorithm details will be presented later on in this paper.

3.3 The Single-Plant Model (SPM)

The single plant model, considered in the present section, is the optimization problem regarding the behavior of a single generation for the whole optimization interval. SPM can be viewed either as a sub-problem of SGM (corresponding to a "dual iteration"), as remarked previously, or as a stand-alone problem of particular interest for the planner. If it is assumed that constraints (3, 4, 5, 6) must be satisfied for each subinterval i, k instead of for each year, or, conversely, if the Lagrange multipliers λ_i and μ_i can be estimated as:

$$\lambda_i = \frac{\rho_{ik}^{gc}}{(1+r)^{(i-1)}} \quad \mu_i = \frac{\rho_{ik}^{CO_2}}{(1+r)^{(i-1)}} \quad i = 1, \ldots, N_y \tag{13}$$

By substituting (13) in the objective function (12), the following modified expression is obtained.

$$\begin{aligned}
&\sum_{i=1}^{N_y}\sum_{k=1}^{N_s}\frac{1}{(1+r)^{(i-1)}}\Bigg[\sum_{j\in N_{tot}^{new}} w_{jik}(\rho_{ik}^{e}E_{jik}^{new} - C_j(E_{jik}^{new}) - I_{jik}) + \\
&+\sum_{j\in N_{tot}^{ex}}\left[\rho_{ik}^{e}E_{jik} - C_j(E_{jik})\right] + \rho_{ik}^{gc} + \sum_{j\in N_{ren}}(E_{jik} + w_{jik}E_{jik}^{new}) + \\
&-\rho_{ik}^{gc}\sum_{j\in N_{therm}}\eta_i(E_{jik} + w_{jik}E_{jik}^{new}) - \rho_{ik}^{CO_2}\sum_{j\in N_{therm}}\varepsilon_j^{CO_2}(E_{jik} + w_{jik}E_{jik}^{new})\Bigg]
\end{aligned} \tag{14}$$

From (14), it is apparent that considering the effect of trading green certificates is that of increasing the income obtained by renewable energy plants and correspondingly decreasing that obtained from fossil fueled plants.

By suitably collecting terms in (14) and by exchanging the order of the summations, the objective function (14) presents a separable structure, i.e. it can be considered as the sum of the revenues separately obtained by each generator plant.

In particular, for an existing generation unit exploiting renewable energy, the SPM objective function becomes as follows:

$$\sum_{i=1}^{N_y}\sum_{k=1}^{N_s}\frac{1}{(1+r)^{(i-1)}}\left[(\rho_{ik}^{e} + \rho_{ik}^{gc})E_{jik} - C_j(E_{jik})\right] \tag{15}$$

For a new generation unit exploiting renewable energy, the SPM objective function becomes as follows:

$$\sum_{i=1}^{N_y}\sum_{k=1}^{N_s}\frac{1}{(1+r)^{(i-1)}}w_{jik}\left[(\rho_{ik}^{e} + \rho_{ik}^{gc})E_{jik} - C_j(E_{jik} - I_{jik})\right] \tag{16}$$

For an existing generation unit using fossil fuel, the SPM objective function becomes as follows:

$$\sum_{i=1}^{N_y}\sum_{k=1}^{N_s}\frac{1}{(1+r)^{(i-1)}}\left[\left(\rho_{ik}^{e}-\rho_{ik}^{gc}\eta_i-\rho_{jk}^{CO_2}\varepsilon_j^{CO_2}\right)E_{jik}-C_j\left(E_{jik}\right)\right] \tag{17}$$

and, for a new generation unit:

$$\sum_{i=1}^{N_y}\sum_{k=1}^{N_s}\frac{1}{(1+r)^{(i-1)}}w_{jik}\left[\left(\rho_{ik}^{e}-\rho_{ik}^{gc}\eta_i-\rho_{ik}^{CO_2}\varepsilon_j^{CO_2}\right)E_{jik}-C_j\left(E_{jik}\right)-I_{jik}\right] \tag{18}$$

In all cases the generation limits (7, 8, 9, 10) have to be taken into account appropriately.

It must be remarked that the maximization of the SPM objective function for an existing unit (expressions (15) and (17)) does not involve consideration of integer variables; in this case the computation of the optimal solution can be carried out analytically for one sub-interval at a time.

For new generation units (objective functions (16) and (18)), consideration of the integer variables w_{jik} demands for more sophisticated optimization methodologies such as implicit enumeration techniques, nonlinear branch-and-bound and dynamic programming.

3.3.1 Solution of SPM for New Power Unit by Dynamic Programming

For the SPM problem, the decision variables are the integer variables w_{jik} which must be selected along the time to achieve the maximization of (16) or (18) over the whole planning horizon.

However, for what regards the installation of a new power plant, some further constraints have to be accounted for. In particular, operation cannot start before construction has been completed (that is after some specified construction time T_c) and the power unit must remain operative for a minimum time interval T_d (power unit life cycle) before de-commissioning.

This amounts to formulate the SPM (one optimization problem for each new power plant) in the following form:

$$max\sum_{i=1}^{N_y}\sum_{k=1}^{N_s}F\left(w_{jik},E_{jik}\right)$$

subject to:

$$\begin{aligned}&0\le w_{jik}E_{jik}^{new}\le w_{jik}E_{jik}^{max,new}\quad i=1,\ldots,N_y\ \ k=1,\ldots,N_s\\&\left(X_j^{cons}(i,k)-T_c\right)\left(w_{jik}-w_{jik-1}\right)\ge 0\ \ i=1,\ldots,N_y\ \ k=1,\ldots,N_s\end{aligned} \tag{19}$$

$$\left(X_j^{deco}(i, k) - T_d\right)\left(w_{jik-1} - w_{jik}\right) \geq 0 \quad i = 1, \ldots, N_y \quad k = 1, \ldots, N_s \tag{20}$$

Function $F(w_{jik}, E_{jik})$ is the contribution to the objective function relative to a single sub-interval i, k and is defined consistently to (16) or (18) above. $X_j^{cons}(i, k)$ and $X_j^{deco}(i, k)$ [26] are functions (counter) providing the number of sub-intervals during which the power unit j has been under construction or, respectively, in operation at each sub-interval i, k. Constraints (19) prevent the decision variable w_{jik} to switch from 0 to 1 in sub-interval i, k unless a sufficient time (T_c) has elapsed to allow the construction of the power plant. Similarly, constraints (20) prevent the decision variable w_{jik} to switch from 1 to 0 in sub-interval **[Trial mode]** until after a minimum time of operation (T_d).

The optimization problem can be defined as that of finding the optimal sequence of the decision variables w_{jik} together with the values of the corresponding energies E_{jik}.

Dynamic programming was proposed in the late 1950s by Dr. Richard Bellman and his co-workers as a means to solve multi-stage decision problems arising in many fields of science and engineering.

According to Bellman's theory, a sequence of values of the variables w_{jik} and E_{jik} is defined as a "policy"; a subset of values of w_{jik} and E_{jik} pertaining to consecutive sub-intervals is defined as a "sub-policy". The theorem of optimality of dynamic programming states that an *optimal policy must contain only optimal sub-policies* or, more explicitly. *a policy is optimal if, whatever the preceding decisions may have been, the decisions still to be taken constitute an optimal policy when the result of the previous decisions is included.*

At each stage of the problem, i.e. for each time sub-interval i, k, the integer variables w_{jik} may take only one of the two possible states: namely 0 or 1.The situation is shown pictorially in Fig. 1 where only four subsequent subintervals are shown for ease of presentation.

According to the formulation of the objective functions (16) and (18), no cost is incurred, but also no revenue can be obtained if $w_{jik+1} = 0$, i.e. when the power unit is not installed and operating.

In contrast, if $w_{jik+1} = 1$, the objective function can be increased by the (hopefully positive) amount F(E^*_{jik}), with E^*_{jik} being the revenue maximizing value of electric energy produced by the power unit in sub-interval i, k.

At the next stage i, k + 1, the power unit considered may conserve its current status or it may change it. No cost has to be paid if the transition does not occur, that

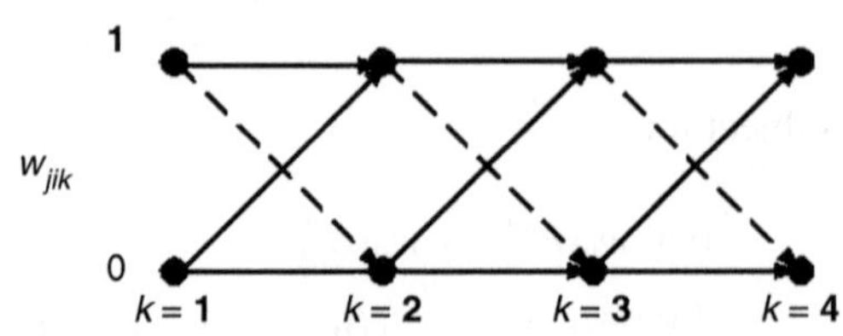

Fig. 1 Decision graph for the w_{jik} variables (four sub-intervals)

is if $w_{jik} = w_{jik+1} = 0$ or if $w_{jik} = w_{jik+1} = 1$. If, however, a transition occurs (when this it is possible), the corresponding transition cost has to be accounted for. If the unit is installed and starts operating in sub-interval $k + 1$ ($w_{jik} = 0$ and $w_{jik+1} = 1$) the transition cost will be taken equal to the present value of the installation cost. If, on the contrary, the unit is de-committed in sub-interval $k + 1$ ($w_{jik} = 1$ and $w_{jik+1} = 0$) the transition cost will be assumed equal to (present value of) the cost of decommissioning. Transition costs have to be properly subtracted from the income relative to the preceding stage; as a result, optimal sub-policies reaching the installation or de-commissioning status can be found in this way, at least with reference to the first two sub-intervals k and $k + 1$. By iterating the procedure just described, the whole planning horizon can be easily covered.

3.4 Further Features of SGM Model and Its Solution

Applying the Lagrangian Relaxation method to the solution of the SGM problem would imply including a lot of additional data into the model. In particular, it would be required to know the consistency of the generation mix presently available to the GENCO with sufficient detail, together with a detailed plan of the near future generation expansion planning (decommissioning of existing plants, expected start-up of already planned units, etc.).

3.4.1 Lagrange Multiplier Update Formula

One of the key problems in the implementation of a successful optimization code based on the Lagrangian Relaxation method is the choice of an efficient strategy for updating the Lagrange multipliers associated to the relaxed coupling constraints. This update has to be carried out at each iteration of the whole procedure, i.e. after solving both the dual and the primal sub-problems and it is recognized that a poor choice of the update scheme may result in a poor convergence of the method or even in a complete failure of the iterative procedure to converge to the optimum. In Fig. 2 we show the flowchart of Lagrangian Relaxation method.

For the problem at hand we are lucky we can exploit the structure of the SGM to find both a good initial estimate of the multipliers guaranteeing a fast convergence and an efficient update formula.

Indeed, from the KKT first order optimality condition applied to the SGM, we obtain a nice estimate of the Lagrange multipliers λ_i and μ_i corresponding to the coupling constraints on the compulsory renewable energy quota (constraints (3) in the model) and on allowed amount of CO_2 emission (constraints (5) in the model). In fact from the first order optimality conditions, we get:

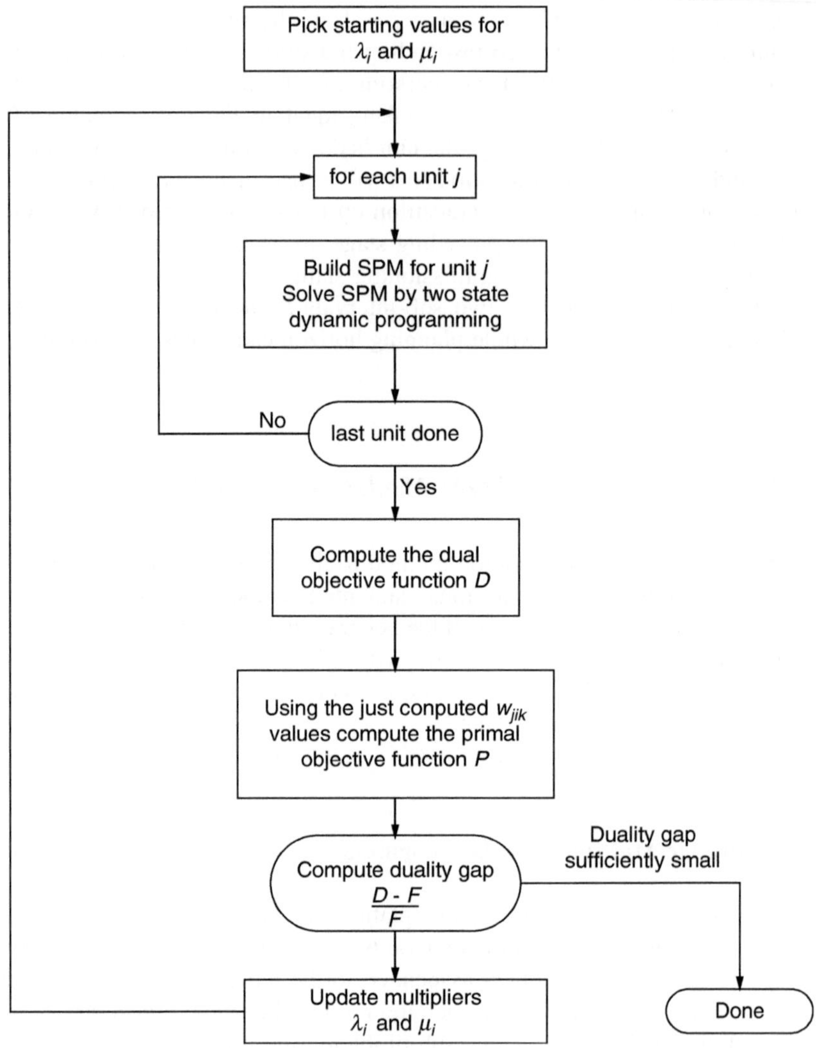

Fig. 2 Flow chart of the Lagrangian Relaxation procedure applied to SGM

$$\frac{\rho_{ik}^{gc}}{(1+r)^{(i-1)}} - \lambda_i = 0 \quad \frac{\rho_{ik}^{CO_2}}{(1+r)^{(i-1)}} - \mu_i = 0 \quad i = 1, \dots, N_y \tag{21}$$

which allows the multipliers corresponding to the two constraints can be estimated as:

$$\frac{\rho_{ik}^{gc}}{(1+r)^{(i-1)}} = \lambda_i \quad \frac{\rho_{ik}^{CO_2}}{(1+r)^{(i-1)}} = \mu_i \quad i = 1, \dots, N_y \tag{22}$$

The estimate turns out to be optimal if the expected prices of green certificates ρ_{ik}^{gc} and of CO_2 emission credits $\rho_{ik}^{CO_2}$ are held constant during each year of the planning horizon. If this is not the case, Eqs. 22 give only an approximation (for example, ρ_{ik}^{gc} and $\rho_{ik}^{CO_2}$ may be replaced by the corresponding yearly averages) and the multipliers λ_i and μ_i will need to be updated.

In the present implementation, the traditional subgradient formula for multiplier update was slightly changed with the aim of speeding up the convergence rate of the Lagrange Relaxation procedure. It is recalled that, according to the subgradient technique, the multipliers have to be updated as follows:

$$\lambda_i^{k+1} = \lambda_i^k + \frac{dq}{d\lambda_i^k}\alpha; \quad \mu_i^{k+1} = \mu_i^k + \frac{dq}{d\mu_i^k}\alpha \tag{23}$$

where q is the objective function of the dual optimization sub-problem and α a suitably chosen scaling parameter. It must be noted that the two derivatives $\frac{dq}{d\lambda_i^k}$, $\frac{dq}{d\mu_i^k}$ are equal to the right-hand-sides of the constraints (3) and (5), i.e. they are a measure of the violations of the corresponding constraints.

In our implementation of the Lagrangian Relaxation method, we have exploited the fundamental relation (22) to develop a faster update scheme of the multipliers. In fact, owing to the sensitivity nature of the Lagrange multipliers, also confirmed by Eq. 22, we know that the correct value of the multipliers will be close to the estimate (23); so the updated values of multipliers will be searched in the vicinity of the (present value) of the prices of green certificates and of CO_2 emission credits.

The adopted updating scheme is as follows:

$$\lambda_i^{k+1} = \lambda_i^k + 0.02\left[\frac{\rho_{ik}^{gc}}{(1+r)^{(i-1)}} - \lambda_i^k\right] \quad \textit{if} \quad \frac{dq}{d\lambda_i^k} \geq 0$$

$$\lambda_i^{k+1} = \lambda_i^k - 0.01\left[\frac{\rho_{ik}^{gc}}{(1+r)^{(i-1)}} - \lambda_i^k\right] \quad \textit{if} \quad \frac{dq}{d\lambda_i^k} < 0$$

$$\mu_i^{k+1} = \mu_i^k + 0.02\left[\frac{\rho_{ik}^{gc}}{(1+r)^{(i-1)}} - \mu_i^k\right] \quad \textit{if} \quad \frac{dq}{d\mu_i^k} \geq 0$$

$$\mu_i^{k+1} = \mu_i^k - 0.01\left[\frac{\rho_{ik}^{gc}}{(1+r)^{(i-1)}} - \mu_i^k\right] \quad \textit{if} \quad \frac{dq}{d\mu_i^k} < 0$$

3.4.2 Limit on Investment

It is possible to reduce the computation burden of the LR algorithm by exploiting some characteristics of the SGM model. When more than one generation unit of the same type are considered for planning, the solution of the single plant model would have to be unnecessarily repeated for all candidate plants. To avoid such wasted effort the (integer) number of units (exploiting the same technology) n_l that may be installed, can be found by solving the following integer linear programming (ILP) problem:

$$max \sum_{l=1}^{T} n_l NPV_l$$

subject to:

$$\sum_{l=1}^{T} n_l I_l \leq I^{max}$$
$$0 \leq n_l \leq n_l^{max} \quad l = 1, \ldots, T$$

NPV_l is the net present value achieved by a single unit exploiting a given generation technology out of a set of T possible choices (say coal, nuclear, hydro and so on). I_l is the investment cost to be paid for installing a single unit exploiting the (lth) technology, I_{max} is the total amount of money the GENCO is willing to invest in generation expansion and n_l^{max} is the maximum number of units (using the (lth) technology) available for planning.

Solving the ILP problem requires that the values of NPV_l and I_l are available for all generation technologies considered. For this reason the solution of the ILP problem is carry out at the end of "dual problem" (the NPV_l is obtained extracting from objective function (14) the terms pertinent to technology l, after optimization performed by "dual problem") and before of the "primal problem" (see Fig. 2). In fact after the "dual problem" we have the inputs for a suitable solution of the above ILP problem and its outcomes are needed to formulation of "primal problem" in order to enforce constraints (3) and (5).

3.5 Monte Carlo Simulation

To take into account the impact of energy and fuel price volatility on the long term generation planning of a generation company we use a Monte Carlo procedure. The proposed model maximizes the Net Present Value (NPV) that can be obtained by a given generation expansion plan, taking into account the risk related to energy market and fuel supplying. For the generation planning problem considered, the main risk factors are: energy price (peak and off-peak), the fuel cost, the renewable resources incentives and the emission credits cost. By means a MC simulation it has been possible to simulate different scenarios with different values of the factors mentioned above and for each scenario, to maximize the considered objective function.

It is known that the price dynamics of the financial product, at the first approximation, can be described from the geometric Brownian motion equation [30]. Analytically, it can be postulated that the price dynamics of the product follows this equation:

$$dS(t) = S(t)\mu dt + S(t)\sigma dW(t) \tag{24}$$

where $S(t)$ represents the price value of the product at time t, μ and σ are, respectively, the annual percentage drift and the annual volatility, while $W(t)$ is the Wiener process realization. A detailed mathematic description of a process like (24) is outside of the aims of this work; however to understand the temporal trend of motion described to (24), it is needed to remember that the Wiener process is characterized by three properties:

- $W(0) = 0$;
- $W(t)$ is wherever continuous in probability;
- $W(t)$ have independent updates, whose distribution is normal with zero average value and variance $(t - s)$

In the basic model, here considered, both parameters μ and σ are constant in the time.

By means traditional techniques of stochastic calculation, it is possible show that the differential stochastic equation described in (24) has the following solution:

$$S(t) = S(0)exp\left(\left(\mu - \frac{\sigma^2}{2}\right)t + \sigma W(t)\right) \tag{25}$$

where $S(0)$ is the initial value of the considered price. Taking into account that (25) represents a stochastic variable with log-normal distribution, the variable $lnS(t)/S(0)$ has lognormal distribution, with average $(\mu - \sigma^2/2)$ and variance $\sigma^2 t$. It's important to note that the quantity $lnS(t)/S(0)$ represents the logarithmic efficiency of product, the so called continuously compound return.

The price commodity has the trend similar to that of a financial product (a share for instance); in the following we will adopt this assumption. To simulate the determined number of trends that describe different possibilities commodity price evolution, it is needed to implement an MC algorithm. Then, it must simulate a determined number of realization of the discrete version of (24). A discrete standard form:

$$S(t + \Delta t) = S(t)(1 + \mu \Delta t + \sigma \varepsilon \Delta t) \tag{26}$$

where ε is a stochastic variable with standard normal distribution.

In the sequel we have simulated 5,000 scenarios, each of them utilizing weekly observations for a time horizon of 30 years. Then, in our case the update $\Delta t = 1/52$, where 52 is the number of week in the year. We have eight different discrete stochastic equations, one for each considered commodity price: nuclear fuel, biomass, green certificates, CO_2 emission credits, coal, natural gas, pick and off-pick energy. We have assumed that the price of nuclear fuel, biomass, green certificate and CO_2 emission credit have independent dynamics with respect to the other commodities. Thus these four commodities have four independent discrete stochastic equations like (26), where the four random component ε_1, ε_2, ε_3 and ε_4 are extracted from four normal standard distribution. On the contrary, we assume that the price dynamic of coal, natural gas, pick and off-pick energy are correlated. In particular:

$$Corr\left(ln\frac{S_i(t+\Delta t)}{S_i(t)}, ln\frac{S_j(t+\Delta t)}{S_j(t)}\right) = Corr(\varepsilon_i, \varepsilon_j)$$

It is possible to estimate the correlations among the stochastic component of the equations for the price trend of coal, natural gas and peak, off-peak energy, from the data available of the considered commodities. In particular, the correlations are estimated using the observation of the monthly average price of the four quantities starting from January 2005 until March 2008. The result may be inaccurate due to the lack of available observations, but is useful for a preliminary parameter calibration. The variables ε_5, ε_6, ε_7 and ε_8 are extracted from a multivariate normal distribution whose correlation matrix is given by:

$$\sum = \begin{array}{ccccc} \ldots & P_{peak} & P_{off-peak} & P_{gas} & P_{coal} \\ P_{peak} & 1 & 0.193 & 0.136 & -0.015 \\ P_{off-peak} & 0.193 & 1 & 0.149 & 0.049 \\ P_{gas} & 0.136 & 0.149 & 1 & 0.069 \\ P_{coal} & -0.015 & 0.049 & 0.069 & 1 \end{array}$$

The above variables are calculated with this formula:

$$[\varepsilon] = L[N]$$

where $[\varepsilon]$ is a random variable vector, L is a lower triangular matrix obtained from the Cholesky factorization of the matrix Σ and $[N]$ is a vector of independently random variables normally distributed.

Analyzing the historical trend of the commodities using the available data, it is possible to note that the deterministic component of commodities price in each of the equation like (26), is very small (10^{-2}). Then we fix μ_i for all variables i $= 1 \ldots 8$.

For what concern the calibration of parameters σ, using the available monthly observations, an estimation of yearly volatility can be obtained by means the formula $\sigma_i = sd_i/\sqrt{1/12}$ where sd_i is the is the standard deviation of the historical trend of the commodity performance $i = 1 \ldots 8$. It can be noted that the factor $\sqrt{1/12}$ reflects the fact that the estimation is done from the monthly observation. It is important to see that we have only monthly data, for a time horizon not particularly long. For this, the value of estimated yearly volatility might result in being too high. We have reduced in suitable way the estimation so obtained for the energy price (both pick and off-pick). The values used for the σ parameter and for price at time t_0 are (Table 1):

Table 1 Scenario variables

Scenario variable	Unit of measurement	Yearly volatility	Value at time t_0
Peak energy price	[euro/MWh]	0.2	110
Off-peak energy price	[euro/MWh]	0.2	60
GC price	[euro/MWh]	0.2	80
CO_2 emission right	[euro/t]	0.325	15
Gas price	[euro/m^3]	0.195	0.21
Goal price	[euro/t]	0.166	0.07
Nuclear fuel price	[euro/t]	0.33	1,200
Biomass fuel price	[euro/t]	0.2	0.10

3.6 *Application of Monte Carlo Method to SGM*

The generation expansion planning and Monte Carlo procedures were implemented in the Matlab programming language exploiting the vectorization capabilities of this computation environment wherever possible. The simulations were carried out on a 2.66 GHz personal computer based on the Intel Core 2 Duo processor on which Matlab 7.5 was installed. The technologies tested in the solution of SGM problem are shown in Table 4; also the rated power of planned generation units and the upper bound on the generation capacity to be installed are given, according to the above described generation planning procedure. Coal, combined cycle and nuclear plants are intended to carry out base load service; mini hydro, with a rated power of up to 1 MW, is included on the assumption that larger and more economically convenient hydro plants have been already exploited and are possibly included in the existing generation park. The data pertaining to these and to the remaining renewable source generation units are shown in Tables 2 and 3, and are consistent with those that can be found in the technical literature [27].

The generation expansion problem of an hypothetical GENCO has been simulated for a 5,000 Monte Carlo scenarios. For each scenario, the LR procedure returns the optimal generation planning maximizing the NPV. The limit of maximum power available for new installation is an indicative value and can be violated, because the ILP optimization is done considering the number of units for each technology, given by the division between the maximum power limit and the unit rated power, rounded to the nearest integer greater than or equal to it. The investment limit was set to 10,000 Meuro. The total generation capacity of the existing units is shown in the first column of Table 5, with reference to each available technology.

In Fig. 3 the probability density of the NPV is represented. The frequency chart has been fitted by means of a lognormal distribution.

The mean value of the NPV represents the more probable profit that the GENCO can achieve, investing the allocated capital in the generation expansion plan provided by Table 5 and Fig. 4.

By comparing the expected value of power to be installed to the maximum power available for installation, the most profitable technologies can be detected. In fact, technologies such as off-shore wind power, biomass, solar thermodynamic and solar photovoltaic have an expected value much lower than the imposed maximum

Table 2 Plants technical data part 1

Type	Investment cost	O&M cost (fix.)	O&M cost (var.)	Decom. cost	Fuel cost	Effic.	Loss
Measurement unit	Meuro/MW	Meuro/ MWyr	euro/MWh	Meuro/MW	euro/kg euro/m^3	%	%
Coal	1	0.02	0.17	0.02	0.07	44	4.5
C. cycle	0.47	0.0095	0.025	0.0094	0.21	56	2.5
Nuclear	1.5	0.07	0.46	0.225	1,200	34	5
Hydro	3	0	10	0	–	–	3
Wind	1.75	0.02	0	0.017	–	–	3
Wind off-shore	2.40	0	15	0.017	–	–	3
Biomass	2.2	0	70	0	0.1	35	3
Geotherm.	3.5	0	11	0	–	–	3
Solar TRM	3	0	50	0	–	–	3
Waste	5	0	40	0	0	25	3

Table 3 Plants technical data part 2

Type	Construction time	Life	WACC	Hours of use
Measurement unit	Years	Years	%	Hours
Coal	4	25	6.8	7,884
C. cycle	2	25	6.8	7,884
Nuclear	7	50	7.8	7,884
Hydro	1.5	40	8.7	3,500
Wind	1.5	20	6.5	2,215
Wind off-shore	2	20	7	2,215
Biomass	1.5	15	9	6,000
Geotherm.	3	15	6.5	7,500
Solar TRM	2	25	5.2	3,500
Waste	1.5	20	9	6,100

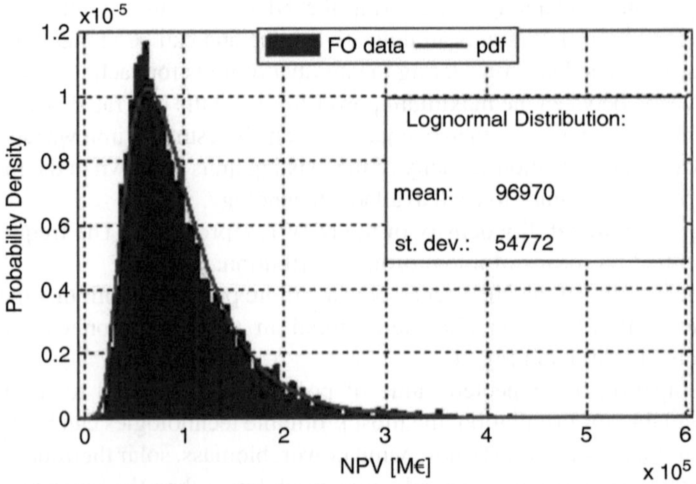

Fig. 3 Probability density function of NPV resulting from simulations

Table 4 Unit rated power and maximum power available for planning

Generation technology	Unit of measurement	Unit rated power	Maximum power
Coal	[MW]	600	2,000
Combined cycle	[MW]	800	2,000
Nuclear	[MW]	1,200	2,000
Mini hydro	[MW]	1	1,000
On-shore wind farm	[MW]	100	300
Off-shore wind farm	[MW]	100	200
Biomass (biogas)	[MW]	20	200
Waste to energy plant	[MW]	50	200
Geothermal	[MW]	40	200
Solar (thermodynamic)	[MW]	10	200
Solar (photovoltaic)	[MW]	10	200

Table 5 Total power existing units and results of MC simulation applied on SGM

Generation technology	Unit of measurement	Existing units total power	Expected value
Coal	[MW]	6,000	1,868
Combined	[MW]	8,000	2,088
Nuclear	[MW]	0	2,055
Mini hydro	[MW]	3,000	196
On-shore wind farm	[MW]	240	234
Off-shore wind farm	[MW]	0	68
Biomass (biogas)	[MW]	0	80
Waste to energy plant	[MW]	0	118
Geothermal	[MW]	650	195
Solar (thermodynamic)	[MW]	0	68
Solar (photovoltaic)	[MW]	0	58

limit and this means that these technologies are competitive in a small number of simulated scenarios. Instead, other technologies are competitive in a larger number of scenarios, guaranteeing a positive NPV.

Figure 4 shows the mix composition in percentage terms. The percentage values are referred at an amount of 7,028 MW that represents the sum of expected value listed in Table 5 (obtained from Monte Carlo simulation). These results show the generation park composition allowing an NPV average value of 96,970 Meuro, considering a investment limit of 10,000 Meuro. Figure 4 shows more clearly what we said about the different technologies competitiveness. In fact, from the point of view of maximizing the GENCO profits (and taking into account the incentive level), the procedure suggests to plan a generation mix constitutes especially of conventional generation technologies (86% of whole generation technologies are shared in approximately equal way among coal, combined cycle and nuclear), with a little quantity of renewable resources (14% shared among different resources). This last consideration leads to the conclusion that actual incentives level is not enough to reach the aims proposed from European Council (directive 2009/28/CE [28]) and Italian Government (2007 Position Paper [29]).

To deal with the previous tests, exploiting the Monte Carlo procedure, we have relaxed the constraint (4) and (6) representing the upper and lower bounds on

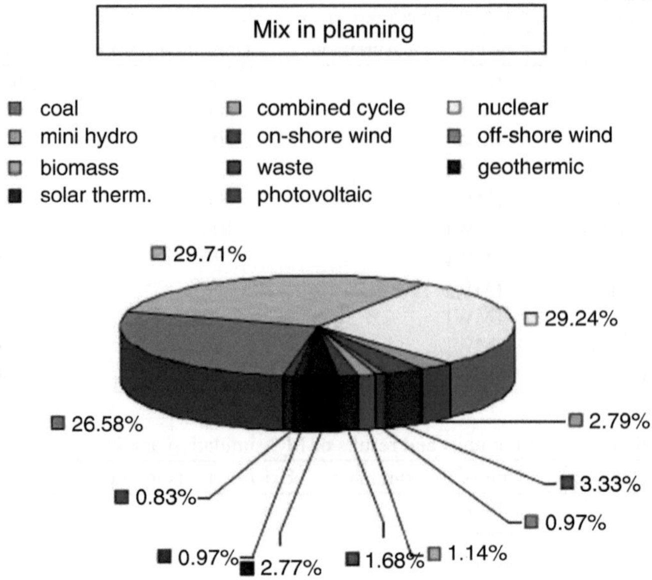

Fig. 4 Mix composition

green certificates and emission rights that GENCO could buy or sell at the respective markets. This is done for the following reasons:

- In Italy the offer and demand of green certificates is guaranteed by GSE (Gestore Servizi Energetici);
- To bound CO_2 emission rights sold or bought, would need to make difficult hypothesis on the worldwide demand and offer of EUAs (European Union Allocations), ERUs (Emission Reduction Units) and CERs (Certified Emission Reductions);
- Enforcing constraints (4) and (6), to reach the solution of Lagrangian Relaxation (LR) many iterations are needed, increasing the computational effort unsustainably (indeed the LR is solved for thousands Monte Carlo scenarios).

However in the next section we will present some tests on LR without Monte Carlo simulation, in order to give at the reader some ideas how LR works when constraints (6) are enforced.

3.7 *Tests on Deterministic SGM*

In the following tests we consider only the conventional portfolio, then the constraints on green certificates market are not considered. To consider all remaining constraints of SGM, it is necessary to make some hypothesis on existing generation mix and its decommissioning

Figure 5 shows the GENCO existing generation park at the time t_0 (and the decommissioning due to life-cycle time), composed by generation plants fed by coal (red line, existing power at t_0 is 6,000 MW), Natural Gas Combined Cycle (NGCC in purple line, existing power at t_0 is 8,000 MW), and traditional plants fed by oil or gas (blue line, existing power at t_0 is 12,000 MW).

The Generation technologies available for planning are listed below:

- Three slots of 3,300 MW of plants feed by coal that could be planned immediately, after 3 and 5 years respectively (taking into account building time);
- Four slots of 2,500 MW of Natural Gas Combined Cycle that could be planned immediately, after 2, 4 and 6 years respectively (taking into account building time);
- One slot of 10,000 MW of Nuclear plant that could be planned immediately (taking into account building time).

The nuclear plants are planned immediately because they do not affect the constraints (5) on maximum emission limit.

Another hypothesis concerns the total GENCO emission limit (given by National Allocation Plan) and the emission credits that the GENCO could buy at the market. These assumptions are as follows:

- The total free allowances start at about 45 $\times$ 10^6 CO_2 tons, decrease linearly in 5 years at about 33 $\times$ 10^6 CO_2 tons and then is kept constant for the whole remaining planning period;
- The maximum emission rights purchasable at the market are kept constant for the first 4 years at 12 $\times$ 10^6 CO_2 tons and at 8 $\times$ 10^6 CO_2 tons for the whole remaining planning period.

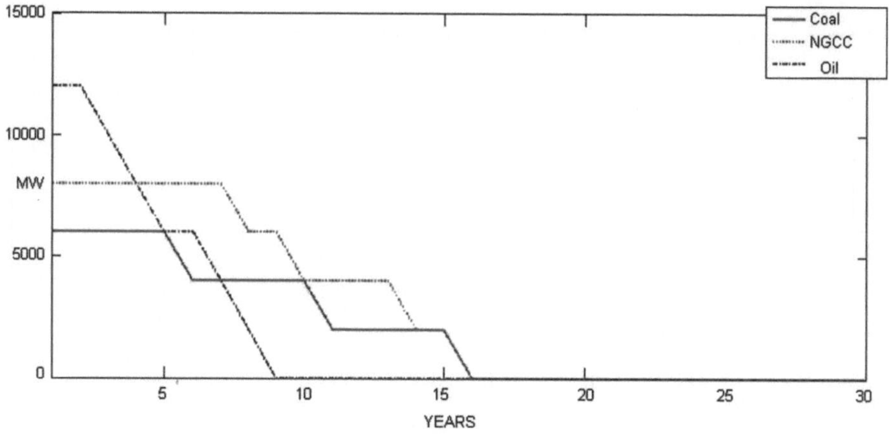

Fig. 5 Decommissioning of existing generation mix

The scenario variables are kept constant (however actualized by means of suitable discount rate) and their values are listed in the Table 6.

To converge, the LR algorithm needs five iterations during which the Lagrangian Multipliers (LMs) on constraint (5) (one for each year of the planning period), corresponding to the modified CO_2 tons prices (euro/CO_2 tons), assume the values reported in Fig. 6.

The mix modification during the convergence is shown in Fig. 7.

Figure 6 shows that the LMs assume oscillating values during the iterations; Indeed, in the first iteration, the procedure starts with LMs set at the actualized CO_2 tons prices, while in the following iterations the LMs are modified by the primal iteration depending from generation technologies chosen in the dual iteration (Fig. 7). For instance when the previous iteration has low values of LMs, the dual problem will choose a mix composed especially of power plants fed by coal (very economical competitive), vice versa the mix found by dual problem will be in prevalence structured by nuclear power plants. The generation mix composition converges (and thus LMs values) after five iterations to an intermediate situation

Table 6 Scenario variables

Scenario variable	Unit of measurement	Yearly volatility
Peak energy price	[euro/MWh]	120
Off-peak energy price	[euro/MWh]	70
CO_2 emission right	[euro/t]	15
Gas price	[euro/m^3]	0.21
Goal price	[euro/t]	0.07
Nuclear fuel price	[euro/t]	1,200
Biomass fuel price	[euro/t]	0.1

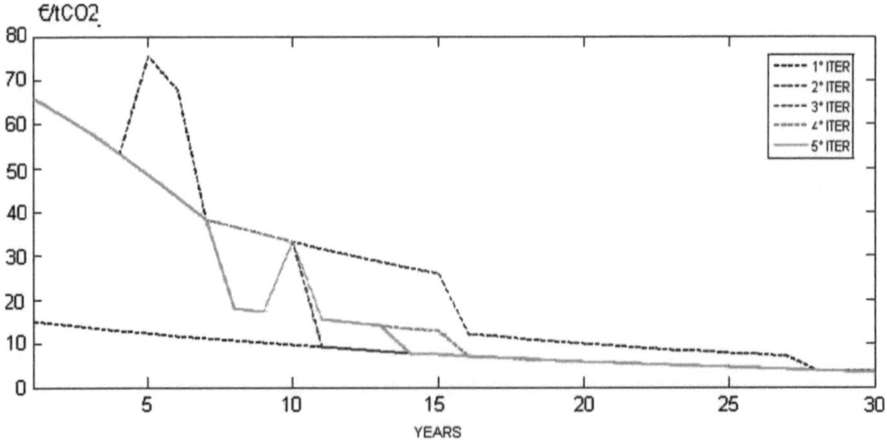

Fig. 6 Lagrangian Multipliers values during the five iterations

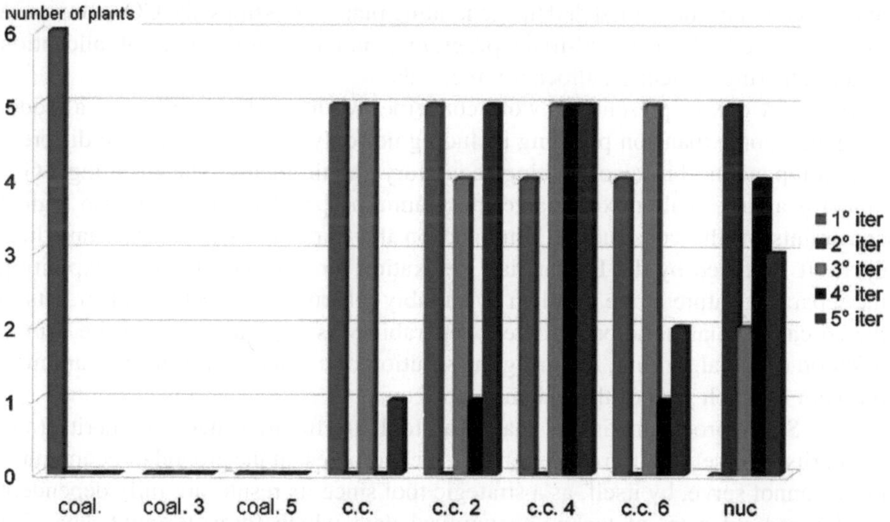

Fig. 7 Mix modification occurring during iteration

between those assumed in the first and second iterations. Considering the mix composition in the first and last iteration we can note:

- In the first one, due to the low CO_2 tons prices (equal to LMs), the mix is completely composed by power plants fed by coal and natural gas combined cycle (c.c. in Fig. 7), while nuclear plants are not considered;
- In the last iteration, the CO_2 tons prices are higher than the first one, thus the generation mix composition is dominated by nuclear power plants and NGCCs.

In the last iteration we note that NGCC can be planned immediately (c.c. in Fig. 7) and after 6 years (c.c. 6 in Fig. 7) are, to the procedure, less competitive than the others. For the first kind of NGCC, this is due especially for LMs high price in the first 10 years (caused by contemporaneous decommissioning of existing generation units and commissioning of new power plants). For the second kind of NGCC (and also for plants fed by coal that could be planned after 3 and 5 years), the low competitiveness is due to the fact that, for a 30 year planning horizon, these plants have less time (to exploit) than the others to refund the investments.

4 Conclusion

In this work we have focused on the GHG emissions impact on Power Generation Expansion Planning problem, giving to the reader a broad overview of the models and the solution methods used in the literature to make these types of analysis. In particular, in the literature summary, it has been considered the way in which many technicians have modeled the generating plants emissions, the CO_2 allowance exchange methods (cap-and-trade programs) and the emission right allocation (grandfathering or auction allocation methods).

Further we have presented a work concerned with the development of a model for generation expansion planning including not only investment and the different cost components, but also existing regulatory requirements. The resulting SGM model is a large scale mixed integer programming problem including also logical constraints on the construction time and on the duration of generation unit life-cycle. It is solved by the Lagrangian Relaxation method which allows exploiting the separable nature of the problem by suitably relaxing the coupling constraints in the so-called dual iteration. Problem separability is exploited also in the primal iteration of the algorithm, requiring the solution of a standard linear programming problem for each year of the planning horizon.

The SGM procedure is an analytical tool, useful to gauge the merits (and de-merits) of a set of different generation technologies on the ground of economics, but it cannot serve, by itself, as a strategic tool since its results are fully dependent on the particular set of techno-economical data which form its input data. For this reason the SGM problem was complemented with a Monte Carlo simulation environment which automatically generates the SGM problem scenario variables thus taking into account the uncertainties in fuel costs, price volatilities and economic risks. The generation expansion planning and Monte Carlo simulation procedures were implemented in the Matlab programming language and tested with reference to different scenarios pertaining to cost and regulation policies.

References

1. Villar CM (2009) Carbon windfall profits in the power sector. Policies to mitigate redistributive effect. EEM. doi: 10.1109/EEM.2009.5311428
2. Ellerman AD, Joskow PL (2008) The European Union's Emissions Trading System in perspective. Pew Center on Global Climate Change, Boston
3. Warwick JM, Morris A, Wilcoxen PJ, Cai Y (2009) Consequences of alternative U.S. cap-and-trade policies: controlling both emissions and costs. The Brookings Institution, Washington, DC
4. Paltsev S, Reilly J, Jacoby H, Gurgel A, Metcalf G, Sokolov A, Holak J (2008) Assessment of US GHG cap-and-trade proposals. Clim Policy 8(4):395–420
5. Jia NX, Yokojama R, Zhou YC, Kozu A (2000) An effective DP solution for optimal generation expansion planning under new environment. In: IEEE PowerCon conference. doi: 10.1109/ICPST.2000.900028
6. Jia NX, Yokojama R, Zhou YC (2001) Advanced DP based method for power expansion planning in the deregulated market. Power Tech. doi: 10.1109/PTC.2001.964602
7. Sun R, Chen L, Song Y, Sun Y (2008) Generation Expansion Planning based on multi-area reliability exponential analytic model and emission control. TD IEEE-PES. doi: 10.1109/TDC.2008.4517104
8. Most D, Genoese M, Esser A, Rentz O (2008) European electricity and emission market modeling—the design of emission allocation plans and its effects on investment planning. EEM. doi: 10.1109/EEM.2008.4579087

9. Neogi S, Pradhan A, Sinha A, Chowdhury S, Chowdhury SP, Gaunt CT (2005) Optimizing generation Portfolio for emission control in Europe's utilities. Ind Eng Chem Res. doi:10.1021/ie049766o
10. Hanshim H, Douglas P, Elkamel A, Croiset E (2009) Optimization model for energy planning with CO_2 emission considerations. EEM. doi: 10.1109/EEM.2009.5207154
11. Muis ZA, Hashim H, Manan ZA, Taha FM (2010) Optimal electricity generation mix with carbon dioxide constraint. Faculty of Chemical and Natural Resources Engineering, University Teknologi Malaysia. Available via DIALOG. http://www.fkkksa.utm.my/prospect/images//igceszarina20301108.pdf. Accessed on January 10, 2010
12. Yamgar S, Nandaa G, Srivastavaa SC, Singh SN, Guptab P, Paulb D, Shrestha RM (2004) Implications of energy tax on generation expansion plan and GHG emission: a case study on Indian power sector. PowerCon, vol 2, pp 1734–1739. doi: 10.1109/ICPST.2004.1460283
13. Rahman S, de Castro A (1994) A framework for incorporating environmental factors into the utility planning process. IEEE Trans Power Syst 9(1). doi:10.1109/59.317591
14. Chappin EJL, Dijkema GPJ (2007) An agent based model of the system of electricity production systems: exploring the impact of CO_2 emission-trading. SoSE IEEE. doi: 10.1109/SYSOSE.2007.4304239
15. Liu T, Li F, Tian C, Zhang H, Cao R, Ding W (2008) Modeling emission limits in product planning. IEEE/SOLI IEEE. doi: 10.1109/SOLI.2008.4682811
16. Meza JLC, Yildirim MB, Masud ASM (2007) A model for the multiperiod multiobjective power generation expansion problem. IEEE Trans Power Syst 22. doi:10.1109/TPWRS.2007.895178
17. Henggeler CA, Gomes AM, Brito IS (2004) A multiple objective mixed integer linear programming model for power generation expansion planning. Energy 29(4):613–627. doi:10.1016/j.energy.2003.10.012
18. Zhan T, Chen S, Tsay M, Kang M, Tsai J, Liao B (2009) Optimal generation expansion planning strategy for the utility with IPPs participation and considering Green House gas mitigation. ICIEA. doi: 10.1109/ICIEA.2009.5138736
19. Nualhong D, Chusanapiputt S, Jantarang S, Pungprasert V (2005) Generation expansion planning including biomass energy sources with global environmental consideration using improved Tabu Search. TENCON. doi: 10.1109/TENCON.2005.300990
20. Shabani A, Hosseini H, Karegar HK, Jalilzadeh S (2008) Optimal generation expansion planning in IPP presence with HCGA. PECon. doi: 10.1109/PECON.2008.4762461
21. Sirikum J, Techanitisawad A, Kachitvichyanukul V (2007) A new efficient GA-Benders' decomposition method: for power generation expansion planning with emission controls. IEEE Trans Power Syst. doi:10.1109/TPWRS.2007.901092
22. Shayenghi H, Pirayeshnegab A, Jalili A, Shayanfar HA (2009) Application of PSO technique for GEP in restructured power systems. Energy Convers Manage 50(9):2127–2135. doi:10.1016/j.enconman.2009.04.018
23. Choi J, Jeong S, Bo S, Park J, El-Keib A, Watada J (2007) CO_2, SO_x and NO_x emissions constrained multi-criteria-best generation mix using fuzzy set theory. In: Conference on large engineering systems. doi: 10.1109/LESCPE.2007.4437364
24. Genesi C, Granelli GP, Marannino P, Montagna M, Rossi S, Siviero I, Desiata L, Gentile G (2009) Impact of renewable energy quotas and emission trade on generation planning. IEEE PowerTech. doi:10.1109/PTC.2009.5281880
25. Genesi C, Marannino P, Montagna M, Rossi S, Siviero I, Desiata L, Gentile G (2009) Risk management in long term generation planning. EEM. doi: 10.1109/EEM.2009.5207136
26. Granelli GP, Marannino P, Montagna M, Zanellini F (2006) Monte Carlo based unit commitment procedures for the deregulated market environment. Electr Power Energy Syst 28:712–722. doi:10.1016/j.ijepes.2006.03.011
27. Desiata L, D'Alberti S (2007) The role of nuclear source in the optimization of base-load generation portfolio (in Italian). Energia 2:74–82

28. Department for European Affairs (2009) On promotion of renewable resources energy use, Brussels, 26 Mar 2009
29. The Presidency of the Council of Ministers (2007) Italian Government Position Paper, Rome, 10 Sep 2007
30. Black F, Scholes M (1973) The Pricing of Options and Corporate Liabilities. Journal of Political Economy 81(3):637–654

Forecasting CO_2 Prices in the EU ETS

Orvika Rosnes, Anne-Franziska Sinner, and Berit Tennbakk

Abstract The paper focuses on the market dynamics of the EU Emissions Trading Scheme (EU ETS), the cap-and-trade system implemented to reduce CO_2 emissions from electricity and heat plus some industrial sectors in the EU. An overview of carbon market models is presented and an analysis based on a particular model is used to illustrate some of the main price drivers in the market. The model results indicate that it is crucial to capture short and long term fuel switching in electricity generation and electricity demand response in order to forecast EU Allowance (EUA) prices. In addition, the impact of other policy measures is significant, e.g., support to renewable energy and compensation for industries at risk of carbon leakage. The applied model, together with the available data and market characteristics, imply that the dynamics of emission reductions from the heating sector is poorly understood, including the combined impact of the ETS and other policy measures such as the renewable energy target.

Keywords EU ETS • CO_2 prices • Modelling • Fuel switching • Abatements

---

The article was written while the authors worked for Econ Pöyry, now Pöyry Management Consulting Norway (www.poyry.com).

O. Rosnes (✉)
Statistics Norway, Research Department, PO Box 8131, Dep, 0033, Oslo, Norway
e-mail: orvika.rosnes@ssb.no

A.-F. Sinner
KPMG LLP, 55 Second Street, Suite 1400, San Francisco, CA 94105, USA
e-mail: asinner@kpmg.com

B. Tennbakk
THEMA Consulting Group, Øvre Vollgate 6, 0158 Oslo, Norway
e-mail: berit.tennbakk@t-cg.no

Q.P. Zheng et al. (eds.), *Handbook of CO_2 in Power Systems*, Energy Systems,
DOI 10.1007/978-3-642-27431-2_11, © Springer-Verlag Berlin Heidelberg 2012

1 Introduction

The EU Emissions Trading Scheme (ETS) is the cornerstone of the EU's strategy to combat climate change and the largest market for emission allowances in the world. The scheme was launched in January 2005 and covers today approximately 10,000 industry installations in the 27 EU Member States and three members of the European Economic Area (Norway, Iceland and Liechtenstein). The sectors included in the scheme are the energy industry (power utilities, combined heat and power utilities, district heating) and emission-intensive manufacturing industry such as iron and steel, pulp and paper, chemical and mineral industries (glass, cement and lime). The ETS sectors are collectively responsible for close to half of the EU's emissions of CO_2 and 40% of its total greenhouse gas emissions.

The duration of the ETS can be divided into different phases. The first phase was the introduction phase of 2005–2007, intended as a learning period for participants and regulators. The second phase coincides with the Kyoto commitment period of 2008–2012. The third phase is the period from 2013 to 2020, corresponding to the adopted targets for 2020 as part of the EU's climate and energy package.

Following a major review of the EU ETS in 2006–2007, an amended EU ETS Directive [1] entered into force in April 2009. The amended directive covers and determines the regulations for the EU ETS second and third trading periods. From 2012, the scheme will be further extended to cover the aviation industry. From 2013 onwards, additional sectors and gases within the mineral, metal and chemical industries as well as new sectors, like aluminum, hydrogen and CO_2 capture, will be covered by the scheme. The total annual average allocation in the first phase was 2,058 Mt CO_2 per year, in the second phase 1,931 Mt CO_2 per year and in the third phase 1,846 Mt CO_2 per year.

The purpose of the EU ETS is to reduce greenhouse gas emissions in the eligible sectors in an efficient manner, both in the short term and in the long term. To this end, the price of EU Allowances (EUAs)[1] for emissions is instrumental: Short term price levels induce efficient emission reductions via reduced industry production, electricity and heat demand, and fuel switching in energy production, whereas expected future price levels affect investment decisions and the future costs of emissions mitigation. Forecasts of EUA prices are therefore crucial inputs for investment decisions. The EUA price has varied substantially since the implementation of the scheme, ranging between 30 € per ton CO_2 and 1 € per ton CO_2. In the aftermath of the financial crisis in the fall of 2008, prices seem to have settled around 15 € per ton CO_2. Forecasts of the CO_2 price in the third phase, based on the EU 2020 targets, range between 22 and 47 € per ton CO_2 in Capros et al. [2] depending on the specific scenario, whereas Stankeviciute et al. [3] estimate the price range to be 25–57 € per ton CO_2 (2005-€). Hence, the range of estimates is substantial.

[1] One EUA represents the right to emit 1 metric ton (*t*) of CO_2 or the equivalent amount of another greenhouse gas.

The determinants of the price and the reasons for the price fluctuations are many. In the first year of the trading scheme, there was abundant uncertainty about a number of features, including the total number of available allowances, and a lack of market data. The EUA price seemed to react adequately to new market information such as changes in fuel prices and cuts in national allocation plans, although price fluctuations may have been too sensitive to changes in market conditions. It was not until the first compliance date, in April 2006, that a complete picture of actual emissions and the market balance emerged. The data showed that the market was unexpectedly over-supplied and prices plummeted, effectively reaching zero towards the end of the scheme. The main determinants of price developments in the first trading period have been attributed to excess supply, but also lower than expected demand for allowances from manufacturing industries (due to lower than expected industry production) and unusual conditions in the power markets influenced the allowance market. Significantly, the trading period was very short, only 3 years, there was a lack of information and banking of allowances from phase 1 to subsequent periods was not permitted.[2] See also Ellerman and Buchner [4] and Convery and Redmond [5] for an overview over the first phase of EU ETS.

This chapter provides a thorough examination of the main price drivers in the EU ETS, and presents a numerical model designed to forecast future EUA prices, as well as examples of scenario and sensitivity analyses based on model simulations.

2 Price Drivers in the EU Emissions Trading Scheme

As any other market, the EU ETS market equilibrium is determined by supply and demand. The commodity supplied and consumed can be thought of as abatements, i.e., emission reductions, although the commodity traded is EUA, i.e., emission allowances. A simplified representation of the (short-term) market equilibrium is shown in Fig. 1.

The demand for abatements is determined by the so-called “gap”. The gap is the difference between *baseline emissions*, i.e., the emission level associated with an EUA price of zero, and the *emissions cap*. Market supply is equal to the marginal cost of emission reductions, the marginal abatement cost curve (MAC). Market equilibrium and the resulting EUA price are found where the demand for abatements intersects with the MAC curve. Forecasting the EUA price then is a question of estimating demand and supply for abatements.

[2] The EU obligation according to the Kyoto protocol is associated with emissions in 2008–2012, and emission reductions in previous years could not be carried forward as compliance with the Kyoto obligation.

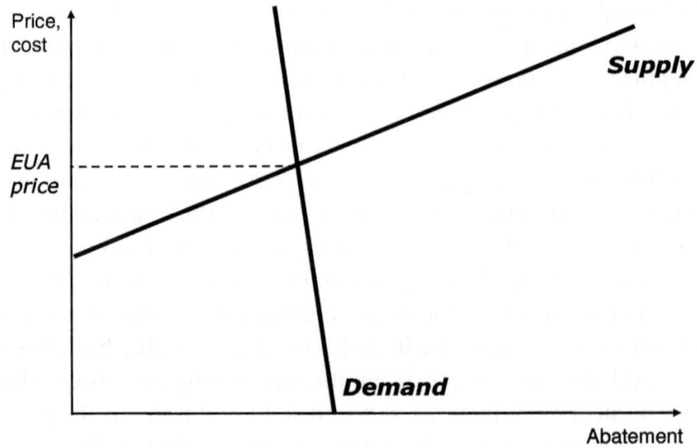

Fig. 1 Supply and demand for abatements in the EU ETS

2.1 Demand Side Drivers

The size of the gap depends on the difference between the cap and baseline emissions. The stricter the cap, i.e., the fewer emission allowances are issued, the larger is the gap and hence the demand for abatements. In Fig. 1 the demand curve shifts to the right if the cap is reduced or baseline emissions increased.

The slope of the demand curve is determined by the price elasticity of demand: increases in CO_2 costs are passed through to market prices for goods and commodities, which in turn reduces demand for electricity and heat, as well as production from the ETS manufacturing industries. Hence, the price elasticity of demand for abatements depends on the price elasticity for heat and electricity and the price elasticity facing industries included in the ETS.

The price elasticity of industrial demand for abatements is thus affected by the extent to which the relevant industries are competitors in a global market or not, and to what extent their competitors in other parts of the world are subject to similar carbon costs. This is the phenomenon known as *carbon leakage*: Migration of industry production from the ETS area in order to avoid exposure to the cost of emitting CO_2. The exposure of European industry to carbon leakage depends on climate policies in the rest of the world and the potency of EU policy measures to mitigate carbon leakage.

2.1.1 Determinants of Baseline Emissions

Baseline emissions are basically affected by the activity level in the economies. Increased economic growth affects the demand for allowances since higher growth generally implies that more goods and services are produced, which in turn requires additional resources including electricity and heat, shifting the demand curve to the right.

Baseline emissions from the power and heat sector depend on the generation mix and the fuels used for generation. Oil, gas and coal emit CO_2, whereas renewable energy sources (RES) and nuclear power do not. This means that if the share of RES and nuclear energy in electricity generation increases, baseline emissions are reduced. Hence, increased investments in RES shift the demand curve in the carbon market to the left.

Fuel prices affect baseline emissions as well. In the power market, the merit order[3] of power generation plants is determined by the marginal cost of generation. In the European power mix, coal and gas power are common fuels and often the fuel used in marginal generation capacity. Other power plants based on renewable and nuclear energy are much less flexible and have low short term marginal cost. This means that relatively flexible coal and gas fired capacity usually determine (hourly) market prices. The merit order of gas and coal power plants depends on efficiency of the plants and on relative fuel prices. When gas prices are low compared to coal prices, more gas power is producing, and vice versa when gas prices are relatively high. This dynamic applies to the long-term development of generation capacity as well: The lower the gas price relative to coal, the more investments in gas power generation.

Since the carbon content of gas is significantly lower than the carbon content of coal, baseline emissions from power generation are lower when the price of gas is low relative to coal. Hence, an increase in the gas price shifts the demand curve for allowances to the right, and a reduction shifts the demand curve to the left.

2.2 *Supply Side Drivers*

The gap between baseline emissions and the number of issued allowances can be closed by reducing emissions from heat and power generation, or manufacturing industries included in the ETS, or by importing credits from the global carbon market. The slope of the supply curve is determined by the marginal cost of reducing emissions from these sources.

2.2.1 Fuel Switching in Power Production

Fuel switching in the power sector represented more than 60% of the total technical abatement options in the first trading period of the EU ETS. The effect goes via the merit order curve. Figure 2 illustrates how fuel switching alters the electricity merit

[3] The merit order curve is the short term electricity supply curve where the generation units are ordered according to short-term marginal cost.

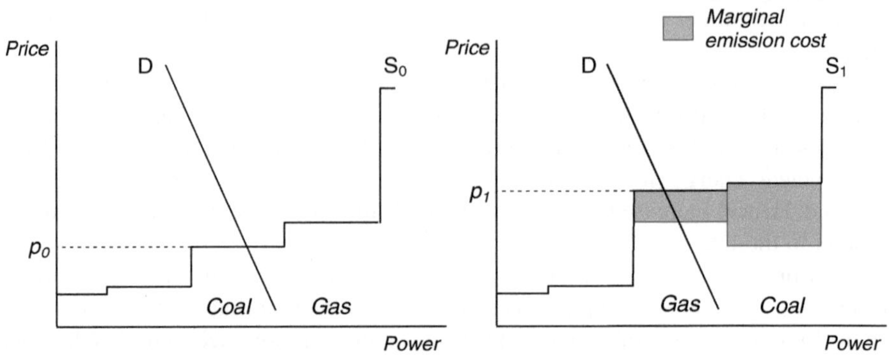

Fig. 2 Fuel switching in electricity generation merit order curve

order curve. The panel on the left shows the market equilibrium in a given hour when generators do not need to buy allowances in order to emit CO_2. Then the short term marginal cost of coal power generation is lower than the marginal cost of gas power generation. With electricity demand as indicated by the downward sloping line, the electricity price p_0 equals the marginal cost of coal power generation, and the gas power plant does not produce.

The panel on the right shows the situation when the generators have to buy allowances for their CO_2 emissions. Since coal power generation emits approximately twice as much CO_2 per kWh as gas power generation, the cost increase for coal power generation is twice as high as for gas power generation. Given the assumed allowance price, gas power becomes cheaper than coal power, and the two plants exchange place in the merit order curve. In the new equilibrium, gas power plant is producing and coal power is not, hence effectively reducing CO_2 emissions. In addition, demand for electricity is reduced due to higher electricity price (p_1), reducing the emissions even more.

In the long term, CO_2 costs will affect investments in power and heat generation as well. Introduction of a CO_2 cost changes the competitiveness between coal and gas power plants in the same manner as changes in the gas price relative to the coal price. If the allowance price is sufficiently high, investments may even switch to emission-free technologies.[4]

[4] Although carbon costs increase power prices, thereby making electricity generation based on renewables more competitive, this is not sufficient to reach the targets set out by the EU renewables directive [6]. With the current cost levels and fuel prices, it will be necessary to subsidize investments or generation in most renewable energy sources in order to reach the ambitious targets set out by the EU directive. This means that the realized investments in renewables depend to a larger degree on policies and support mechanisms (i.e., the implementation of the RES directive in the member states) than on the allowance price levels.

2.2.2 Abatements in Manufacturing Industries

Investments in new technologies in the industrial sector may reduce process emissions or increase energy efficiency and thereby reduce overall emissions. This in turn will imply a lower allowance price, since fewer allowances will be demanded. However, the potential for cheap abatements outside the energy sector (power and heat) is limited. We assess that emission reductions of about 200 Mt CO_2 per year can be achieved in the manufacturing industries from 2009 to 2020 for costs below 30 € per ton CO_2. This is based on the EU Commission's impact assessment [7] and our own estimates.

2.2.3 Credit Import and Banking

The gap may partly be closed by imports of credits from the so-called flexible mechanisms under the Kyoto Protocol. These are (mainly) Certified Emission Reduction units (CERs) from abatement projects in developing countries that do not have a binding emissions reduction target under the Kyoto protocol, and Emission Reduction Units (ERUs) from abatement projects in other countries with a binding commitment under the Kyoto Protocol.[5]

CER and ERU prices depend on international supply and demand, and as such on the adoption of emissions trading schemes outside of Europe, but also on the management and policies adopted by, and the efficiency of, the UN Framework Convention on Climate Change (UNFCCC), which is the body providing guidelines for and the approval of such credits. Imports to the EU ETS are, however, restricted. The newest version of the EU ETS Directive limits the use of CERs and ERUs for EU ETS installations to 11% of the 2008–2012 allocation.

From phase 2 onwards, allowances from one trading phase can be banked and used for compliance in the next phase (such banking was not permitted from phase 1 to phase 2). This means that unused EUAs have a future value, and if that value is expected to be greater than the current price plus interest rate, the EUAs will be banked. Banking of credits effectively reduces the emissions cap in a given trading phase.

3 Modelling the EU Emissions Trading Scheme

Numerous studies have tried to quantify the costs of climate change and combating climate change, and several numerical models have been developed during the past two decades. The first numerical models to analyse the economics of climate

[5] CERs are derived via the Clean Development Mechanism (CDM) and ERUs via Joint Implementation (JI) projects. Both types of projects are subject to strict approval procedures decided by the UNFCCC. One CER or ERU permits the holder to emit one metric ton of CO_2 equivalent.

change were DICE (Dynamic Integrated Model of Climate and the Economy; [8]) and RICE, a regional version of DICE [9].

Several other modelling teams made efforts to quantify the costs of climate change and various modelling approaches have been applied: general equilibrium and partial equilibrium models, top-down and bottom-up approaches, global and regionally focused studies, and estimates of marginal abatement cost curves by econometric modelling.

The aim of the early models was primarily to assess the optimal trade-off between mitigation and adaptation to climate change; in other words – what is the cost of mitigation measures, whether and how we should respond to climate change. Later, as the climate policies began to take shape, the analyses of more specific targets (e.g., the Kyoto targets, the EU "bubble") came into focus.

The Energy Modeling Forum (EMF)[6] at Stanford University has made considerable efforts to bring forth and analyse the differences of the various models. Weyant [11] documents EMF-16, and compares results of 13 different models that quantify the costs of reaching the Kyoto targets. Alternative technology strategies were studied in EMF-19 [12].

Springer [13] gives a comprehensive overview over a range of models, comparing their methods and results. He finds that 13 models estimate the permit price in full global trading scenario between 1 and 22 $ per ton CO_2 (2000-$). If only Annex 2 countries[7] are considered, the price range is 3–74 $ per ton CO_2. This wide range of costs reflects the many uncertainties involved: future GDP and population growth, technical developments, elasticities, policy scenarios and baseline emissions. The estimates are necessarily more uncertain, the more long-term the studies.

The early models focused on CO_2 emissions. More recently, other GHGs have been included in the models and in de la Chesnaye and Weyant [14] the focus is extended to multi-GHG studies.

Hourcade et al. [15] give an overview over models that try to unify the bottom-up and top-down approaches in hybrid modelling of energy-environment policies.

Due to the global nature of climate change, most of the models are global (even though some countries are modelled in greater detail than others) and long-term. However, as the EU ETS was introduced, analyses of the regional market became pertinent. Moreover, the short-term impacts became important: the impact of today's events on the carbon price next year or 2 years from now became interesting

[6] EMF was started in 1976 as an organized forum for discussion and evaluation of important energy and environmental issues. It operates around organized working groups that focus their work by comparing the results of different economic, market and planning models [10].

[7] Annex 2 refers to the developed countries which are Parties under the UN Framework Convention on Climate Change: Australia, Austria, Belgium, Canada, Denmark, European Economic Community, Finland, France, Germany, Greece, Iceland, Ireland, Italy, Japan, Luxembourg, Netherlands, New Zealand, Norway, Portugal, Spain, Sweden, Switzerland, United Kingdom of Great Britain and Northern Ireland, and the United States of America.

(as opposed to decades and 50–100 years into the future). It was necessary to analyse the impacts – both the price and the outcome on emissions – of the regional market. Global models have been used to analyse the EU ETS, for instance DART [16] and POLES [17]. However, a few models have been developed that focus on the EU ETS: Econ Pöyry's Carbon Market Model (detailed below) and PRIMES [18]. PRIMES has been used for impact assessments by the European Commission, e.g., in European Commission [1]. This paper presents results based on Econ Pöyry's carbon market model that is specifically designed to analyse the EU ETS market.

3.1 *Econ Pöyry's Carbon Market Model*

Econ Pöyry's Carbon Market Model is a numerical partial equilibrium model for the EU ETS. Econ Pöyry's Carbon Market Model includes emissions from all ETS sectors and covers the 27 EU Member States plus Norway and Switzerland.[8] The model is a combination of a bottom-up and top-down model.

The model finds equilibrium for the EU ETS allowance market and the European electricity market simultaneously. For the allowance market, the allowance price is found by equilibrating supply and demand of allowances in the EU ETS market for the whole trading period (e.g., 2008–2012). For the electricity market, the model determines for each country electricity prices, electricity consumption, production and investments to meet the electricity demand, reservoir levels for hydropower capacity, and trade between the countries throughout each year.

Figure 3 illustrates the model. A technical description of the model is included in the appendix.

3.1.1 The Allowance Market

For allowance market, the model captures the fundamental supply and demand functions in the EU ETS market. Emissions from the EU ETS industries – energy (electricity and heat production), pulp and paper, metal, mineral and chemical industries, refineries and aviation – are "matched" with the cap.

Allowance price is found by balancing supply and demand. Total supply is given by the cap, i.e., the amount of allowances allocated to the participants. Total demand for allowances is determined by activity level in the sectors. The gap between baseline emissions and the cap determines the required abatements.

[8] Switzerland does not participate in the EU ETS, but is included since Swiss hydropower is essential for the European power market; the Swiss power system should therefore be modelled endogenously.

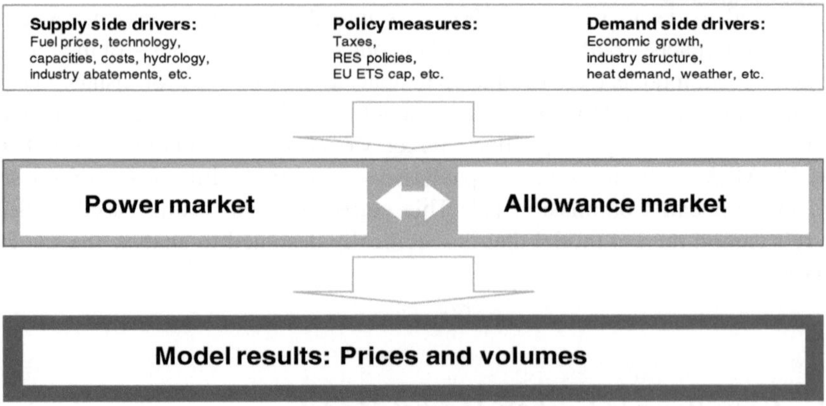

Fig. 3 Simplified overview of Econ Pöyry's Carbon Market Model

Which abatements are actually carried out depends on the costs of reducing emissions in the different ETS sectors.

As a result of this optimisation, the model finds one equilibrium price for the whole emissions trading period (e.g., 2008–2012). Since allowances can in practice be borrowed from the next year during the period, it is the total emissions and the total amount of allowances during the entire period that determine the price. High emissions in one year cause many allowances to be used in that year, leaving fewer allowances available for the next year, thus implicitly raising the price of allowances in the future. This mechanism equalizes the price in all years. With perfect information and no uncertainty (as in the model), there is no fluctuation in allowance price.

Emissions and abatements from electricity generation and industrial production are determined endogenously, while emissions from heat generation are included exogenously. Heat demand is determined by temperatures and economic growth. Supply comes from combined heat and power (CHP) units (that produce both power and heat) and district heating units.

Emissions and Abatements in Manufacturing Industries

In addition to energy (electricity and heat) generation, other emission-intensive industries are participants in the EU ETS market: metal industries (iron and steel), pulp and paper, mineral industries (cement and lime, glass), chemical industries, refineries and aviation.[9]

[9] The model set-up is very flexible in terms of accommodating new sectors.

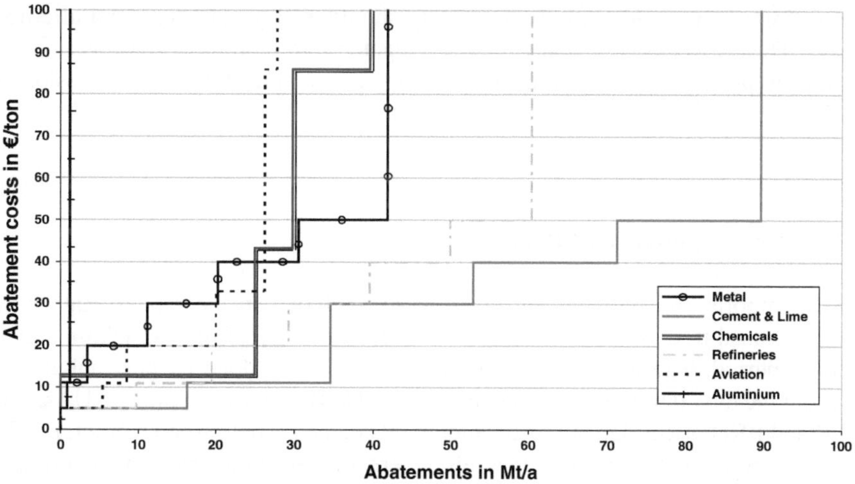

Fig. 4 Emission reduction costs and potentials in the manufacturing industries per 2010 (Source: European Commission [7] and the authors' estimates)

The starting point is the assumption that emissions are equal to the emissions in baseline case. Emissions from industries in the baseline case depend on the activity level, which is in turn a function of global economic growth and domestic conditions.

Industry abatements are determined endogenously in the model. The assumed abatement potential in each industry is limited by a potential pool of activities. Potentials and costs in the different manufacturing industries are shown in Fig. 4.

Some industries (iron and steel, pulp and paper, and chemical industry) are also power intensive, and as such important for the outcome of power market. In other words, there is an important interaction between the allowance market and electricity market.

3.1.2 The Electricity Market

The electricity market module includes demand functions for up to five demand segments in each country, described by income and price elasticities and initial demand levels. Supply is represented by generation capacities and corresponding marginal costs, i.e., the merit order curves for each country. In the short term, investments are exogenous (taking into account actual developments only); in the longer term, investments are determined endogenously in the model (both the amount and technology choice). Cross-border trade takes place according to price differences and transmission capacities.

The model takes into account the diurnal and seasonal variation that characterizes electricity demand and hydropower inflow. The basic time unit is a week, consisting of five load periods (night, day shoulder, day, day peak and afternoon peak). The 52 weeks of a year can be aggregated into months or seasons.

Power prices, consumption, production, investments, trade and emissions in each period and each country are calculated.

3.1.3 Credit Import and Banking

Credit imports and banking are not directly included in the model. Rather, the model is used to derive a price curve for different levels of the cap within a given trading period. The analysis of import and banking is done ex post, taking into account the uncertainties pertaining to CER and ERU prices and future EUA prices. CER and ERU prices depend on developments in the global carbon market, i.e., UN policies, regional trading schemes elsewhere in the world, capacity building in government institutions in non-annex 1 countries, etc. Banking depends on expectations for future EUA prices which depend on developments in the global climate policy negotiations as well as future EU policies. These aspects are not captured by the model.

4 Model Results: EU Allowance Prices in 2020

In order to forecast future carbon prices in a cap-and-trade system like the EU ETS, an adequate market model is required and a number of basic assumptions must be made about the underlying market conditions and policy framework developments. We use Econ Pöyry's Carbon Market Model to forecast EU allowance prices for 2020. We start by discussing the most important assumptions, before we turn to the model results. Sensitivity analyses are presented in the next chapter.

4.1 Base Scenario Assumptions

The Base scenario represents a reference case for the later assessment of the EUA price sensitivity on different market drivers. It describes the "most likely" EU ETS market development until 2020, based on current and proposed EU policy regulations, an assessment of the global carbon market outlook and expected future developments in the market drivers. The most important assumptions that are used in the Base scenario are described in the following paragraphs.

4.1.1 Climate Policies in the EU

The EU ETS is merely one instrument in the EU climate policy package adopted in 2008–2009. The climate policy package sets out targets for energy efficiency, renewables deployment and emission reductions for the EU Member States in 2020 [7]. The main policy goals are:

- Reduce EU's GHG emissions in 2020 by 20% compared to 1990, and by 30% in case of a sufficiently binding global climate agreement.
- Increase the share of RES to 20% of total energy consumption compared to 2005 levels.
- Improve energy efficiency by 20% compared to a baseline trajectory.

The following model analysis has been based on the above-mentioned 2020 European policy targets.

4.1.2 Rules and Regulations Regarding the EU ETS

In general, the Base scenario applies the EU's target of 20% emissions reductions by 2020 (compared to 1990-levels). The assumptions about allocation and trading rules in 2020 are based on European Commission [1]. This regulation clarifies allocation and auctioning rules. However, due to some outstanding definitions and specifications, uncertainties still remain. These uncertainties are dealt with by defining simple and European-wide assumptions to quantify the total allocation, the auctioning volume and the import volume of external credits.

The Cap

The Base scenario assumes that existing ETS sectors must reduce emissions 21% by 2020 compared to 2005 levels.[10] CO_2 emissions from aviation, CO_2 and N_2O emissions from some chemical processes and GHG emissions from capturing, transporting and storing CO_2 will be included in the ETS. In our analysis we assume that the new sectors get the same reduction target as existing ETS sectors, i.e., 21% reduction compared to 2005 emissions.

The EU Commission determines a single EU-wide cap and allowances will be allocated on the basis of harmonized rules. Consequently, the overall allocation volume is based on the verified 2005 emissions of installations already included in the scheme and will decrease in a linear manner by 1.74% of average 2008–2012 allowances each year counted from 2010.

[10] Observe that the 20% reduction target applies to all sectors in the EU economy and refers to 1990 emission levels, whereas the ETS only applies to some sectors and refers to 2005 verified emissions for these sectors.

Table 1 Model assumptions on EU ETS allocation volume in 2020, Mt CO_2

	2005 emissions	2020 allocation (−21%)
Current ETS sectors	2,141	1,692
New sectors	253	200
Aviation	156	
Chemicals	85	
Aluminum	12	
Total	2,394	1,892

Table 2 Fuel price assumptions (2008 real prices)

	2020
Coal ($/t)	65.0
Natural gas (€/MWh)	25.0
Oil ($/barrel)	61.5

With these assumptions, the total cap on emissions will be 1,892 Mt CO_2 in 2020. Table 1 summarizes the assumptions behind this estimate.

4.1.3 Fuel Prices

Table 2 lists the assumed 2020 fuel prices.

4.1.4 Renewable Electricity Generation

We assume that renewable electricity production capacity will increase, although we assume that the EU renewables target will not be met by 2020. The EU has estimated that the 2020 target for renewable energies implies that 34% of European electricity demand is covered by renewable sources in 2020 [6]. This is a very ambitious target, and in our calculations we assume that only 80% of the targeted growth in renewable generation volume is achieved by 2020. Hence, compliance with the EU's 2020 renewables targets will be delayed so that the targets are not fulfilled until sometime between 2025 and 2030.

4.1.5 Credit Imports

A maximum credit import limit of 1.6 billion allowances over the period 2008–2020 is assumed. This volume corresponds to the EU Commission's estimated total import volume. The proposal in the draft ETS Directive is that no further imports of CERs/ERUs will be allowed unless an international post-Kyoto agreement enters into force. CERs/ERUs that are not used for compliance in the 2008–2012 trading period may be carried over to the third trading period. On average, the credit import may be 116 allowances per year.

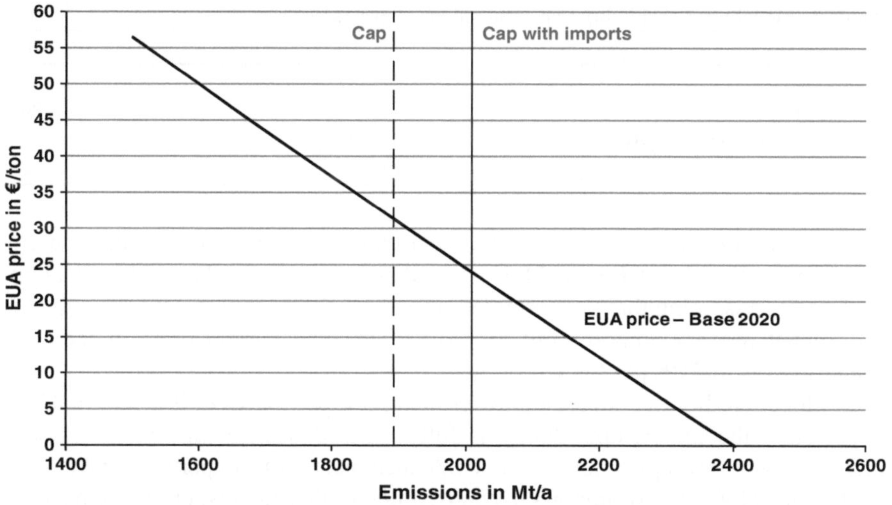

Fig. 5 Allowance prices in the Base scenario in 2020

4.2 Base Scenario Results

Using Econ Pöyry's Carbon Market Model, baseline emission levels are calculated for 2020.

With the assumptions for the Base scenario as described above, baseline emissions from the ETS sectors are estimated to 2,402 Mt CO_2 in 2020. The baseline describes the situation if emissions are not capped. The 2020 cap in the Base scenario is expected to be 1,892 Mt CO_2, which implies a gap of 510 Mt CO_2 compared to baseline emissions, or about 20% of the total allocation.

By imposing a restriction on emissions (the cap), the model finds a solution which complies with the cap. Running the model with different caps, we find the EU allowance price curve. The EU allowance price curve can be interpreted as the marginal abatement cost curve for the whole EU ETS, reflecting the cost of meeting the different potential caps and including demand side response.

If the gap of 510 Mt CO_2 is to be closed by abatements in the ETS sectors alone, the marginal abatement cost, and hence the market price for EUAs, would be 32 € per ton CO_2. This is where the vertical dotted line intersects with the EUA price curve on Fig. 5. The figure depicts the allowance price for different levels of emissions. Emissions (in million tons) are measured on the horizontal axis and the corresponding carbon price in € per ton CO_2 on the vertical axis. The higher the allowed emissions are, the lower is the allowance price. If the allowed emissions equal baseline emissions, the allowance price is zero. This is the point where the price line crosses the horizontal axis. Please observe that the price curve is not linear, although for the particular input parameters employed here, the price curve appears to be linear in the figure.

Table 3 Model results with different assumptions (€ per ton CO_2)

	2020
Base case	32
Low gas price	26
Less renewable electricity (65% of RES target is achieved)	42
More renewable electricity (compliance with RES target)	22

However, the price is considerably lower than 32 € per ton CO_2 if the opportunity to import credits (CERs and ERUs) is taken into account (the solid vertical line on Fig. 5). Assuming that the permitted imports in 2020 are 116 Mt CO_2 (equal to the annual average import limit in 2013–2020), the carbon price in the Base scenario falls to 25 € per ton CO_2.

If global credit price levels are lower than 25 € per ton CO_2, it is reasonable to expect that the import opportunity is fully utilized. Hence, as long as external credits present lower price levels than EUA abatement costs, imports will take place.

The actual amount of imports depends, among other factors, on the demand for credits elsewhere and on the factors that determine the EUA marginal abatement costs. Non-delivery risks, at least for primary CERs, are still expected to be higher than for EUAs; hence CERs might be traded at a slight discount.

4.2.1 Caveat

With the chosen approach, it is implicitly assumed that the year 2020 is representative for the ETS third trading period from 2013 to 2020. Hence it is assumed that the carbon market balance and marginal abatements cost in 2020 represent the average of the whole trading period. This is clearly a simplification. In reality, the EUA price should be determined by the market balance over the whole trading period if we assume that there are no restrictions on banking from one year to another, taking the market interest rate into account. Our model finds the equilibrium solution for one year at the time, and does not handle market dynamics over a time span of several years. However, checking the market balance in intermediate years (e.g., 2015) using the derived price prognosis for 2020, we find that the 2020 model simulation seems to be quite representative for the third trading phase.

5 Sensitivity Analyses

In this chapter, we discuss how the EUA price is affected by changes in important input assumptions, namely relative fuel prices and subsidies to renewable electricity generation. Table 3 summarizes the model results in the sensitivity analyses. Here we compare the results to the Base scenario without credit imports, i.e., to the price of 32 € per ton CO_2. The magnitude of the price changes should, however, be quite similar if we take the same amount of imports into account in the sensitivities as well.

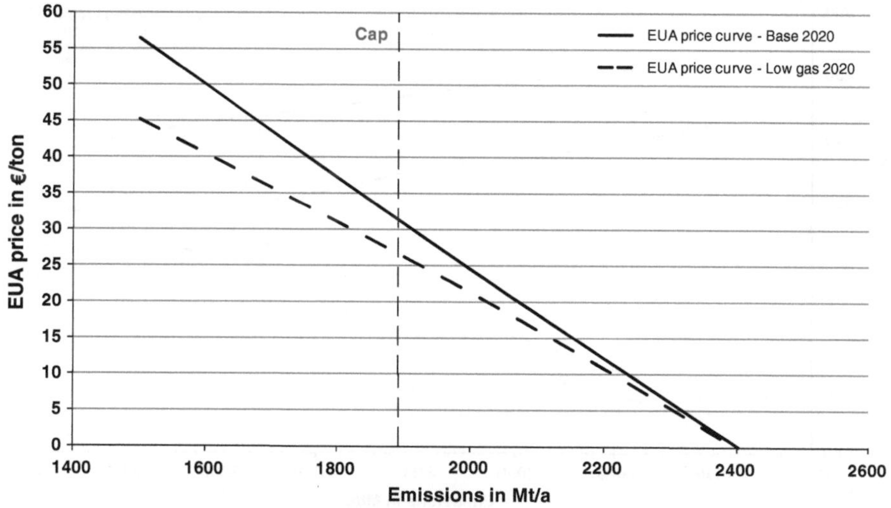

Fig. 6 Allowance prices with different gas prices

5.1 Fuel Prices

The EUA price is very sensitive to assumptions about relative fuel prices (especially gas price relative to coal price) since the fuel prices strongly impact emission levels. Fuel switching in the power sector represented more than 60% of the technical abatement options in the ETS first trading period.

In general, lower gas prices impact the carbon price via the effect on baseline emissions and via the effect on the slope of the EUA price curve. The slope of the EUA price curve becomes less steep because fuel switching becomes less expensive. Baseline emissions are altered by two effects that work in opposite directions: On the one hand, electricity demand increases due to lower gas prices. On the other hand, baseline emissions are reduced since gas becomes more competitive towards coal, so the increased demand can be met by a generation mix with lower emissions.

In the *Low gas price scenario* we assume a lower gas price (21 €/MWh instead of 25 €/MWh) in order to analyse the EUA price sensitivity. At a lower gas price level, abatements through fuel switching from coal to gas are cheaper than in the Base scenario. Consequently, the EUA price level drops from 32 to 26 € per ton CO_2 (disregarding credit imports and banking).

Figure 6 shows the abatement cost curves in the two scenarios. Baseline emissions are almost identical in both scenarios. At very loose caps, lower gas price levels generally result in a higher power demand, but emission levels do not increase since the increased demand is met by higher gas power generation, while coal power generation is reduced. When the cap is tightened, the carbon price

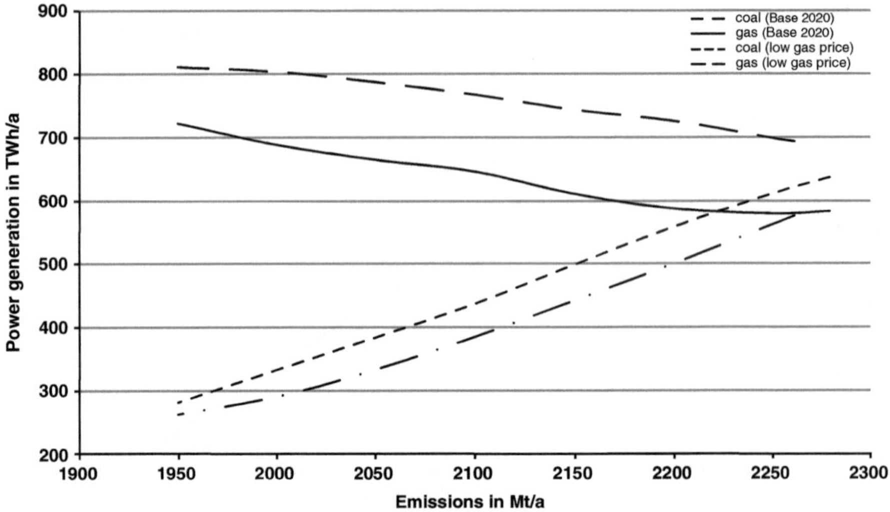

Fig. 7 Fuel switching with different gas prices

generally increases. But the price increases more in the Base scenario due to the higher gas price: With a higher (relative) gas price, a higher EUA price is necessary in order to induce the needed fuel switching from coal to gas.

Baseline emissions are lower in the Low gas price scenario than in the Base scenario because there is more gas power generation and less coal power generation. Fuel switching could be both less and more expensive than in the Base case, but since the gap is smaller, less fuel switching is required, and the EUA price is lower for all levels of the cap.

In Fig. 7 we explore fuel switching from coal to gas in the two scenarios. The figure shows the levels of gas and coal power generation for the different scenarios associated with different levels of the cap. Starting from the situation without a cap (on the right hand side of the figure), fossil fuel generation is reduced via reduced power demand as a cap is imposed and gradually tightened. As the cap gets tighter, fuel switching is induced, i.e., coal power generation starts to decrease and gas power generation starts to increase. We also see that the initial level of gas power generation is higher and coal generation lower in the Low gas price scenario than in the Base scenario. Even in this scenario the share of gas in the fuel mix increases at the expense of coal as the emissions cap tightens. The EUA price curve in Fig. 6 shows, however, that fuel switching comes at a lower cost in the Low gas price scenario than in the Base scenario.

In other words, when the cap gets tighter, gas power generation increases and coal power generation decreases. However, the slopes of the (increasing) gas curve and the (decreasing) coal curve are not equal; the reduction in coal power generation is larger than the increase in gas power generation. This is due to demand response: at tighter caps and higher price levels, power demand decreases.

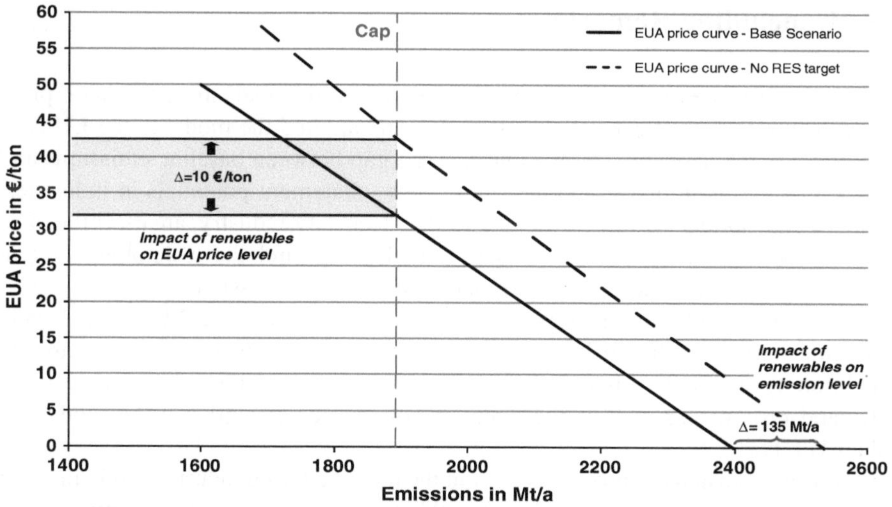

Fig. 8 Allowance prices with different levels of renewable electricity generation

5.2 *Renewable Electricity Generation*

In general, the deployment of RES for electricity generation significantly reduces greenhouse gas emissions from power generation, and consequently allowance and power prices as well. A sensitivity analysis suggests that 80% compliance with the renewable target (Base scenario) induces about 25% of the abatements required to comply with the cap in the EU ETS, i.e., baseline emissions are reduced by a volume corresponding to 25% of the total gap. Increased deployment of renewables constitutes a significant downside for the EUA price. For instance, the EUA price is reduced about 10 € per ton CO_2 if the target for renewables is complied with.

Figure 8 compares two scenarios: the *Base scenario*, in which 80% of the required investments in RES are achieved by 2020, and a *No RES target scenario*, in which the investments in renewable electricity generation follow a business as usual development. In the latter case, no additional policies are assumed to stimulate investments in RES; consequently, the only 65% of the RES target is achieved.

Comparing the baseline emission levels (2,534 Mt CO_2 per year in the No RES target scenario and 2,402 Mt CO_2 per year in the Base scenario), we find that the higher deployment of renewables reduces annual CO_2 emissions from the ETS sectors by approximately 135 Mt CO_2. Furthermore, achieving 80% of the targeted growth in renewable energy reduces the (internal) marginal abatement costs in the EU ETS sectors by a range of 10 € per ton CO_2 on average, depending on the considered cap.

Further, the sensitivity analysis indicates that the effect on the overall emissions and the future EUA price is sensitive both to the stringency of the carbon target and to the assumed RES electricity generation volume, but relatively insensitive to differences in the distribution of RES electricity technologies.

6 Concluding Remarks

We have presented a numerical model approach to the analysis of the main price drivers in a cap-and-trade emissions market and applied the model to the EU ETS market. On the demand side, we find that the gap between baseline emissions and the cap is the main driver. On the supply side, abatement potentials in industries, particularly options for fuel switching in the electricity industry, are essential. The general activity level in the economies, fuel prices and the technology mix in electricity generation are key determinants of baseline emissions.

Numerical model simulations show that price formation in the EU cap-and-trade system is strongly driven by the cost of fuel switching in the power sector, in addition to demand response. Our analysis shows that it is crucial to capture demand response as well as the dynamics of the power market when modelling the EU ETS.

However, abatement potentials exist in the manufacturing industry and in the heat sector as well, although one might argue that the main determinants of emissions in these sectors are incentivized by forces outside the ETS, that is, emissions are not primarily determined by the allowance price. For EU manufacturing industries, the activity level, and hence the demand for allowances and electricity, is to a large extent driven by European allowance and power prices *relative to* global emission costs and power prices. Relative prices, in turn, depend on global climate policies and developments in other regions. Currently, the EU is working on a scheme to compensate EU industries at risk of carbon leakage, therefore, supply and demand for allowances from the manufacturing industries are likely to be heavily influenced by additional policy measures in the third trading phase.

Emissions from district heating and industrial CHPs are similarly driven by renewables policies (compliance with the RES target) and developments in industrial activity. Nevertheless, there is a need for a better understanding – and modelling – of the underlying dynamics of the heat market and how emissions here are affected by the combined impact of the renewables target and policies, and the EU ETS. Energy system models, such as PRIMES and MARKAL, do capture some of these trade-offs, but they only capture the heat market in a rather crude fashion. Heat market dynamics are complicated to model as it is not really one market, but rather local markets with different regulations, alternatives and demand patterns.

A Appendix: Econ Pöyry's Carbon Market Model

A.1 Overview

Econ Pöyry's Carbon Market Model – a numerical model for the European carbon and power market – was used for the simulations presented above. This appendix documents the model.

Econ Pöyry's Carbon Market Model is a partial equilibrium model that solves for the EU ETS allowance market and the European electricity market simultaneously. The model covers EU-27 countries plus Norway and Switzerland.[11] The model is a combination of a bottom-up and top-down model.

For the *allowance market*, the allowance price is found by equilibrating supply and demand of allowances in the EU ETS market for the whole trading period (e.g., 2008–2012). For the *electricity market*, the model determines for each country electricity prices throughout each year, electricity consumption, production and investments to meet the electricity demand, reservoir levels for hydropower capacity, and trade between the countries.

The model determines a *long-run equilibrium* in the sense that investments in new power generation capacity are determined endogenously. Optimising behaviour of producers and consumers is assumed within the confines of restrictions that describe the actual market conditions. The restrictions include excise taxes and regulatory mark-ups that describe government interventions. There is no uncertainty in the model.

The model solves for wholesale (spot market) prices and end user prices of all agents. There are no explicit forward markets or long term contracts in the model. We interpret the model to contain implicit contracts: Since we assume rational expectations and full certainty, spot market prices over the contract period will equal implicit contract prices.

The model is developed in the GAMS programming language [19].

A.2 Market for Carbon Allowances

For allowance market, the model captures the fundamental supply and demand functions in the EU ETS market as well as in the linked Kyoto project-based mechanisms. Allowance price is found by balancing supply and demand. Total supply of allowances is given by the amount of allowances allocated to the participants. Total demand for allowances is determined by activity level in the sectors that are subject to the EU ETS: energy (electricity and heat production), pulp and paper, metal, mineral and chemical industries, refineries and aviation. These are grouped into *electricity*, *heat* and *other industries* in the model. The gap between business as usual (BAU) emissions[12] and the cap given by the total amount

[11] Norway participates in the EU ETS. Switzerland does not participate in the EU ETS, but is included since Swiss hydropower is essential for the European power market; the Swiss power system should therefore be modelled endogenously.

[12] Business as usual emissions is what emissions would have been without restrictions, given the economic activity level in the industries. The activity level is a function of general economic conditions and temperatures.

of allowances determines the required abatements. Which abatements are actually carried out depends on the costs of reducing emissions in the different ETS sectors.

CO_2 emissions stem from power generation, heat generation and industrial activities in other sectors i:

$$E = \sum_l E_{l,\mathrm{El}} + \sum_l E_{l,\mathrm{Heat}} + \sum_i \left(\sum_l E_{l,i}^{\mathrm{BAU}} - \sum_j A_{i,j} \right) \tag{1}$$

Equation 1 defines CO_2 emissions from different sources. The first line (right hand side) specifies emissions ($E_{l,\mathrm{El}}$) from power production in each country l. These stem from use of fossil fuels both in existing and new power plants. The power sector will be discussed in detail below (Sect. A.3).

The second line of Eq. 1 is emissions from heat generation. Heat demand is determined by temperatures and economic growth. Supply comes from combined heat and power (CHP) units and district heating units and is specified by capacities and costs in these units. The heat market is not modelled explicitly; instead, emissions from heat production ($E_{l,\ \mathrm{Heat}}$) in each country are included exogenously.

The third line of Eq. 1 defines emissions from other industries i that are part of the EU ETS: metal industries, pulp and paper, minerals, chemical industries, refineries and aviation. Emissions from industries in the business as usual case depend on the activity level, which is in turn a function of global economic growth and domestic conditions. Some of them (iron and steel, aluminium, pulp and paper, and chemical industry) are also power intensive, and as such important for the outcome of power market. Others (e.g., minerals) are not power intensive. The starting point is that emissions are equal to the emissions in BAU case (E^{BAU}). However, various abatement activities j can be carried out to reduce emissions. Whether and which abatements are carried out depends on the marginal abatement costs ϖ and the allowance price. Hence, A is an endogenous variable. The abatement potential in each industry is limited by a pool of potential activities:

$$A_{i,j} \leq A_{i,j}^{\max} \tag{2}$$

Emissions from all the installations in the EU ETS – grouped into power, heat and other industries in the model – are matched with the cap. In other words, total emissions from these sectors must not exceed the total amount of allowances:

$$E \leq \overline{E} + \widehat{E} \tag{3}$$

If the total emissions exceed the cap, a penalty ϑ must be paid on the excess emissions $\widehat{E}$ (in addition to the allowances that must be purchased in the market).

Since the EU ETS is a system where allowances can be traded, it is the total cap that is relevant, not individual countries' caps. Therefore, Eq. 3 applies to the whole EU ETS area, not to different countries and installations separately.

As a result of this optimization, the model finds one equilibrium price for allowances for the whole emission trading period (e.g., 2008–2012). Since allowances can in practice be borrowed from the next year during the period, it is the total emissions and the total amount of allowances during the entire period that determine the price. High emissions in one year cause many allowances to be used in that year, leaving fewer allowances available for the next year, thus implicitly raising the price of allowances in the future. This mechanism equalizes the price in all years. With perfect information and no uncertainty (as in the model), there is no fluctuation in allowance price.

A.3 *Electricity Market*

For electricity market, it is important whether demand occurs during day or night, or during winter or summer. Hence, the electricity market module has a considerably finer time resolution than the allowance market module. The basic time unit in the electricity market module is a week that is divided into five load periods (night, day shoulder, day, day peak and afternoon peak). However, the weeks can be aggregated into months or seasons. The weeks (or months) are further aggregated into years, as the seasonal variation plays a key role in electricity demand and supply. The model can be run for one year or for several consecutive years.

A.3.1 Demand for Electricity

There are up to five end users of electricity in each country: *households, services, industry (not including process industry), process industry (including pulp and paper) and electric boilers*. Not all end users are present in all countries (e.g., electric boilers are specific to Norway and Sweden only).

End users maximize utility functions $W_{l,e} = \Sigma_m W_{m,l,e}(D_{m,l,e})$ of electricity demand (D) of end user e of country l in every period m. The outcome is log-linear electricity demand functions that are written in levels as

$$D_{m,l,e} = \left[\alpha_{m,l,e}(\xi_{l,e})^{\gamma_{l,e}}\pi^{\kappa_{l,e}}\right]P_{m,l,e}^{\beta_{l,e}} \tag{4}$$

Electricity demand is a function of the end user price of electricity (P), the oil price (π) and income level (ξ). The first part of Eq. 4, inside the square brackets, is a constant for our purposes. α is used to calibrate the model to actual price and consumption in the base month. It also includes an exogenous seasonal variation pattern, which is especially strong in Norway and Sweden, due to the large amount of electricity used for heating in these countries. In other countries, electricity demand for air condition contributes to the seasonal pattern. γ, κ and β are the income elasticity, oil price elasticity and electricity price elasticity, respectively.

Different price and income elasticities are assigned to each sector. The electricity price elasticity varies between −0.2 and −0.8. Electric boilers are assumed to have a fairly elastic demand, as they are able to change between electricity and oil on a short notice. Demand in process industries is price inelastic.

The price elasticity refers to *end user price of electricity* (P) that includes the *spot price* (p) of electricity, the *distribution tariffs* (τ^n), the *excise tax on electricity* (τ^e), the *mark-up* (τ^o) to the supplier and VAT(τ^m).

$$P_{m,l,e} = (\tilde{p}_{m,l} + \tau^n_{l,e} + \tau^e_{l,e} + \tau^o_{l,e})(1 + \tau^m_{l,e}) \tag{5}$$

Consumption during a period m is divided between the different load periods a according to given ratios. Accordingly, consumers respond only to an average price level of a period (e.g., month). The spot price p of a period m is an aggregate of spot prices over load periods[13]:

$$\tilde{p}_{m,l} = \Sigma_a \lambda_{l,a} p_{m,l,a} \tag{6}$$

λ is the share of consumption assigned to each load period.

A.3.2 Supply of Electricity

The supply side is represented by producers with different technologies and capacities at their disposal. The supply curve (merit order curve) for each country is the sum of the marginal cost curves of producers in the country. The marginal cost curve is dynamic, since investments can expand production capacity.

The existing and potential power plants are grouped into categories (cost classes) according to *technology* (hydropower, thermal power incl. combined heat and power, nuclear power, wind power, etc.), *fuel* (coal, gas, oil, biomass, etc.) and *plant efficiency*. Altogether, there are more than 70 different production categories (k). Exogenous short run marginal costs (δ) and emission factors (η) are assigned to each production category k. The marginal costs reflect fuel cost, operating cost and emission cost.

Production X is determined endogenously (exceptions are discussed below). Producers maximise profits:

$$\Pi_{m,a,l,k} = p_{m,a,l}\, X_{m,a,l,k} - \delta_{l,k} X_{m,a,l,k} \tag{7}$$

[13] The model calculates the spot price in each load period a and each period m.

Profit, Π, is income less short-run costs. Each period and production category requires a separate profit decision in each country. The outcome of maximising profit *in isolation* is a zero–one decision: no production ($X_{m,a,l,k} = 0$) if costs exceed the output price ($\delta_{k,l} > p_{m,a,k,l}$), and maximum production ($X_{m,a,l,k} = X^{\max}_{m,a,l,k}$) if the output price exceeds costs ($p_{m,a,k,l} > \delta_{k,l}$). Since prices vary with time, a technology may be operated part of the time.

The next step is to constrain production during those periods where the price exceeds costs – in other words to determine $X^{\max}_{m,a,l,k}$. The basic constraint in the model is physical production capacity, but *available* production capacity is further constrained by requirements for reserve capacity, maintenance and seasonal variation. Also, the minimum production level may be higher than zero (for instance for nuclear power plants). Hence, the minimum and maximum production for each production category in each country, month and load period is restricted to:

$$\forall m, a, l, k : X^{\min}_{m,a,l,k} \leq X_{m,a,l,k} \leq X^{\max}_{m,a,l,k} \tag{8}$$

Additionally, the whole power system of a country is subject to a reserve requirement, ε, that further limits available capacity:

$$\forall m, a, l : \sum_k X_{m,a,l,k} \leq (1 - \varepsilon) \sum_k \mu_a X^{\max}_{m,a,l,k} \tag{9}$$

To understand the combined effects of the constraints of Eqs. 8 and 9, it is useful to note that the capacity requirement, Eq. 8 gives one constraint for every technology k, area l, month m and load period a; while the reserve requirement, Eq. 9 gives one constraint for every area l, month m and load period a.

Investments and refurbishments

The physical production capacity can be expanded by investing in new capacity and refurbishing existing, outdated capacity. Refurbishing prolongs the lifetime of an existing power plant and is a cheaper option than building a new power plant. Unit costs are defined for investments and refurbishments, ι and ς, respectively. The total volumes of investments ι and refurbishments J in each production category and country are limited:

$$\forall k, l : I_{k,l} \leq I^{\max}_{k,l} \tag{10}$$

$$\forall k, l : J_{k,l} \leq J^{\max}_{k,l} \tag{11}$$

Total capacity in each plant category in each country is therefore the sum of original installed capacity ($X^{\max\ \mathrm{Start}}_{m,a,k,l}$), investments and refurbishments:

$$\forall m, a, k, l : X_{m,a,k,l}^{\max \text{ Start}} + I_{k,l} + J_{k,l} = X_{m,a,k,l}^{\max} \tag{12}$$

Hydropower

In addition to production capacity, hydropower production is further restricted by water inflow and reservoir capacity. Three different types of hydropower plants are modelled: plants with reservoirs, run-of-river and pumped storage.

The short term marginal costs of hydropower are very low and the constraints given by Eqs. 8 and 9 are quite lax. But the alternative cost of producing hydropower *now* is the price one can obtain in the *future*. Thus, hydropower production depends both on the short term marginal cost and on the value of the stored water, which equals the expected future price less the expected future cost. Reservoirs are used in order to optimise production over a year: water can be stored in the reservoirs for production in later periods. Because of the flexibility in the hydropower production, the total supply curve of a period is not fixed by assumption. The reservoir level R of a country l in a period m depends on the inflow θ and hydropower production level of the country in the period:

$$\forall m, l : R_{(m+1),l} \leq R_{m,l} + \theta_{m,l} - \sum_{a} X_{m,a,l,\text{hydro}} \tag{13}$$

There is one reservoir per country. Bottlenecks within a country are thus not modelled. There is a minimum requirement on reservoir capacity that generates the alternative cost of producing now:

$$\forall m, l : R_{m,l}^{min} \leq R_{m,l} \leq R_{m,l}^{\max} \tag{14}$$

The minimum reservoir level is typically 15%, the maximum is 95% of total reservoir capacity. The reason is that it is unlikely that all the reservoirs in a country are empty (full) at the same time.

In contrast to plants with reservoirs, *run-of-river plants* cannot store water and, hence, transfer production from low-price to high-price periods. Instead, production must occur when the water is there. Production in these plants is exogenously determined and is assumed to follow a seasonal profile.

A third type of hydropower plant, *pumped storage plants*, can use the same water over and over again. The water is used for electricity production during high-price periods (e.g., day) and the water is pumped from lower reservoir to higher reservoir during low-price periods (e.g., night). Two additional constraints limit production in pumped storage plants. Firstly, production in pumped storage plant requires consumption of electricity, Q, and there is an energy loss ϕ related to that:

$$\forall m, l : \phi \sum_{a} Q_{m,a,l} = \sum_{a} X_{m,a,\text{Pumped Storage},l} \tag{15}$$

Secondly, the same equipment cannot be used for production and consumption at the same time:

$$\forall m, l, a : Q_{m,a,l} + X_{m,a,\text{Pumped Storage},l} \leq X^{\max}_{m,a,\text{Pumped Storage},l} \qquad (16)$$

In an electricity market with significant hydro production, the amount of precipitation is important. We assume that precipitation is "normal" throughout the year. Normal precipitation is defined as a situation with median historic precipitation. Since we assume full certainty, precipitation is known: everybody knows at the beginning of the year that precipitation is going to be normal (or wet or dry) throughout the whole year. Sensitivities are used to analyse the impact of dry or wet years.

Exogenous production

Production in some technologies is not determined by profit-maximising behaviour, that is, production does not depend on power prices. For instance, power production in combined heat and power (CHP) plants is in reality determined by heat demand; power is merely a by-product. Heat demand and production are assumed to follow a seasonal profile. They can vary with temperature, but are assumed price inelastic. Wind power can only be produced when there is enough wind, and given enough wind, marginal costs are negligible. Production in these power plants is therefore included exogenously in the model.

Emissions

CO_2 emissions stem from use of fossil fuels both in existing and new power plants. Emissions from electricity sector $E_{1,\text{El}}$ depend on the production level and emission factor η_k in each plant category:

$$E_{l,\text{El}} = \sum_{m,a,k} \eta_k X_{m,a,l,k} \qquad (17)$$

A.3.3 Trade

Both trade between the model countries (i.e., the countries that are endogenous in the model) and with the surrounding countries is included. There is no trade within a country, i.e. a country is seen as an entity (except Denmark that is modelled as two regions).

Trade between the model countries occurs when prices, adjusted for costs related to transmission losses, differ between the countries. Trade occurs until the price differences are eliminated or until the transmission capacity is fully utilised.

Trade between the model countries is described by the variable T. $T_{m,a,l,s}$ means exports from country l to country s. There is a transmission loss ($\sigma^{t}_{l,s}$) on this trade, to the effect that available imports in country s are reduced compared with the original exports. Available imports are $T_{m,a,l,s} / (1 + \sigma^{t}_{l,s})$.[14]

Imports $U_{m,a,l,c}$ and exports $V_{m,a,l,c}$ to outside countries c are either modelled as fixed, based on historical levels and assumptions about future development, or price-dependent. In the latter case, prices in the countries outside the model (π^{i} for imports and π^{e} for exports) are constant, i.e. not influenced by trade. Nevertheless, π^{i} and π^{e} reflect daily and seasonal variation.

Capacity constraints on transmission lines are in effect on both types of trade.

A.3.4 Electricity Market Equilibrium

The various parts of the market – consumption, production and trade – are tied together in the equations for electricity flow equilibrium. These equations say that demand in any period and any country equals supply:

$$\begin{aligned}\forall m,a,l: & \sum_{e} \lambda_{l,a}(1+\sigma^{d}_{l})D_{m,l,e} + Q_{m,a,l} + \sum_{s} T_{m,a,l,s} + \sum_{c} V_{m,a,l,c} \\ & \leq \sum_{k} X_{m,a,l,k} + \sum_{k} \Gamma_{m,a,l,k} + \sum_{s} \frac{T_{m,a,s,l}}{1+\sigma^{t}_{l,s}} + \sum_{c} U_{m,a,l,c} \end{aligned} \tag{18}$$

To start at the right hand side of the equation: available supply equals domestic endogenous production ($X_{m,a,l,k}$) plus domestic exogenous production ($\Gamma_{m,a,l,k}$); plus imports from other model countries after "melting" $\left(\frac{T_{m,a,s,l}}{1+\sigma^{t}_{l,s}}\right)$; plus imports from non-model countries ($U_{m,a,l,s}$). Demand can be no larger than this supply. Demand (LHS) consists of consumption demand of the load period ($\lambda_{l,a}(1 + \sigma^{d}_{l})$ $D_{m,l,e}$), including domestic transmission losses (δ^{d}_{l}); plus power use by pumped storage plants ($Q_{m,a,l}$); plus exports to model countries ($T_{m,a,l,s}$); plus exports to non-model countries ($V_{m,a,l,c}$).

A.4 *Solution for Power and Allowance Market*

The solution to the model is found by maximizing utility (that is, the areas under the demand curves of Eq. 4) less private economic costs including short run marginal costs, taxes and tariffs, investment and refurbishment costs, abatement costs of CO_2

[14] This modelling approach is sometimes called the iceberg approach, since losses are modelled as if the export melts on its way to becoming import. In electricity trade, the iceberg approach is especially pertinent.

emissions and penalty for non-compliance, if emissions exceed the cap.[15] The maximization is constrained by the various technical and economical constraints of the market, including distribution and transmission losses. In other words, the idea is to *mimic private behaviour*. This implies to maximise

$$
\begin{aligned}
W = \sum_{m,a,l} \Bigg\{ & \sum_{e} W_{m,l,e}(D_{m,l,e}) - \sum_{k} \delta_{l,k} X_{m,a,l,k} - \sum_{e} \left(\tau^{n}_{l,e} + \tau^{e}_{l,e} + \tau^{o}_{l,e}\right) D_{m,l,e} \\
& - \sum_{e} \left(\tilde{p}_{m,l} + \tau^{n}_{l,e} + \tau^{e}_{l,e} + \tau^{o}_{l,e}\right) \tau^{m}_{l,e} D_{m,l,e} \\
& + \sum_{c} \pi^{e}_{c,a} V_{m,c,l,a} - \sum_{c} \pi^{i}_{c,a} U_{m,c,l,a} - \sum_{s} \tau^{t}_{l,s} T_{m,a,l,s} \Bigg\} \\
& - \sum_{k,l} \iota_{k,l} I_{k,l} - \sum_{k,l} \varsigma_{k,l} J_{k,l} \\
& - \sum_{i,j} \varpi_{i,j} A_{i,j} - \vartheta \widehat{E}
\end{aligned} \tag{19}
$$

The maximisation is carried out subject to the electricity market equilibrium condition (18) that ties the quantity variables together; the constraints on electricity production, Eqs. 8, 9, 10, 11, 12, 13, 14, 15 and 16; the capacity constraints on trade; the constraints related to CO_2 emissions and allowances, Eqs. 1, 2, 3 and 17; and formally even the equations that determine prices in terms of consumption, Eqs. 4, 5 and 6.

The maximisation of Eq. 19 determines all quantity variables of the system: consumption ($D_{m,l,e}$), production ($X_{m,a,l,k}$), investments $I_{k,l}$, refurbishments $J_{k,l}$ and trade ($T_{m,a,l,s}$, $U_{m,c,l,a}$ and $V_{m,c,l,a}$), emissions (E) and abatements (A_i) in different sectors and formally even the producer price ($p_{m,l}$) and allowance price.

A.5 Parameters and Variables

Endogenous variables *are written in Latin letters.*

D is electricity consumption
X is electricity production
T is trade within the model countries (EU-27 plus Norway and Switzerland)
U is imports to the model region
V is exports out of the model region
Q is demand in pumped storage plants

(continued)

[15] We ignore the income effects in this equation and calculate ordinary consumer's surplus. Consumer's surplus is an average of equivalent variation and compensating variation, see, e.g., [20].

R is reservoirs
I is investments
J is refurbishments
P is end user price
p is wholesale (spot market) price
E is CO_2 emissions
$\widehat{E}$ is the excess emissions
A is abatements (in EU ETS industries other than power and heat)

Parameters *are written (mostly) in Greek letters.*

α is the constant term that defines the starting point of the model (found by a combination of observed end-user prices and consumption). α also includes seasonal variation
ξ is income level (GDP for services, industry and electric boilers, and private income for households)
π is oil price
γ is income elasticity
β is electricity price elasticity
κ is oil price elasticity
τ^n is distribution cost
τ^e is excise tax on electricity
τ^o is mark-up on spot price to different end-users
τ^m is VAT
τ^t is border tariff
δ is marginal production costs
η is emission factor (CO_2 emissions per unit of output)
σ^d is transmission losses within a country
σ^t is transmission losses on international interconnectors
θ is total inflow to the reservoirs
ε is minimum reserve requirement
λ is the share of consumption assigned to each load period, $\Sigma_a \lambda_{l,a} = 1.0$,
μ is the relative length of each load period, $\Sigma_a \mu_a = 1.0$
$X^{\max}$ is maximum production
$X^{\min}$ is minimum production
$X^{\max\mathrm{Start}}$ is maximum production, based only on today's installed capacity (i.e., without additional investments)
Γ is exogenous production
$I^{\max}$ is investment potential
$J^{\max}$ is refurbishment potential
ι is unit cost of investment
ς is unit cost of refurbishment
ϕ is efficiency factor of pumped storage
π^{e} is price for exports to non-model countries (e.g., Russia)
π^{i} is price for imports from non-model countries (e.g., Russia)
ϖ is abatement cost in the EU ETS industrial sectors (other than power and heat)
E^{BAU} is CO_2 emissions from industrial sectors in the business-as-usual case, before abatements,
E_{Heat} is CO_2 emissions from heat production
$\bar{E}$ is the total emission cap

(continued)

ϑ is penalty if the emission cap is exceeded
A^{max} is the maximum abatement potential in the industrial sectors

Indices *are written as footscripts in lower case Latin letters.*

$m \in \{1,...,13\}$ is a 4-week period ("month")
$a \in$ {*night, day shoulder, day, day peak and afternoon peak*} is load period
$e \in$ {*households, services, process industry (including pulp and paper), other industry, electric boilers*} is end-user category
$k \in$ {*hydro, nuclear, wind, CHP, coal condensing, CCGT, ...*} is production technology and cost category
$i \in$ {*iron and steel, aluminium, pulp and paper, chemical industry, mineral industry, refineries, aviation*} is sector in allowance market (except electricity and heat)
$j \in \{1,...,25\}$ is potential abatement activity
$l,s \in$ {*EU-27, Norway, Switzerland*} is country in the model region
$c \in$ {*Russia, ...*} is trade partner outside the non-model region

References

1. European Commission (2009) Directive 2009/29/EC of the European Parliament and of the Council of 23 April 2009 amending Directive 2003/87/EC so as to improve and extend the greenhouse gas emission allowance trading scheme of the Community. Official Journal of the European Union, 5 June 2009, pp 63–87
2. Capros P, Mantzos L, Papandreou V, Tasios N (2008) Model-based analysis of the 2008 EU Policy Package on climate change and renewables. E3M Lab, National Technical University, Athens
3. Stankeviciute L, Kitious A, Criqui P (2008) The fundamentals of the future international emissions trading system. Energy Policy 36(11):4272–4286
4. Ellerman AD, Buchner BK (2007) The European Union Emissions Trading Scheme: origins, allocation, and early results. Rev Environ Econ Policy 1(1):66–87
5. Convery FJ, Redmond L (2007) Market and price developments in the European Union emissions trading scheme. Rev Environ Econ Policy 1(1):88–111
6. European Commission (2009) Directive 2009/28/EC of the European Parliament and of the Council of 23 April 2009 on the promotion of the use of energy from renewable sources and amending and subsequently repealing Directives 2001/77/EC and 2003/30/EC. Official Journal of the European Union, 5 June 2009, pp 1–47
7. European Commission (2008) Commission staff working document. Impact assessment. Document accompanying the Package of implementation measures for the EU's objectives on climate change and renewable energy for 2020. SEC (2008) 85/3. Commission of the European Communities, Brussels, 23 Jan 2008
8. Nordhaus WD (1993) Rolling the 'dice': an optimal transition path for controlling greenhouse gases. Resour Energy Econ 15:27–50
9. Nordhaus WD, Boyer JG (1999) Requiem for Kyoto: an economic analysis of the Kyoto Protocol. In: Weyant J (ed) The costs of the Kyoto Protocol: a multi-model evaluation. Energy J (Special issue):93–130
10. Weyant JP, de la Chesnaye FC, Blanford GJ (2006) Overview over EMF-21: multigas mitigation and climate policy. In: de la Chesnaye FC, Weyant JP (eds) Multi-greenhouse gas mitigation and climate policy. Energy J (Special issue 3):1–32

11. Weyant JP (ed) (1999) The costs of the Kyoto Protocol: a multi-model evaluation. Energy J (Special issue)
12. Weyant JP (2004) EMF-19 Alternative technology strategies for climate change policy. Energy Econ 26(4):501–755
13. Springer U (2003) The market for tradable GHG permits under the Kyoto Protocol: a survey of model studies. Energy Econ 25:527–551
14. de la Chesnaye FC, Weyant JP (eds) (2006) Multi-Greenhouse gas mitigation and climate policy. Energy J (Special issue 3)
15. Hourcade J-C, Jaccard M, Bataille C, Ghersi F (eds) (2006) Hybrid modeling of energy-environment policies: reconciling bottom-up and top-down. Energy J (Special issue 2)
16. Klepper G, Peterson S (2006) Emissions trading, CDM, JI, and more: the climate strategy of the EU. Energy J 27(2):1–26
17. European Commission (1996) POLES 2.2. European Commission DG XII. EUR17358 EN
18. Capros P (2005) The PRIMES energy system model. Summary description. National Technical University of Athens, European Commission Joule-III Programme, Athens
19. Brooke A, Kendrick D, Meeraus A, Raman R (1998) GAMS – a user's guide. GAMS Development Corporation, Washington DC
20. Varian HR (1992) Microeconomic analysis, 3rd edn. Norton, New York

Portfolio Optimization of Power Generation Assets

Reinhard Madlener

Abstract In this chapter I provide an overview of the theoretical and applied literature dealing with mean-variance portfolio analysis used to study the efficiency of portfolios of power generation assets. The relevant literature focuses on the risk-mitigating benefits of technological diversification vis-a-vis single-technology analysis with conventional levelized cost analysis, to varying degrees taking into account real-world constraints. Part of the cutting-edge research deals with the benefits that accrue from intra-technology diversification and geographical dispersion. Some studies also take into account country-specific differences in national regulatory framework conditions and local resource potentials (esp. in the case of wind power). Other research has focused on technical and system-related aspects, such as load dispatch and portfolio restrictions, e.g., arising from grid constraints and the intermittent nature of many renewable energy sources. Complementary approaches to mean-variance portfolio analysis, such as real options analysis and fuzzy modeling, as well as alternative measures of risk (e.g. Value at Risk – VaR, Conditional Value at Risk – CVaR, and (semi-) mean absolute deviation), are briefly discussed as well, thus acknowledging some of the most important recent developments in this research area that have not been reviewed elsewhere yet.

Keywords Mean variance portfolio theory • Fuzzy sets • VaR • Real options • CAPM • Uncertainty

---

R. Madlener (✉)
Institute for Future Energy Consumer Needs and Behavior (FCN), School of Business and Economics/E.ON Energy Research Center, RWTH Aachen University, Aachen, Germany
e-mail: RMadlener@eonerc.rwth-aachen.de

Q.P. Zheng et al. (eds.), *Handbook of CO_2 in Power Systems*, Energy Systems,
DOI 10.1007/978-3-642-27431-2_12, © Springer-Verlag Berlin Heidelberg 2012

1 Introduction

The use of portfolio theory Markowitz, 1952 & 1959 [55, 56] has proven useful for providing new insights to decision- and policy-makers regarding the possible role and potential benefits of renewable energy technology for power generation and for determining optimal degrees of technological diversity of an energy system, and also the link between energy diversity and security of energy supply.

The purpose of this chapter is to survey some of the relevant literature, and to review some of the more recent work that has taken new research avenues. Overall, the field of research has been expanding rapidly in recent years. In this respect I have tried to add a number of recently published articles. Much of this literature was not yet covered, for instance, in the compilation of manuscripts edited by Morgan Bazilian and Fabien Roques in 2008 as a tribute to Shimon Awerbuch, who very much popularized the mean-variance portfolio optimization approach in the energy literature.

In liberalized, competitive electricity markets power generators face many risks, including those related to changes in fuel input and electricity prices, and more recently also the price of CO_2, but also of changes in demand, unforeseen unit outages, volatile renewable energy sources, regulatory regime changes etc. This renders risk management as paramount, i.e. the process of achieving a desired risk – return balance by strategic action, and helps to avoid inefficient portfolios of (real) assets.

Conventional approaches, such as the levelized cost methodology, do neither adequately incorporate risk and uncertainty (but instead typically assume some practitioner's discount rates), and thus need to be supplemented with approaches that are able to do that. In particular, expected future costs and revenue streams are subject to important risks that cannot be properly valued by just adding some (arbitrary) risk premium into the calculations. Moreover, the benefits of diversifying away some of the risk is ignored in such analysis.

Much of the theoretical and applied research on portfolio optimization is driven by tailoring the return and risk of an investment venture to the particular preferences of an investor. While there is less debate about the measure of (expected) return, there has been much more discussion about the most adequate modeling and measurement of an investment's risk. Most of the literature in many areas of decision-making has followed the fundamental view brought up by Markowitz [55] that any decision under uncertainty may be studied in terms of risk and return. At the same time, it has been recognized very early (e.g. already by Markowitz [56]) that an adequate estimation of risks induced by uncertainties cannot always be achieved by two-sided risk measures such as the variance.

Mean-variance portfolio analysis (MVPA) is complementary to other approaches adopted from the finance literature (such as the Capital Asset Pricing Model – CAPM, Real options analysis – ROA etc.). MVPA is usually static only (one-period analysis), whereas dynamic approaches are still few and far between. Principles of finance theory [55], applied to the energy domain, purporting the idea of efficient portfolios (which implies for a risk-averse investor to accept a minimum risk for a given return, or put differently, a maximum return for a given or accepted level of risk).

Still, a lot of further research is needed to account for the special characteristics of power generation (i.e. real, in contrast to financial) assets, and the energy system more broadly speaking (e.g. with regard to grid characteristics or technically feasible technology combinations including energy storage options). For instance, grid constraints ought to be taken into account as well as load dispatching (e.g. [14, 29]), scheduling trades [17], and maximum share of intermittent generation [19], among others.

The remainder of this chapter is organized as follows. Section 2 introduces the main methodology used. Section 3 contains a discussion and categorization of the literature reviewed. Section 4 addresses some methodological issues, while Sect. 5 concludes.

2 Methodology

In this section we briefly describe three major methodologies to investigate optimal portfolios, starting with Markowitz' mean-variance portfolio theory, followed by the less well known semi-mean absolute deviation model and a fuzzy version of it. For obvious reasons, the aim is merely to give an idea of how these models look like, rather than a thorough treatment which can be found elsewhere in the literature.

2.1 Mean-Variance Portfolio Analysis

The mean-variance portfolio analysis (MVPA) approach [55] shows how rational investors can reap the risk-reducing benefits of diversification, and make their portfolios efficient (i.e. an investor should not take more risk than necessary for a given level of expected return). According to this probabilistic methodology it is not enough to look at the expected return, and the risk of one security in isolation, especially when investing in more than one asset. Diversification is the basic ingredient of Markowitz's MVPA, and the relationship between expected returns and risks (the basic portfolio characteristics) constitutes an integral part of this approach.

In modern portfolio theory, however, some strong assumptions are made about the investors and markets, which have led to criticism (e.g. [22, 23]). In particular, the investor is assumed to be risk-averse, return-loving, and acting perfectly rational. Regarding the market, it is assumed that each class of asset is sufficiently diversified to eliminate any non-market risk.

The mean-variance portfolio selection model can be defined as follows:

$$V(R_p) \to \min \tag{1}$$

s.t.

$$E(R_p) = \sum_{i=1}^{n} R_i x_i \quad \text{with} \quad 0 \leq x_i \leq x_{i,\max} \quad \text{and} \quad \sum_{i=1}^{n} x_i = 1 \tag{2}$$

where $E(R_p)$ denotes the portfolio's expected return, R_i the expected return of individual assets i, x_i the share of asset i in the portfolio, and $V(R_p)$ the portfolio risk. The risk of the portfolio can be specified as

$$V(R_p) = \sqrt{\sum_{i=1}^{n} x_i^2 \sigma_i^2 + 2 \sum_{i=1}^{n-1} \sum_{j=i+1}^{n} x_i x_j \sigma_i \sigma_j \rho_{ij}}, \tag{3}$$

where σ_i^2 is the variance of component asset i, σ_i the volatility of asset i, and ρ_{ij} the correlation coefficient between asset i and j. Figure 1 provides a graphical illustration of the mean-variance portfolio model. The dots represent portfolios, whereas the envelope curve (upper part) is the efficient frontier that represents all portfolios that are efficient (for a certain expected return no more risk than necessary, or for a certain risk level no lesser return than possible).

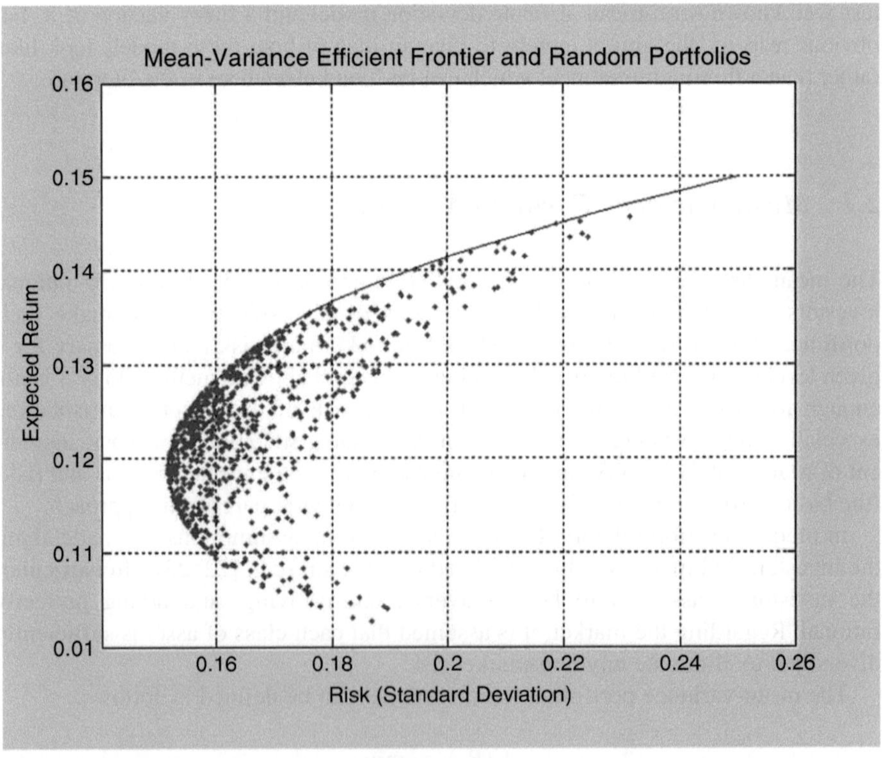

Fig. 1 Mean-variance portfolio model (illustration) Source: MathWorks

2.2 Semi-Mean Absolute Deviation Model

Konno and Yamazaki [41] proposed the (linear) mean-absolute deviation (MAD) model, a mean-lower partial risk model that constitutes an interesting alternative to the classic mean-variance portfolio model. Apart from the linearity of the problem, which can be solved more easily than a quadratic one, another advantage of the MAD model is that it can be used to tackle large-scale problems where a dense covariance matrix occurs, and that no particular return distribution is needed. Konno and Koshizuka [39] review some important properties of the MAD portfolio optimization model, arguing that it is theoretically and computationally superior to the MVP model.

Formally, the portfolio rate of return in the MAD model is defined very much like the MVP selection problem of Markowitz. Specifically, the definition of portfolio risk $w_p$is given as the expected value of the mean absolute deviation between the realization of the portfolio's rate of return and its expected value, formulated as (cf. [41]):

$$w_p = \mathrm{E}\left[\left|\sum_{i=1}^{n}(R_i x_i) - \mathrm{E}\left[\sum_{i=1}^{n}(R_i x_i)\right]\right|\right] \tag{4}$$

With this definition of risk, a two-dimensional portfolio optimization model can be specified as:

$$\mathrm{E}\left[\left|\sum_{i=1}^{n}(R_i x_i) - \mathrm{E}\left[\sum_{i=1}^{n}(R_i x_i)\right]\right\}\right|\right] \rightarrow \min \tag{5}$$

$$\sum_{i=1}^{n} E(R_i)x_i \rightarrow \max \tag{6}$$

$$\sum_{i=1}^{n} x_i = 1,\ x_i \geq 0 \tag{7}$$

From this, a semi-mean absolute deviation (SMAD) portfolio optimization model can be derived, where the risk measure is defined by the mean absolute deviation of the portfolio's rate of return *below the average* (for an investor only the downside risk is relevant):

$$\mathrm{E}\left[\left|\min\left\{0, \sum_{i=1}^{n}(R_i x_i) - E\left[\sum_{i=1}^{n}(R_i x_i)\right]\right\}\right|\right] \rightarrow \min \tag{8}$$

$$\sum_{i=1}^{n} \mathrm{E}(R_i)x_i \rightarrow \max \tag{9}$$

$$\sum_{i=1}^{n} x_i = 1,\ x_i \geq 0 \tag{10}$$

2.3 Fuzzy Semi-Mean Absolute Deviation Model

More recent work on portfolio optimization of power generation assets has increasingly focused on alternative measures of risk, and thus on alternatives to the Markowitz method (for a detailed discussion see [28]). VaR and CVaR approaches have received some attention, but also mean-absolute deviation (MAD) and semi-mean absolute deviation (SMAD) models. Finally, in combination with fuzzy sets theory, MAD and SMAD have also been combined, leading to a fuzzy semi-mean absolute deviation (FSMAD) approach [27].

The mean-absolute deviation (MAD) model is an alternative to the classic mean-variance model in that the measure of risk (standard deviation) is replaced by the absolute deviation. This methodology was proposed by Konno and Yamazaki [41] and Konno and Koshizuka [39], respectively. In this alternative approach a new definition of risk is used. Again, as for the SMAD model, because positive deviations of investments from the expected mean value are beneficial, only negative deviations are defined as risks.

Considering fuzzy theory, where investors' aspiration level of return and risk are represented by adequate membership functions, the fuzzy semi-mean absolute deviation (FSMAD) portfolio selection model was proposed

$$\Lambda \rightarrow \max \tag{11}$$

s.t.

$$\alpha_R \sum_{i=1}^{n} R_i x_i - \Lambda \geq \alpha_R R_M \tag{12}$$

$$\alpha_w \frac{1}{T} \sum_{t=1}^{T} d_t + \Lambda \geq \alpha_w w_M \quad \Lambda \geq 0 \tag{13}$$

$$d_t + \sum_{i=1}^{n} (R_{it} - R_i) x_i \geq 0 \, d_t \geq 0 \tag{14}$$

$$\sum_{i=1}^{n} x_i = 1 \quad 0 \leq x_i \leq x_{i,\max} \tag{15}$$

where $\Lambda = \log(\lambda/1 - \lambda)$ (λ denotes the value of membership function for expected return and risk[1]). Parameters R_M and w_M are the mid-points where the membership function value λ is equal to 0.5, and parameters α_R and α_w determine the shape of these membership functions[2]. The variable dt is a time variable connected with the semi-mean absolute deviation, R_{it} represents the return value for asset i at time t.

---

[1] A formal presentation of these membership functions is provided in Glensk and Madlener [27].

[2] More information about this model and other fuzzy portfolio optimization problems proposed and analyzed can be found in Glensk and Madlener [27].

3 Literature Review

3.1 Early Studies

The first widely recognized application of mean-variance portfolio optimization in the field of energy utilities was done by Bar-Lev and Katz [7]. By using Markowitz's diversification, the authors derive an efficient frontier for a fossil fuel mix, and investigate various investment opportunities by the variation of cost and risk assumptions. The authors conclude that the regulated utilities investigated are, in general, efficiently diversified.

Humphreys and McClain [34] apply MVP theory to show what energy mix should be adopted in the US in the light of a national goal to reduce the risks from unanticipated energy price shocks to the national economy. They argue that the electric utility industry has moved towards higher production efficiency since the 1980s and that the move of the industry towards higher use of natural gas in the 1990s might have been motivated by seeking higher returns on energy investments.

3.2 More Recent Studies

The use of MVPA in the energy domain has been revived, and popularized with a special focus on the beneficial role renewables can play in power generation portfolios, by the financial economist and energy analyst Shimon Awerbuch.

Awerbuch [3], for the example of the photovoltaics (PV) technology, shows the benefits of approaches such as mean-variance portfolio optimization and the Capital Asset Pricing Model (CAPM) to properly reveal the "true value" of low-risk, passive and capital-intensive technologies. He calls for responsibility to broaden the analytic horizon of policy-makers to include new valuation models and concepts that are able to properly reflect the unique risk-return attributes of new technologies such as PV.

Awerbuch and Berger [4] are the first to apply MVP theory on generation portfolios in liberalized power markets, putting their focus on the generation portfolio of the European Union (EU-15). In contrast to Bar-Lev and Katz, they make use of a more detailed portfolio model that also reflects the risks of relevant generation cost streams, such as construction period costs, operation and maintenance (O&M) costs, and fuel costs. On the basis of this model, they show portfolio effects of different generation mixes. Their results show that the existing EU generation portfolio is slightly suboptimal from a risk-return perspective. Specifically, by adding more wind power or comparable renewable energy technologies to the conventional mix, the overall efficiency of the portfolio could be improved.

Awerbuch [5] investigates the potential of mitigating fossil fuel price volatility by portfolio optimization. He shows that renewable energy technologies, apart from their potentials to enhance energy security, can reduce the portfolio-generating cost, even if they are more costly. This result seems counterintuitive at first sight, but is of course perfectly consistent with finance theory. It also illustrates the importance of explicitly accounting for the benefits of risk diversification, rather than to evaluate alternative technologies separately to determine their relative merits. In Awerbuch et al. [7] the authors investigate the prospects of wind and other renewables for the case of Scotland.

Doherty et al. [18] employ MVP theory and Integrated Resource Planning (IRP) to investigate the role of wind power generation in future generation portfolios, accounting for the characteristics of load and wind, generation adequacy and plant utilization. Sensitivity analysis is done for the discount rate, carbon tax and fuel prices. Wind power is found to have a significant role to play in the future power generation portfolio of Ireland. In a related paper Doherty et al. [19] investigate how the generation portfolio of Ireland might evolve until 2020. A main focus is again on the cost of carbon and wind energy, but this time also on policy design, with the aim to inform policy-makers about how policy instruments have to be tailored to mitigate risk arising from fuel price volatility by an optimally diversified portfolio.

Krey and Zweifel [43] determine efficient power generation mixes for Switzerland and the U.S. by applying MVP theory. In their investigation, they consider correlated shocks in electricity generation costs by usage of the seemingly unrelated regression (SUR) estimation method (cf. Sect. 4.1). In doing so, the systematic component of the covariance matrix is eliminated in the case of cost changes, and thus a time-invariant correlation matrix can be achieved. They come to the conclusion that the maximum expected return from the portfolio in Switzerland could be increased if capacity were shifted from run-of-river and storage hydro power plants to more nuclear and solar power. In two related papers [44, 45] the authors expand the analysis to the current user view (expected return expressed as kWh/US$) and the investor view (expected return expressed as *changes* in kWh/US$), respectively, thus accounting for the different views and preferences of the actors involved in the market, and the different recommendations for efficient portfolios that may result.

Borchert and Schemm [11] apply Markowitz's portfolio theory in a different context, viz. for the evaluation of investment options in wind power. In particular, as target variables for the investigation of German wind power projects, they take the expected return of the investment, and the Conditional Value at Risk (CVaR) as a risk parameter. They conclude that the current promotion scheme for electricity from renewables in Germany (based on guaranteed feed-in tariffs) only leads to minor diversification effects.

Drake and Hubacek [20], by using a risk-return portfolio approach, investigate the potential to reduce the variability of wind power in the UK by geographical dispersion of the location of wind farm sites. The investigation is for two wind scenarios obtained from wind speed data. The analysis is done for a single location (a centralized wind farm with 2.7 GW_{el} of installed capacity), compared to a situation where this capacity is spread across four different locations. The authors conclude that reductions in wind power variability in the order of 36% can be achieved as a result of spatially distributing wind power capacity.

Jansen and Beurskens [35] report on results from applying MVP theory to the future power generation technology portfolio in the Netherlands in 2030 (for two base case generation mixes/scenarios), with a particular emphasis on the role of renewables. They find that diversification into renewable generation technologies may allow for substantial risk reduction at no extra cost, and that offshore wind can reduce the cost risk of the Dutch generation portfolio, and have a dampening effect on Dutch electricity prices in 2030.

Kienzle and Andersson [38] apply MVP theory to portfolios generating multiple forms of energy, such as electricity, heating or cooling power, providing an extension of MVP analysis to an arbitrary number of output energies. Instead of using means and variances derived from historical time series data, the authors use a set of energy scenarios. The model developed is applied to a portfolio of distributed electricity and heat generation technologies. In a related paper [24], the authors show how integrated planning for multi-energy networks that is based on MVP theory can be used to determine an efficient mix of power generation that also incorporates siting decisions and risk preferences.

Kotsan and Douglas [42] use MVP analysis to study the locational value of power generation assets. Risk-averse investors maximize the expected value of the weighted average locational marginal price (LMP) in their portfolio, and to minimize its variance. For a simulated bus network (IEEE 68-type) similar to the New York – New England system the authors find that network topology is an important determinant of the investment decision, since congested power lines make it difficult to supply power to certain nodes during peak-load hours.

Madlener and Wenk [54] investigate the future development of the generation portfolio in Switzerland, and the electric utility AXPO in particular, and aim at identifying efficient investment options for the electricity supply sector. By applying MVP theory, their study covers power generation technologies currently in operation as well as new options, such as new renewables (wind, photovoltaics) and combined-cycle gas turbines, and retrofits of run-of-river and storage hydro power plants. In doing so, they explicitly account for differences in construction lead times and (asymmetric) distributions in the stochastic variables, explicitly differentiate between base-load and peak-load technologies, and compare the lifetime-adjusted average return and the return-specific variance as target variables. Their results show that the current production portfolio for base-load in Switzerland is very close to the efficient frontier, and leaves only little leeway for further optimization, whereas the peak-load portfolio still allows for some improvement from a return-risk perspective.

Gotham et al. [29], applying MVP theory, focus on the heterogeneous ability of fuels/technologies to adapt to varying loads, thus accounting for the fact that load cycles can play critical roles in fuel selection. The authors propose a load decomposition approach that allows to "retain the convenience and elegance of the mean-variance approach". The MVP model variant is applied to data for the state of Indiana (U.S.), and three load factors distinguished (base, cycling, and peak). Furthermore, the usefulness of the model for evaluating changes in optimal fuel mixes as a consequence of policy change is shown.

Muñoz et al. [57] introduce a model for the optimal investment portfolio containing various types of renewables (wind, photovoltaics, mini-hydro, and thermo-electrical). The authors apply the model, which consists of a system of linear equations representing all available technologies/energies, to various case studies implemented in Spain within the Renewable Energies Plan (2005–2010), and also investigate the impact of different types of financing.

Sunderkötter and Weber [68] propose an analytical approach that combines elements from peak-load pricing and MVP theory to derive optimal power generation technologies in the presence of uncertainty in the fuel prices. In a static optimization model, the authors derive analytically derived optimality conditions, which contrast to the existing numerical methods, and apply the proposed framework for the case of Germany.

Westner and Madlener [72] apply MVP theory for analyzing the prospects of four different cogeneration technologies in Germany. They take the ambitious policy target of a doubling of the share of cogeneration from 12.5% to 25% into account as well as the new regulatory framework that contains guaranteed feed-in tariffs for electricity from cogeneration plants, additional revenues from heat sales, specific operational features, and specific rules concerning the allocation of CO_2 allowances. The investigation is done on the basis of fuel cost and electricity and heat price projections taken from the *Leitstudie 2008* [58]. The authors conclude (1) that combined-cycle gas turbine (CCGT) and engine-CHP are the most attractive technologies; (2) that additional capacity will likely be installed in the industrial sector; and (3) that it is unlikely that the political target of a 25% share of cogeneration can be achieved. In related research, the authors take a different perspective [73]. They investigate the benefits of regional diversification of cogeneration in Europe, analyzing the role in cogeneration portfolios of the CCGT- and engine-CHP technology when implemented in four different EU member states (France, Germany, Italy, and the UK) with very different promotion schemes and other framework conditions in place. The results provide evidence that the returns of investments in cogeneration differ markedly across the countries studied, but also with respect to the support scheme and the technology concerned. Specifically, regional diversification of CCGT-CHP does not contribute to reducing the portfolio's risk, whereas engine-CHP can have such a risk-mitigating effect.

Zhu and Fan [75] apply MVP theory to the medium-term (2020) plans of China regarding four different power generation technologies (coal, natural gas, nuclear, and wind) and study both the consequences of CO_2^- emission constrained scenarios and the policy implications of the analysis. The authors find that the planned 2020 power generation portfolio is more risky than the 2005 portfolio, and that an increase in the share of nuclear power could effectively reduce the portfolio's risk. They also conclude that, due to the higher generation cost, stronger political support will be needed to promote the development of power generation from renewable energy sources.

Delarue et al. [14] introduce a MVP model that explicitly distinguishes between installed capacity (generation), electricity (energy), and the actual instantaneous power delivery. In this way the ramp-up and shut-down limits of conventional power plants as well as the variability of wind power are properly accounted for in the portfolio-guided investment optimization. The model itself is specified as a quadratically constrained programming model. The results indicate that wind power can indeed lower the financial risk related to generation costs, but in their opinion to smaller levels than typically reported in the literature. The requirement of sufficient reserve capacity is emphasized to enable the level of wind power deployment that is desired from a portfolio optimization perspective.

Lang and Madlener [49] analyze the impact of credit risk mitigation by means of margining (i.e. the mandatory deposit of collaterals by outside counterparties with the clearing company of the futures exchange) on the value of power plants in a MVP optimization framework. The authors find that the explicit consideration of the margining requirements stipulated by the European Commodity Clearing AG (ECC) can markedly change the composition of efficient portfolios.

Madlener and Glensk [52] present results of a study where MVP theory is applied to power generation assets in three European countries (Germany, Sweden, and the UK). In particular, the authors first determine the efficiency frontier for both existing and prospective future portfolios. Further, they analyze whether recently announced investment plans of E.ON are likely to enhance or deteriorate the efficiency of the company's power generation portfolios. As return-risk criteria they use the expected net present value per unit of installed capacity and the corresponding standard deviation. The restricted portfolio optimization model reveals fairly well-diversified generation mixes in all three markets analyzed. New renewable energy technologies tend to have a positive impact on the existing portfolios, while the contrary is found for new investments in conventional technologies.

Roques et al. [66] analyze the relevance of geographical diversification of wind farms in Europe with a MVP modeling approach, using historical wind power production data from five EU member countries (Austria, Denmark, France, Germany, and Spain). For unconstrained portfolios they find that countries with the best wind energy resources, or whose sizes contribute to smoothing out the country's output variability, dominate the optimal portfolios. In the constrained optimization, they take account of wind resource potentials and transmission constraints, finding that the efficiency gains from geographical diversification are reduced, but nevertheless still considerable relative to both actual and projected 2020 portfolios. The authors conclude from the study that an increased cross-border interconnection capacity, greater coordination of renewable energy support policies in Europe, and locational incentives implemented in renewable support mechanisms and electricity market designs, and suggest the development of a mechanism for renewables credits trading as a remedy to cure the misalignment between actual wind power portfolios and the theoretically efficient geographic dispersion.

3.3 Fuzzy Approaches

Nguene and Finger [59] use fuzzy set theory to investigate strategic choices in electricity supply for the case of Switzerland. Economic performance is calculated for different time horizons. The authors show that expected profits not only depend on the perception of a market player, but also on the imprecision of the variables used. The methodology is based on five major steps: (1) screening of the variables; (2) modeling of the explanatory variables; (3) definition and construction of strategies; (4) construction of the scenarios; and (5) assessment and evaluation of the strategies.

Glensk and Madlener [27] provide an overview of different fuzzy portfolio selection models in which the rate of return as well as the investor's aspiration levels of portfolio return and risk are treated as fuzzy variables. Portfolio risk is defined as a downside risk. Moreover, the authors introduce a semi-mean absolute deviation (SMAD) portfolio selection model that is then applied to the selection of power generation mixes (using data for the existing and announced new generation assets of energy provider E.ON in Germany).

Madlener et al. [53] investigate how fuzzy portfolio optimization and alternative risk measures can be used to optimize an onshore wind power plant portfolio in Germany consisting of five wind parks. The authors show the parallels between the valuation of equity shares on financial markets and investment planning for projects. The results of the empirical analysis show that the return and risk spans for the semi-mean absolute deviation (SMAD) model are much wider than for the fuzzy model version (FSMAD), while the highest returns are generated by portfolios using the FSMAD model. Furthermore, portfolio optimization based on the FSMAD model offer a smaller choice of portfolio possibilities, which could facilitate the eventual choice of a certain portfolio, reflecting a major advantage often brought forward with respect to fuzzy sets theory-based optimization, viz. to better reflect the preferences of the decision-maker and the real-world conditions.

3.4 Miscellaneous and Hybrid Approaches

Yu [74] introduces a spatial mean-variance mixed integer programming (MIP) model for short-term risk analysis in electricity markets. The spatial character is captured by correlations of geographically separated markets and the consideration of wheeling administration. Additionally, transaction costs and other practical constraints are incorporated in the model. Interestingly, the efficient frontier determined from applying the model in a case study is neither smooth nor concave. Finally, possible extensions of the model to other energy markets are investigated.

Roques [63] investigates the value of operating flexibility and contractual arrangements in the context of technology choice of new market entrants in liberalized electricity markets. For three different contractual risk transfer case studies, combined cycle gas turbines (CCGT), due to their operating flexibility and "self-hedging" merits of the cash flows (high correlation between electricity and gas prices), are found to be particularly attractive for new entrants, compared to coal and nuclear power plants. Roques et al. [64, 65], rather than putting the focus of their investigation on existing generation portfolios, concentrate on investments in new generation technologies. Specifically, they focus on three base-load technologies (gas, coal, and nuclear) and consider the net present value (NPV) and its standard deviation as target variables. One result of their investigation that focuses on the power generation market in the UK is that the high correlation between gas and electricity prices reduces the risks for gas-fired power plants, and makes those portfolios attractive that are dominated by a high gas share. More diversified portfolios, including also nuclear and coal, are only economically feasible in combination with risk-reducing, long-term contracts for power sales. Roques et al. consider in their investigation many cost and technology-specific parameters, whereas the value of long-term investments, like that in nuclear power capacity, is underestimated.

van Zon and Fuss [71], in a stylized model, combine a clay-clay vintage modeling framework with elements from MVP theory for studying the role of fuel price risk and embodied technical change (ETC), thus capturing both dynamic investment considerations and irreversibilities associated with large sunk cost. They find that ETC, combined with an expected future switch toward some alternative technology, can reduce current investment in that technology, while (temporarily) increasing current investment in competing technologies. This enables a rational, risk-averse investor to maximize productivity gains by waiting until technical change has materialized, after which they switch and invest more heavily in the most recent available vintages.

Denault et al. [13] investigate the diversification effect of wind power to reduce the risk of water inflow shortages on hydropower production, usingcopulas to model the dependence between the two energy sources studied. The data used are for the province of Quebec, spanning the period 1958–2003. The results show that for all scenarios considered, any proportion of wind up to30% improves the production deficit risk profile of an all-hydro system.

Pindoriya et al. [60] introduce a multi-objective mean-variance-skewness (MVS) model [40], an extension of the classical mean-variance portfolio model, for the allocation of a power generation portfolio. The non-dominated (Pareto-optimal) solution is derived from a meta-heuristic technique based on multiple-objective particle swarm optimization (MOPSO). They find that the MVP approach leads to suboptimal portfolios in the presence of skewness, and that their approach can provide much better solutions in situations where non-normally distributed assets for trading exist.

Fuss et al. [25] combine portfolio analysis with real options modeling, focusing on the role of bioenergy in climate change and robust long-term climate risk mitigation. In a plant-level analysis, the authors study the impact of uncertainty on investment decision-making, using different socio-economic scenarios for a range of climate stabilization targets. The authors also analyze the impact of alternative risk measures on the conclusions derived.

Hickey et al. [30] review the strengths and limitations of alternative approaches to develop optimal portfolios of electricity supply options (portfolio theory, real options theory, and diversity indices), and a number of constraints that have to be satisfied during optimization. They conclude that from the approaches compared and contrasted, the concept of diversity is the most appropriate one. In an empirical

Table 1 Summary of literature on portfolio optimization of power generation assets

Study (in chronological order)	Approach/es	Country coverage	Special focus
Humphreys and McClain [34]	MVPA[a]	U.S.	Unanticipated energy price shocks
Awerbuch [3]	Engineering-based and finance approach to valuation	Europe, U.S.	Photovoltaics, managerial options
Awerbuch and Berger [4]	Cost minimization	EU-15	EU electricity planning and policy-making
Yu [74]	Spatial MIP[b]	U.S.	Multiple commodity products; NYPP/PJM data
Awerbuch [5]	MVPA[a]	EU, U.S., Mexico	Renewables (wind, geothermal, other), energy security
Doherty et al. [18]	MVPA[a] IRP[c]	Ireland	Wind energy, diversity
Krey and Zweifel [43–45]	Cost minimization, SUR estimation[d]	US, Switzerland	Perspective of investor and of current user
Borchert and Schemm [11]	MVPA[a]	Germany (Dutch wind data)	Offshore/onshore wind power (fictitious portfolio), CVaR
Drake and Hubacek [20]	MVPA[a] (not stated explicitly)	UK	Wind energy, geographical dispersion
Nguene and Finger [59]	Fuzzy sets theory	Switzerland	Strategic choices in electric energy supply
Awerbuch et al. [6]	MVPA[a]	Scotland	Wind and other renewables
Doherty et al. [19]	MVPA[a]	Ireland	Max. share of intermittend gen., isolated system
Huang and Wu [31]	MVPA[a]	Taiwan	Risk-weighted generation costs, electricity supply planning
Jansen and Beurskens [35]	MVPA[a]	Netherlands	Future power generation mixes (2030), renewables, esp. offshore wind
Kienzle and Andersson [38]; Favre-Perrod et al. [24]	MVPA[a]	(generic)	Energy system planning; multi- energy generation & transmission model
Kotsan and Douglas [42]	MVPA[a]	New York – New England	Locational value of generation assets; network topology
Madlener and Wenk [54]	MVPA[a]	Switzerland	Profit maximization; base-load/peak-load; retrofits; non-normally distributed construction lead times
Roques et al. [64, 65]	MVPA[a]	UK	Base-load generation, large electricity generators
Denault et al. [13]	VaR, copulas	Québec	Hydro power, wind power
Gotham et al. [29]	MVPA[a]	US (Indiana)	Load dispatch
Muñoz et al. [57]		Spain	Renewables
Sunderkötter and Weber [68]	MVPA[a], analytical model for risk characterization	Germany	Peak-load pricing
Westner and Madlener [72]	MVPA[a]	Germany	CHP[e], intra-technology diversification

(continued)

Table 1 (continued)

Study (in chronological order)	Approach/es	Country coverage	Special focus
Zhu and Fan [75]	MVPA	China	Evaluation of medium-term policy planning
Delarue et al. [14]	MVPA[a] QCP[f], inclusion of ramping limits	Belgium	Wind versus conventional power plants, dispatchability
Glensk and Madlener [27]	Fuzzy sets theory; SMAD[g], FSMAD[h], MVPA[a]	Germany	Comparative analysis of portfolio selection models and risk measures
Lang and Madlener [49]	MVPA[a]	Germany	Impact of credit risk mitigation and margining
Madlener and Glensk [52]	MVPA[a]	Germany, Sweden, and UK	Utility portfolio optimization
Pindoriya et al. [60]	MVSPA[i], MOPSO[j]	US (PJM), stylized case study	Assets with non-normal return distribution
Roques et al. [66]	MVPA[a]	Austria, Denmark, France, Germany, Spain	Wind power
Westner and Madlener [73]	MVPA[a]	Germany, Italy, France, and UK	Cogeneration, geographical dispersion
Madlener et al. [53]	Fuzzy theory; SMAD	Germany	Onshore wind
Fuss et al. [25]	MVPA[a], Real options theory	(generic)	Renewables and GHG[k] mitigation
Hickey et al. [30]	Diversity study, Shannon-Wiener index	Continental U.S. and the State of Illinois	Electricity supply portfolio

Source: Own compilation
[a]Mean-variance portfolio analysis
[b]Mixed integer programming
[c]Integrated resource planning
[d]Seemingly unrelated regression estimation
[e]Combined-heat-and-power
[f]Quadratically constrained optimization problem
[g]Semi-absolute mean
[h]Fuzzy semi-mean absolute deviation
[i]Mean-variance skewness portfolio analysis
[j]Multi-objective particle swarm optimization
[k]Greenhouse gases

application for both the state of Illinois and continental U.S., they then demonstrate how the Shannon-Wiener index can be applied to assess electricity supply portfolios (Table 1).

Next, we briefly describe selected studies in the energy economics literature that have applied MVP theory, but in the context of portfolios of energy commodities (or derivatives thereof), rather than power generation assets. The studies covered are summarized in Table 2. The idea is to see what can be learned from these studies for the context of power generation assets.

Table 2 Summary of literature on portfolio optimization of energy commodities (derivatives)

Study	Approach	Country coverage	Comments
Bar-Lev and Katz [7]	MVPA[a]	US	Fossil fuel procurement
Liu and Wu [50]	MVPA[a]	US (PJM)	Spot markets versus bilateral contracts (risk-free asset, *n* risky assets)
Huisman et al. [33]	MVPA[a]	Switzerland	Peak-/off-peak forward contracts, electricity portfolio management
Galvani and Plourde [26]	MVPA[a]	Canada	Futures for crude oil, natural gas and unleaded gasoline
de Oliveira et al. [17]	MVPA[a], MDE[b]	Brazil	Scheduling trades, energy contracts

Source: own compilation
[a]Mean-variance portfolio analysis
[b]Mixture design of experiments

Bar-Lev and Katz [7] was already discussed in Sect. 3.1. Liu and Wu [50] consider risk management in competitive electricity markets through diversification by means of a general quadratic portfolio optimization problem. Their approach considers energy allocation between spot markets and bilateral contracts, with a risk-free asset and *n* risky assets. The methodology is illustrated for historical data from the PJM electricity market.

Huisman et al. [33] apply MVP theory to investigate the optimal allocation of peak- and off-peak forward contracts for electricity of a rational electricity purchaser who aims at hedging both price and volumetric risks. In a two-step procedure the authors show, first, that an optimal (and risk attitude-independent) allocation of peak relative to off-peak contracts is sought, and how they benefit from differences in the relative hedging cost efficiency over both contracts. Second, the exact allocation between peak- and off-peak contracts is chosen depending on the risk appetite. In a related earlier paper, Huisman et al. [32] investigate the hedging exposure to electricity price risk in a Value at Risk (VaR) framework, assuming that purchasers of forward contracts minimize expected costs of adoption, respecting an *ex ante* risk limit (defined in terms of the VaR). The results indicate that a risk-loving actor can achieve lower expected costs, compared to a risk-averse actor. Risk-loving behavior is rewarded by lower expected costs, due to gains from the difference in the hedging efficiency of using base-load and peak-load contracts.

Galvani and Plourde [26], in their investigation on Canadian energy commodities, find that futures for crude oil, natural gas and unleaded gasoline fail to enhance the performance of representative energy stocks in terms of return to risk, but can help to decrease the overall level of risk exposure borne by passive equity investors. They find further that futures contracts on energy commodities are valuable to market participants interested in hedging against price fluctuations in energy markets by means of buy-and-hold hedging strategies. Interestingly, they show that this finding is reversed if one takes the perspective of a trader whose interest is better approximated by return-to-risk bearing. Specifically, the authors conclude that return-to-risk-maximizing agents are unlikely to benefit from trading energy futures in addition to energy stocks. Futures for energy commodities are found to fail to offer significant diversification gains with respect to energy stocks once investors adopt simply dynamic trading strategies that are based on readily available pricing information.

de Oliveira et al. (2011) provide a novel approach for adjusting the Conditional Value-at-Risk (CVaR) metric to the mix of contracts on the energy markets, using the method of Mixture Design of Experiments (MDE) [12, 67]. In this experimental strategy, the relevant design factors are treated as proportions in a mixture system that is considered adequate for testing portfolios. In contrast to the common approach of using linear programming, so-called desirability functions are used instead to combine the multi-response, non-linear objective functions for the mean, with the variance of a specific portfolio obtained through MDE. The maximization of the desirability function is implied in the portfolio optimization, and generates the efficient frontier. The paper contributes to the literature by providing a way to include risk aversion in the optimization routine, by assessing contract interaction, and by showing how to lessen the computational effort needed to solve the constrained non-linear optimization problem. The case study is for the energy market in Brazil.

4 Methodological Issues

In this section we briefly discuss three important methodological issues related to (1) the estimation of the covariance matrix; (2) the use of alternative measures of risk; and (3) the use of alternative measures of return for power generation assets.

4.1 Seemingly Unrelated Regression (SUR) Estimation

Krey and Zweifel [43–45] emphasize the use of the Seemingly Unrelated Regression (SUR) estimation technique to address the problem of time-variant correlation matrices and resulting biases. The problem is the following. Portfolio risk σ_p depends on individual standard errors σ_i of asset (or technology) i and the correlations between the returns, ρ_{ij}. Estimates of the covariance matrix (which consists of σ_i and σ_{ij}) should be reasonably time-invariant. When error terms are significantly correlated across the energy sources considered in portfolio optimization, estimation efficiency can be enhanced by exploiting this information, and ordinary least squares (OLS) estimation is inappropriate. SUR estimation allows to estimate the expected returns of all equations (e.g. the number of electricity generation technologies considered) in one regression, taking care of possible correlation between error terms across the equations investigated.

4.2 Alternative Risk Measures

In recent years, the popularity of alternative risk measures, such as the Mean Absolute Deviation (MAD), the Value at Risk (VaR), the expected shortfall or the semi-variance, has been growing. In this subsection, we discuss some recent developments with regard to the inclusion of different types of risk measures in portfolio analysis. A useful review on this topic is Krokhmal [46], among others.

Popular downside risk models are the Lower Partial Moment, or Expected Regret, in stochastic programming also referred to as the Integrated Change Constraint model (e.g. [8, 16, 69, 70]). In finance a widely used measure is the Value-at-Risk (JP [21, 36, 37]). However, minimization of risk by means of VaR measures has often turned out to be methodologically and technically difficult, due to VaR's notorious non-convexity as a function of the decision variables considered. As a consequence, an alternative yet very similar measure that circumvents the shortcoming of the VaR, is the Conditional Value-at-Risk [47, 61]. Many other measures closely related to the CVaR exist, but their discussion is way beyond the scope of this short exposition. In recent years, there have also been attempts to develop axiomatic approaches for the definition and construction of risk measures, such as the concept of coherent risk measures [2, 16], while the theory of risk measures has been extended to spectral measures [1] and deviation measures [62]. Finally, substantial effort has been made to develop risk-return optimization models and risk measures that are consistent with expected utility maximization. Under certain conditions, it has been shown that MVPA is even consistent with the von Neumann-Morgenstern utility theory [48].

Konno and Yamazaki [41] proposed the MAD model as a linear model for portfolio selection, and applied it to the Tokyo Stock Market. The MAD model is an interesting alternative to the quadratic Markowitz model that yields quite similar optimal portfolios. Obviously, a linear model can be solved more easily than a quadratic one. Also, the MAD model can be applied to tackle large-scale problems where a dense covariance matrix can occur. A further advantage is that it does not require any particular type of return distribution but still encompasses all properties of the MVP approach.

4.3 Alternative Return Measures

Many portfolio optimization studies applied to power generation, especially earlier ones, have used some electric output per unit cost measure (e.g. kWh/US$) for the "return" or productivity of the power generation assets studied and of the portfolio overall. In a liberalized electricity market, however, it can be argued that the consideration of cost measures alone is no longer sufficient, as plant operators aim at maximizing profit (i.e. revenue minus cost).

A natural candidate return measure for both existing and new plants is the net present value (NPV), which is based on the discounted cash flow (DCF) technique.

The cash flows estimated for power plants depend on revenues from power generation on the one hand, and fuel, CO_2 emission (if applicable) and operating costs, capital costs, and depreciation on the other hand. An alternative measure to the NPV is the use of the annual return, which in contrast to the NPV is a static measure. Westner and Madlener [72, 73], for instance, have used both expected net present value (NPV) and expected annual returns (in €/kWh_{el}) in order to compare the results with each other. Finally, Drake and Hubacek [20] use average Megawatt per hour (MW/h) as a measure for the portfolio's return, i.e. a non-monetary value.

5 Conclusion

In this chapter we have discussed the literature on portfolio optimization of power generation assets using mean-variance portfolio analysis (MVPA) and related methodologies. Starting from the development of the state-of-the-art in MVPA used in the context of power generation and trading, both in methodology and application, we also described some new frontiers based on fuzzy theory, the inclusion of non-normal distributions, and single-side risk analysis.

Both energy providers and policy-makers are well advised to avoid single-technology investment decisions, and instead to take into account the correlations between risky power generation technologies, thus aiming for efficient portfolios and thus acknowledging the risk-mitigating benefits accruing from diversification of less than perfectly correlated risks. Given that real assets involved in electrical energy supply features a number of special features that need not be dealt with when dealing with financial assets, portfolio optimization for power generation units and power supply contracts tends to be more complex.

The review shows that the existing literature provides several different approaches concerning the application of MVP theory on power generation portfolios. Overall, a growing and relatively recent interest of both theoreticians and practitioners in this set of problems is revealed. The major differences between the research articles discussed can be found in the criteria used to investigate the portfolio performance. We conclude that the application of portfolio optimization to the field of power generation and related issues in energy sector analysis is a fruitful and promising new field for theoretical and empirical research especially in light of the enormous investment needs in the energy sector, and especially in the electricity system, in order to safeguard the sustainable provision of energy in the years to come.

References

1. Acerbi C (2002) Spectral measures of risk: a coherent representation of subjective risk aversion. J Bank Financ 26(7):1487–1503
2. Artzner P, Delbaen F, Eber J-M, Heath D (1999) Coherent measures of risk. Math Financ 9 (3):203–228

3. Awerbuch S (2000) Investing in photovoltaics: risk, accounting and the value of new technology. Energy Policy 28(14):1023–1035
4. Awerbuch S, Berger M (2003) Applying portfolio theory to EU electricity planning and policy-making. IEA/EET Working Paper No. EET/2003/03, IEA/OECD, Paris
5. Awerbuch S (2006) Portfolio-based electricity generation planning: policy implications for renewables and energy security. Mitig Adapt Strat Glob Chang 11(3):693–710
6. Awerbuch S, Jansen JC, Beurskens L (2008) The Role of wind generation in enhancing Scotland's energy diversity and security: a mean-variance portfolio optimization of Scotland's energy mix. In: Bazilian M and Roques F (eds) op cit., Chap 7, pp 139–150
7. Bar-Lev D, Katz S (1976) A portfolio approach to fossil fuel procurement in the electric utility industry. J Financ 31(3):933–947
8. Bawa VS (1975) Optimal rules for ordering uncertain prospects. Rev Financ Stud 2(1):95–121
9. Bazilian M, Roques F (eds) (2008) Analytical methods for energy diversity and security – portfolio optimization in the energy sector: a tribute to the work of Dr. Shimon Awerbuch, 1st edn. Elsevier, Oxford/Amsterdam
10. Bazilian M, Roques F (2008b). Using portfolio theory to value power generation investments. In: Bazilian M and Roques F (eds) op cit., Chap 3, pp 61–68
11. Borchert J, Schemm R (2007) Einsatz der Portfoliotheorie im Asset Allokations-Prozess am Beispiel eines fiktiven Anlageraums von Windkraftstandorten. Z Energie 31(4):311–322
12. Cornell JA (2002) Experiments with mixtures. John Wiley & Sons, New York
13. Denault M, Dupuis D, Couture-Cardinal S (2009) Complementarity of hydro and wind power: improving the risk profile of energy inflows. Energy Policy 37(12):5376–5384
14. Delarue E, De Jonghe C, Belmans R, D'haeseleer W (2010) Applying portfolio theory to the electricity sector: energy versus power. Energy Econ 33(1):12–23
15. Delbaen F (2002) Coherent risk measures on general probability spaces. In: Sandmann K, Schönbucher P (eds) Advances in finance and statistics: essays in honour of dieter sondermann. Springer, Berlin, pp 1–38
16. Dembo RS, Rosen D (1999) The practice of portfolio replication: a practical overview of forward and inverse problems. Ann Oper Res 85:267–284
17. de Oliveira FA, de Paiva AP, Marangon Lima JW, Balestrassi PP, Amaury Mendes RR (2010) Portfolio optimization using mixture design of experiments: scheduling trades within electricity markets. Energy Econ 33(1):24–32
18. Doherty R, Outhred H, O'Malley M (2006) Establishing the role that wind generation May have in future generation portfolios. IEEE T Power Syst 21(3):1415–1422
19. Doherty R, Outhred H, O'Malley M (2008) Generation portfolio analysis for a carbon constrained and uncertain future. In: Bazilian M and Roques F (eds) op cit., Chap 8, pp 151–166
20. Drake B, Hubacek K (2007) What to expect from a greater geographic dispersion of wind farms? – a risk portfolio approach. Energy Policy 35(8):3999–4008
21. Duffie D, Pan J (1997) An overview of value-at-risk. J Derivatives 4:7–49
22. Elton EJ, Gruber MJ (1997) Modern portfolio theory, 1950 to date. J Bank Financ 21 (11–12):1743–1759
23. Fabozzi G, Gupta F, Markowitz H (2002) The legacy of modern portfolio theory. J Invest, Inst Investor (Fall) 11:7–22
24. Favre-Perrod P, Kienzle F, Andersson G (2010) Modeling and design of future multi-energy generation and transmission systems. Eur T Electr Power 20(8):994–1008
25. Fuss S, Szolgayová J, Khabarov N, Obersteiner M (2012) Renewables and climate change mitigation: Irreversible energy investment under uncertainty and portfolio effects. Energy Policy 40:59–68
26. Galvani V, Plourde A (2010) Portfolio diversification in energy markets. Energy Econ 32(2):257–268
27. Glensk B, Madlener R (2010) Fuzzy portfolio optimization for power generation assets. FCN Working Paper No. 10/2010. Institute for Future Energy Consumer Needs and Behavior, RWTH Aachen University

28. Glensk B, Madlener R (2011) Portfolio selection methods and their empirical applicability to real assets in energy markets. FCN Working Paper No. 11/2011. Institute for Future Energy Consumer Needs and Behavior, RWTH Aachen University, May (in prep.)
29. Gotham D, Muthuraman K, Preckel P, Rardin R, Ruangpattana S (2009) A load factor based mean-variance analysis for fuel diversification. Energy Econ 31(2):249–256
30. Hickey EA, Carlson LJ, Loomis D (2010) Issues in the determination of the optimal portfolio of electricity supply options. Energy Policy 38(5):2198–2207
31. Huang Y-H, Wu J-H (2008) A portfolio risk analysis on electricity supply planning. Energy Policy 36(2):627–641
32. Huismann R, Mahieu R, Schlichter F (2007) Hedging exposure to electricity price risk in a value at risk framework. ERIM Report Series Reference No. ERS-2007-013-F&A (available at SSRN: http://ssrn.com/abstract = 968855)
33. Huisman R, Mahieu R, Schlichter F (2009) Electricity portfolio management: optimal peak/off-peak allocations. Energy Econ 31(1):169–174
34. Humphreys H, McClain K (1998) Reducing the impacts of energy price volatility through dynamic portfolio selection. Energy J 19(3):107–131
35. Jansen JC, Beurskens L (2008) Portfolio analysis of the future dutch generating mix. In: Bazilian M and Roques F (eds) op cit., Chap 6, pp 117–138
36. Jorion P (2006) Value at risk: the benchmark for controlling market risk, 3rd edn. McGraw-Hill, Maidenhead
37. Morgan JP (1994) Riskmetrics. JP Morgan, New York
38. Kienzle F, Andersson G (2008) Efficient multi-energy generation portfolios for the future, paper presented at the 4th annual Carnegie Mellon conference on the electricity industry, Pittsburgh, USA, 2008
39. Konno H, Koshizuka T (2005) Mean-absolute deviation model. IIE T 37(10):893–900
40. Konno H, Suzuki K (1995) A mean-variance skewness optimization model. J Oper Res Soc Jpn 38:137–187
41. Konno H, Yamazaki H (1991) Mean-absolute deviation portfolio optimization model and its application to Tokyo stock market. Manage Sci 37(5):519–531
42. Kotsan S, Douglas S (2008) Application of mean-variance analysis to locational value of generation assets. In: Bazilian M and Roques F (eds) op cit., Chap 13, pp 263–274
43. Krey B, Zweifel P (2006) Efficient electricity portfolios for Switzerland and the United States. Working Paper No. 0602. Socioeconomic Institute, University of Zurich
44. Krey B, Zweifel P (2008a) Efficient and secure power for the USA and Switzerland. In: Bazilian M and Roques F (eds) op cit., Chap 10, pp 193–118
45. Krey B, Zweifel P (2008) The impact of liberalization on the scope of efficiency improvement in electricity-generating portfolios for the United States and Switzerland. Z Energie 32(3):203–209
46. Krokhmal P (2007) Higher moment coherent risk measures. Quantitative Finance 7(4): 373–387
47. Krokhmal P, Palmquist J, Uryasev S (2002) Portfolio optimization with conditional value-at-risk objective and constraints. J Risk 4(2):43–68
48. Kroll Y, Levy H, Markowitz HM (1984) Mean-variance versus direct utility maximization. J Financ 39(1):47–61
49. Lang J, Madlener R (2010) Portfolio optimization for power plants: the impact of credit risk mitigation and margining. FCN Working Paper No. 11/2010, Institute for Future Energy Consumer Needs and Behavior, RWTH Aachen University
50. Liu MK, Wu F (2007) Portfolio optimization in electricity markets. Electr Power Syst Res 77 (8):1000–1009
51. Madlener R, Glensk B, Raymond P (2009) Investigation of E.ON's power generation assets by using mean-variance portfolio analysis. FCN Working Paper No. 12/2009. Institute for Future Energy Consumer Needs and Behavior, RWTH Aachen University

52. Madlener R, Glensk B (2010) Portfolio impact of new power generation investments of E.ON in Germany, Sweden and the UK. FCN Working Paper No. 17/2010. Institute for Future Energy Consumer Needs and Behavior, RWTH Aachen University
53. Madlener R, Glensk B, Weber V (2011) Fuzzy portfolio optimization of onshore wind power plants. FCN Working Paper No. 10/2011. Institute for Future Energy Consumer Needs and Behavior, RWTH Aachen University, May
54. Madlener R, Wenk C (2008) Efficient investment portfolios for the swiss electricity supply sector. FCN Working Paper No. 2/2008. Institute for Future Energy Consumer Needs and Behavior, RWTH Aachen University
55. Markowitz H (1952) Portfolio selection. J Financ 7(1):77–91
56. Markowitz HM (1959) Portfolio selection, 1st edn. Wiley & Sons, New York
57. Muñoz JI, Sánchez de la Nieta AA, Contreras J, Bernal-Augustín JL (2009) Optimal investment portfolio in renewable energy: the Spanish case. Energy Policy 37(12):5273–5284
58. Nitsch J (2008) Weiterentwicklung der Ausbaustrategie Erneuerbare Energien – Leitstudie 2008, Bundesministerium für Umwelt, Naturschutz und Reaktorsicherheit, Reihe Umweltpolitik, Berlin
59. Nguene GN, Finger M (2007) A fuzzy-based approach for strategic choices in electric energy supply: the case of a swiss power provider on the eve of electricity market opening. Eng Appl Artif Intell 20(1):37–48
60. Pindoriya NM, Singh SN, Singh SK (2010) Multi-objective mean–variance–skewness model for generation portfolio allocation in electricity markets. Electr Power Syst Res 80 (10):1314–1321
61. Rockafellar RT, Uryasev S (2000) Optimization of conditional value-at-risk. J Risk 2:21–41
62. Rockafellar RT, Uryasev S, Zabarankin M (2006) Generalized deviations in risk analysis. Stoch Financ 10(1):51–74
63. Roques FA (2008) Technology choices for new entrants in liberalised markets: the value of operating flexibility and contractual arrangements. Util Policy 16(4):245–253
64. Roques FA, Newbery DM, Nuttall WJ (2008) Fuel mix diversification in liberalized electricity markets: a mean-variance portfolio theory approach. Energy Econ 30(4):1831–1849
65. Roques FA, Newbery DM, Nuttall WJ (2008b) Portfolio optimization and utilities' investment in liberalized power markets. In: Bazilian M and Roques F (eds) op cit., Chap 11, pp 219–245
66. Roques FA, Hiroux C, Saguan M (2010) Optimal wind power deployment in europe – a portfolio approach. Energy Policy 38(7):3245–3256
67. Scheffe H (1958) Experiments with mixtures. J Roy Stat Soc B 20(2):344–360
68. Sunderkötter M, Weber C (2009) Valuing fuel diversification in optimal investment policies for electricity generation portfolios. EWL Working Paper No. 6. University of Duisburg-Essen, Germany
69. Testuri CE, Uryasev S (2004) On relation between expected regret and conditional value-at-risk. In: Rachev Z (ed) Handbook of computational and numerical methods in finance. Birkhäuser, Boston, pp 361–373
70. Klein Haneveld WK, Streutker MH, van der Vlerk MH (2010) An ALM model for pension funds using integrated chance constraints. Ann Oper Res 177(1):47–62
71. van Zon A, Fuss S (2008) Risk, Embodied technical change and irreversible investment decisions in UK electricity production: an optimum technology portfolio approach. In: Bazilian M and Roques F (eds) op cit., Chap 14, pp 275–304
72. Westner G, Madlener R (2009) Development of cogeneration in Germany: a dynamic portfolio analysis based on the new regulatory framework. FCN Working Paper No. 4/2009. Institute for Future Energy Consumer Needs and Behavior, RWTH Aachen University, November (revised March 2010)
73. Westner G, Madlener R (2010) The benefit of regional diversification of cogeneration investments in europe: a mean-variance portfolio analysis. Energy Policy 38(12):7911–7920
74. Yu Z (2003) A spatial mean-variance MIP model for energy market risk analysis. Energy Econ 25(3):255–268
75. Zhu L, Fan Y (2009) Optimization of China's generating portfolio and policy implications based on portfolio theory. Energy 35(3):1391–1402

Market Clearing Mechanisms for Efficiently Incorporating Renewable Energy and Mitigating CO_2

Golbon Zakeri and Javad Khazaei

Abstract In recent years there has been a move in the majority of industrialized countries to invest in renewable resources for the production of energy. This move has come about as people worldwide are more aware of negative effects of fossil fuel sources of energy on the environment, including the release of green house gases such as CO_2. Utilization of renewable sources of energy, for instance, harnessing wind power in electricity production, is deemed to be reducing the use of fossil fuels and hence results in the reduction of CO_2. Mechanisms that promote and facilitate utilization of renewable sources of energy are being developed. In particular, recently stochastic programming market clearing mechanisms have been suggested that would seemingly allow for a more efficient use of wind energy hence reduction of fossil fuel use, that ultimately would result in a reduction of CO_2. In this paper we will examine the steady state behaviour of participants in an electricity market to fully analyze the hypothesis that the stochastic programming market clearing mechanism is less fossil fuel (and hence CO_2) intensive than a conventional two settlement market through some simple examples.

Keywords Wind • Electricity • Uncertainty • Stochastic program • Dispatch • Game theory • Market power

1 Introduction

The United Nations Conference on Environment and Development (UNCED), informally known as the Earth Summit (held in Rio de Janeiro from 3 to 14 June 1992) produced an international environmental treaty known as the United Nations Framework Convention on Climate Change (UNFCCC). The objective of this

G. Zakeri (✉) • J. Khazaei
Department of Engineering Science, University of Auckland, Auckland, New Zealand
e-mail: g.zakeri@auckland.ac.nz

Q.P. Zheng et al. (eds.), *Handbook of CO_2 in Power Systems*, Energy Systems, DOI 10.1007/978-3-642-27431-2_13, © Springer-Verlag Berlin Heidelberg 2012

treaty as stated in Article 2 of the treaty is "to achieve stabilization of greenhouse gas concentrations in the atmosphere at a level that would prevent dangerous anthropogenic interference with the climate system. Such a level should be achieved within a time-frame sufficient to allow ecosystems to adapt naturally to climate change, to ensure that food production is not threatened and to enable economic development to proceed in a sustainable manner."

To achieve the ends set out in the UNFCCC, the Kyoto Protocol was adopted in December 1997 and entered into force in February 2005. As of November 2009, the Protocol has 187 signatories. Under the Protocol, the so called Annex I countries, a group of 37 industrialized countries, have committed themselves to the reduction of four green house gases (GHG) that include CO_2. It is widely agreed upon that the primary human source of carbon dioxide in the atmosphere is from the use of fossil fuels in energy production (e.g. using coal fired thermal generators) and transport. Deforestation is also another leading source of CO_2 emissions. Therefore, one potentially substantial way to decrease CO_2 emissions is for countries to disincentivize the use of fossil fuels in the production of energy. To this end, many countries and jurisdictions have imposed, or are about to impose, CO_2 related charges (e.g. the European Union, California and New Zealand) and almost all industrialized countries are investing in renewable resources for the production of energy, primarily electricity. Renewable resources such as wind and solar power are volatile as they are dependent on the cooperation of the elements. The wind must blow before the turbines can produce, and it must be sunny in order to utilize solar photovoltaic panels.

This introduces a tradeoff between the volatility of electricity production and emission reduction. To remedy this volatility it is possible to increase the electricity system requirements of dedicated reserve stations (which can be very expensive), or it can be required that volatile renewable generation comes coupled with "firm" production. A third, smarter alternative has been to seek the aid of stochastic programming (see e.g. [1–3]) to design an economic dispatch mechanism that is hoped to perform more effectively in an uncertain environment. Economic dispatch mechanisms are utilized when clearing the market in electricity pool markets (such as various jurisdictions in the US, Australia, New Zealand, and Scandinavia). We will start with a brief description of the conventional dispatch model used to clear electricity pool markets. In Sect. 3 we present examples that motivate the stochastic programming mechanism for market clearing introduced by Pritchard et al. in [1]. We will then review the stochastic programming market clearing mechanism proposed by Pritchard et al.

Introduction of a new market clearing mechanism is likely to result in change of behaviour for the participants in that market. We will examine the effects of the introduction of the stochastic programming market clearing mechanism on the behaviour of participants and ultimately on two performance indicators for our market. These indicators are the overall expected welfare and the expected use of thermal resources. While overall expected welfare is a conventional benchmark for the economic performance of a mechanism, in the context of CO_2 emissions, the expected use of thermal resources, or the expected use of particular kinds of thermal resources, is a more relevant benchmark. We will provide examples to illustrate that mechanisms designed to encourage harnessing renewable resources can lead to counter-intuitive increase in CO_2 emissions due to the volatile nature of renewable resources such as wind.

2 Review of the Conventional Economic Dispatch Model

In this section, we briefly review the economic dispatch model that is currently used in wholesale electricity pool markets. Typically, in a wholesale electricity market, in each period of the day, each generator offers in generation quantities for each of its plants (possibly located at different grid injection points), at certain prices. In its most general form, the generation offers are supply functions (also known as offer curves) denoted $p = C(q)$, where $C(q)$ is the marginal price of producing quantity q. The functions $C(q)$ are required to be increasing. When a generator owns multiple generation units, the efficient generation of a quantity of electricity is achieved by turning on the least costly unit first then moving to more costly units successively (in absence of any constraints). This is the rationale for the requirement that $C(q)$ be an increasing function. It is also the reason why in most electricity spot markets, $C(q)$ is effectively an increasing step function.

The supply offers are collected by the independent system operator (ISO). The ISO also estimates the demand over that period. The ISO then solves a side constrained network optimization problem where the objective is to minimize the total cost of production of electricity. The constraints of this optimization problem reflect that demand must be met at every node of the network, and that physical flow constraints such as transmission line capacities and Kirchhoff's laws must be complied with. Often reactive power modeling is left out of the ISO's dispatch problem and the problem is, in fact, a direct current equivalent load flow model [4, 5]. A general model for the ISO's economic dispatch problem (EDP) is formulated below.

$$
\begin{aligned}
\text{EDP}: \quad \text{minimize} \quad & \textstyle\sum_i \sum_{m\in\mathcal{O}(i)} \int_0^{q_m} C_m(v)dv \\
\text{s.t.} \quad & g_i(u) + \sum_{m\in\mathcal{O}(i)} q_m = D_i, \quad i \in N, \quad [f_i] \\
& q_m \in \phi_m, \quad m \in \mathcal{O}(i), \qquad i \in N, \\
& u \in U.
\end{aligned}
$$

- We use i as the index for the nodes in the transmission grid.
- We use m as the index for the generators.
- $\mathcal{O}(i)$ indicates the set of all generators located at node i.
- Generator m can supply quantity q_m.

- The demand at node i is denoted by D_i.
- ϕ_m indicates the capacity of generator m.
- The components of vector v measure the dispatch of each generator.
- The components of the vector u measure the flow of power in each transmission line with u_{ik} denoting the flow in the directed line from i to k. By convention, we assume $i < k$.
- f_i denotes the nodal price of electricity for node i and is put in brackets to indicate that it is determined by the optimal duals on the node balance constraints for node i. We follow this notation throughout the paper.

Here we require that u lies in the convex set U, which means that each component satisfies the thermal limits on each line, and satisfies loop flow constraints that are required by Kirchhoff's law. The function $g_i(u)$ defines the amount of power arriving at node i for a given choice of u. This notation enables different loss functions to be modeled. For example, if there are no line losses then we obtain

$$g_i(u) = \sum_{k<i} u_{ki} - \sum_{k>i} u_{ik}.$$

With quadratic losses we obtain

$$g_i(u) = \sum_{k<i} u_{ki} - \sum_{k>i} u_{ik} - \sum_{k<i} \frac{1}{2} r_{ki} u_{ki}^2 - \sum_{k>i} \frac{1}{2} r_{ik} u_{ik}^2.$$

As indicated above, one of the functions of the ISO is to set the price. The price of electricity is determined as the shadow price f_i of the node balance constraints above that indicate demand must be met at all nodes. This price is the system cost of meeting one more unit of demand at node i. This method of determining the electricity price is sometimes referred to as locational marginal pricing (LMP). New Zealand and the PJM market in the US are examples of electricity markets with LMP.

In some electricity markets, such as the England and Wales market, all trading takes place at a single node, and the market runs as a single node market. Clearly, this is also the case if the network constraints are inactive, e.g. in periods where the electricity flow is less than the capacity of the line that it flows on, for every line, and when the losses are negligible enough to ignore. In such situation, the optimization problem EDP reduces to the following single node economic dispatch problem (SNEDP).

$$\begin{array}{llll} \text{SNEDP}: & \text{minimize} & \sum_m \int_0^{q_m} C_m(v)dv & \\ & \text{s.t.} & \sum_m q_m = D, & [f] \\ & & q_m \in \phi_m, & \end{array}$$

Here D denotes the total demand and f is the clearing price of electricity.

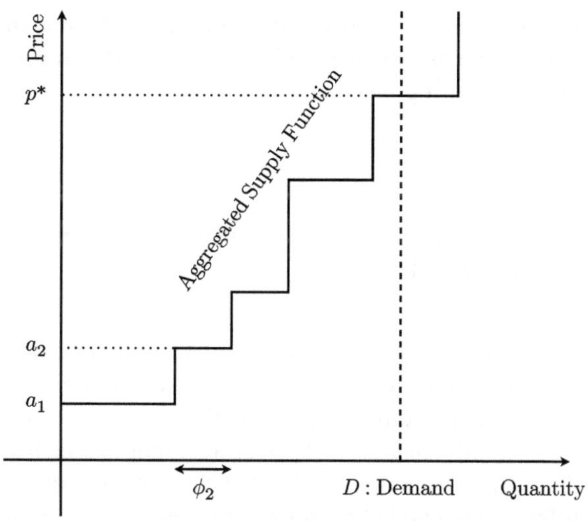

Fig. 1 The simple one-node conventional market clearing mechanism

As alluded to earlier, frequently the supply functions are specified as *offer stacks*. Offer stacks are increasing piecewise constant functions where each piece, often called a tranche, is specified by a quantity and a price. Let a_t and ϕ_t denote the price and the quantity of tranche t respectively. Then the optimization problem SNEDP reduces to simply ordering the available tranches of energy in an increasing order then observing where the total demand D meets this aggregated supply function as depicted in Fig. 1.

For the rest of this paper, we confine our attention to the case of the single node electricity markets. Although the results presented in this paper stand when we have an electricity market over a network (including losses, etc.) it is clearer to see the effects of uncertainty in a single node case. We also bypass anomalies that are introduced through the transmission network in this paper. It is worth noting that Downward [6] examines what can happen to emissions when a CO_2 charge is introduced into an electricity market with a transmission network. Downward starts by proving that when generators offer their production at their true marginal cost, then introducing a CO_2 charge will result in the reduction of emissions. However, when exercise of market power is present, then even in a simple two node network electricity market, with a single transmission line, there are situations in which the introduction of a CO_2 charge can in fact lead to an increase of production of electricity and hence increased emissions. This is because the generators end up with a different cost structure (that now have the embedded CO_2 charges) and their behaviour is different. In Downward's model the introduction of the CO_2 charge results in closer marginal costs for the generators which induces more competition and production in the market. Ultimately, this results in increased CO_2 emissions. This is an example of the general theory of second best that manifests in economics literature frequently dating back to 1950s [7]. Here the introduction of an 'improvement' actually leads to line constraints binding which itself leads to a worse market outcome in steady state. We examine a similar phenomenon in relation to introduction of the stochastic programming market clearing mechanism.

3 Review of the Stochastic Programming Market Clearing Mechanism

In this section, we introduce a stochastic programming economic dispatch model. The central idea here is to view the dispatch problem as having two stages, following the conventional models of two stage stochastic programs [8].

Currently, in wholesale electricity markets such as the NZEM, generators submit their supply functions to the ISO a number of periods prior to the trading period the offers are intended for (in some markets the offers are submitted an entire day ahead of the actual trading period). Before the trading period in question, the ISO solves the conventional dispatch problem EDP using the submitted offers and estimates of demand (for instance, instance expected demand) and proceeds to dispatch generation according to that plan from various generators. During the trading period, various random variables (such as wind farm output and demand) are realized and there will be discrepancies between the planned dispatch of generation, the realized generation (from the wind farms for instance) and actual consumption. Currently, ISOs rely on a dedicated frequency keeping, load following, station to close this difference (at least in short sub-intervals of the trading period prior to a re-dispatch). With the increased volatility introduced through intermittent renewable generation such as wind, this system is unlikely to be sufficient, and it will almost certainly not function efficiently. This will hinder the current policies that preferentially dispatch wind over other forms of generation.

In the stochastic programming mechanism, in stage one, an initial dispatch is computed in advance, with only probabilistic estimates for quantities such as wind generation. This first stage produces a plan of production, which will be modified once the random variables (such as quantity of energy produced from wind) are realized. The second stage then represents "real time" dispatch, i.e. the actual dispatches over a short period (Pritchard et al. intend the duration of this second stage to be somewhere between a minute and an hour, so that it would coincide with a trading period or be a sub-interval of that period). During this second stage, outputs of wind farms take on realizations unknown in the first stage. Adapting to these changes will require modifications to the initial dispatch plan. Therefore, various generators may have to ramp up or down. The formulation of the stochastic program will enable all firms to bid in what are termed "regulations". Through regulation bids, firms indicate whether or not they have the ability to deviate production from the level that was agreed upon in the first stage and how much would this deviation cost. The deviations can be above or below the levels indicated in the first stage production plan therefore the regulation costs are for both up or down deviations when such deviations are possible. The effect of enabling such bids is that the market is opened to all available participants and not restricted to certain designated load following stations. Therefore, it is hoped that through this increased participation, the system can cope with more volatility. Furthermore, as the stochastic programming market clearing mechanism produces the dispatches taking account of the distribution of volatility (for instance in wind production) it will be a more efficient dispatch over a deterministic dispatch model, provided the market participants bid in the same offer curves in both mechanisms.

In the stochastic programming mechanism, the optimization problem aims to minimize expected total cost of generation and regulation in the real time. In addition to the first stage decisions, the stochastic program gives a plan for each of the possible states of uncertainty i.e. scenarios. A general model for the stochastic programming dispatch problem is as follows.

$$
\begin{array}{lll}
\text{SNSP}: & & \\
\min & \sum_{s} \theta_s \sum_{m} \left(\int_0^{q_m + x_{m,s}} C_m(v)dv + \int_0^{x_{m,s}} \rho_m(v)dv \right) & \\
\text{s.t.} & \sum_{m} q_m = Q, & [f] \\
 & Q + \sum_{m} x_{m,s} = D_s, & s \in S, \quad [\theta_s p_s] \\
 & q_m, x_{m,s} \in \phi_m, &
\end{array}
$$

Here we use the same notation as SNEDP for the common parameters between this model and SNEDP. In this model, we have assumed a finite discrete distribution where each scenario is indexed by s. The probability associated with scenario s is denoted θ_s. Let q_m denote the first stage production quantity of generator m; we use $x_{m,s}$ as the deviation in the spot market from q_m if scenario s occurs (this can be positive or negative). The components of vector v measure the dispatch of each generator, and C_m and ρ_m indicate the marginal generation and deviation cost functions of generator m respectively. Note that in this market clearing mechanism, C_m and ρ_m are offered by the generators participating in the market.

Similar to the conventional economic dispatch, SNSP is responsible for determining the nodal prices. The first stage price is determined as the shadow price of the constraint that determines the total generation in the first stage (f). The dual variables p_s, of the constraints that enforce the total generation should meet the demand in the spot market, determine the probability included price of the second stage in scenario s. The payment to generator m, if scenario s happens, is equal to $fq_m + p_s x_{m,s}$. In what follows, we will discuss a simple illustrative example that demonstrates the desired properties of the stochastic programming market clearing mechanism, which are the reasons why they have been advocated in the literature.

Example 1. Consider a simple electricity market over a single node in which there exists a cheap non-thermal (for instance nuclear) and an expensive thermal generator. These generators are labeled G_1 and G_2 respectively and the production cost of G_1 is assumed to be \$20.00 per MW and that of G_2 is assumed to be

\$27.00 per MW. We further assume that our cheap generator G_1 is not easily able to generate more than its pre-dispatch quantity in the spot market. Ramping up or down in spot is comparatively costly for G_1. In contrast, we assume that our thermal generator is flexible and can change its output at a significantly lower cost than G_1 (at \$10.00 vs. \$2.00 for G_2). The generation and ramp costs of these generators are noted in Fig. 2 along with generator capacities (both fixed at \$5.00 MW) and demand. We have assumed that demand is deterministic and fixed at 4 MW. Here the conventional deterministic market clearing will result in a dispatch of 4 MWs from G_1 and nothing from G_2. Total thermal generation is zero in this case.

Now imagine a wind generator is added to the system. This wind generator can produce nothing or two units with the probabilities 0.6 and 0.4 respectively. Figure 3 illustrates this situation.

Consider a market clearing mechanism that dispatches wind generator in all the scenarios (this is following the current New Zealand dispatch policy that prefers the dispatch of wind). Therefore, this example can be represented as Fig. 4 with wind generation taken into account as offsetting demand. As discussed above, the conventional dispatch system solves a pre-dispatch problem for the expected

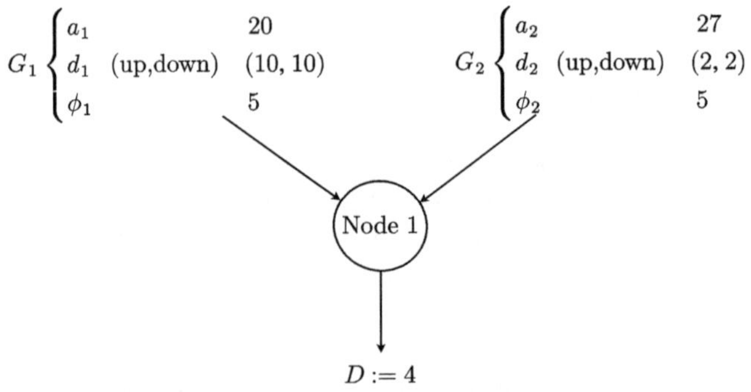

Fig. 2 One-node example market structure

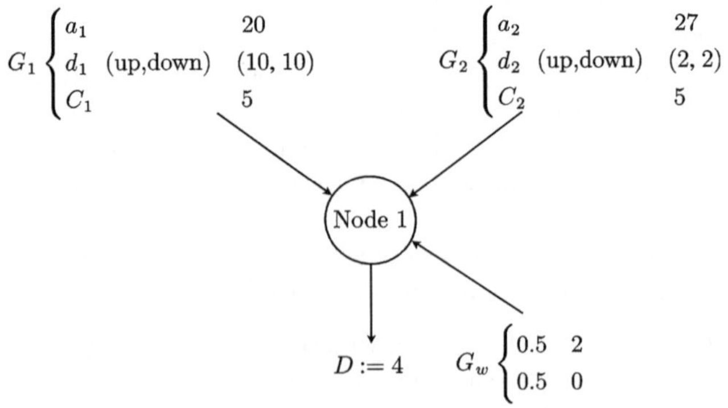

Fig. 3 Market structure after introducing a new wind plant to the electricity market

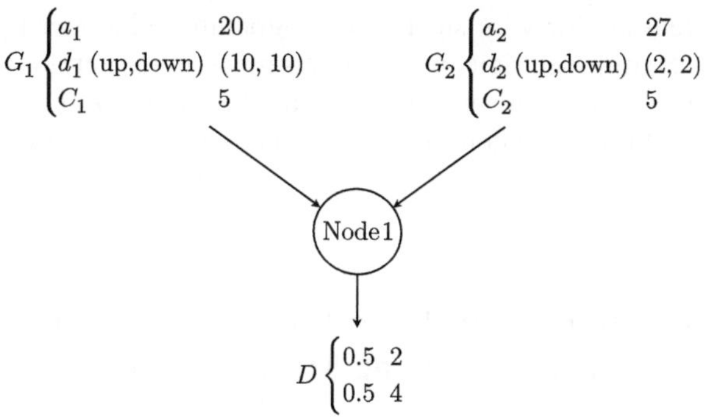

Fig. 4 The equivalent system, after wind introduction

Table 1 Generation quantities and total cost of the conventional market for Example 1

	Pre-dispatch	Scenario 1	Scenario 2	Expected value
G_1	3	3	2	2.6
G_2	0	1	0	0.6
Cost	60	89	50	73.4
Thermal use	0	1	0	0.6

Table 2 Generation quantities and total cost of the stochastic programming market for Example 1

	Pre-dispatch	Scenario 1	Scenario 2	Expected value
G_1	4	4	2	3.2
G_2	0	0	0	0
Cost	80	80	60	72
Thermal use	0	0	0	0

demand (i.e. 3.2 MW). In this case, the pre-dispatch quantity of the nuclear and thermal generators are 3 and 0 MW respectively. When wind generation realizes, the ISO will have to re-dispatch generation. For example, if scenario 2 happens and wind does not blow, one more unit is produced and thermal generator G_2 is dispatched for this amount, as ramp up is expensive for generator G_1. Table 1 contains the dispatch results for this example when the conventional market clearing mechanism is used.

The same problem illustrated in Fig. 3, can be solved using the stochastic programming market clearing mechanism. In this case, the stochastic programming market clearing mechanism pre-dispatches G_1 (the cheaper and cleaner but less flexible generator) 4 MW. G_2 is not pre-dispatched. The second stage decisions in the stochastic programming market clearing determine the adjustments that are needed in each scenario once the volume of generation from wind is realized. Table 2 contains the stochastic programming dispatches along with the corresponding costs. Note that from Tables 1 and 2 it is evident that the stochastic programming approach is more efficient, i.e. it has a lower expected cost. Furthermore, the thermal dispatch is less in the market that is cleared through stochastic programming as compared to the market cleared from the deterministic approach. Both attributes are desirable and reasons for why stochastic programming has been proposed as a mechanism for clearing electricity pool markets involving uncertainty. Nevertheless, introduction of a new mechanism can change the behaviour of the participants in a market, and we will turn our attention to this issue in the next section.

4 Impacts of the Introduction of Stochastic Programming Market Clearing Mechanism in Presence of Market Power

While dispatches acquired through a stochastic programming market clearing mechanism are more efficient than dispatches that are arrived at from the conventional deterministic EDP, the change in the market clearing mechanism may affect the offer behaviour of the market participants. The efficiency gains of the stochastic program over the conventional deterministic EDP are only proven when the offers into both market clearing problems are the same. In order to make the right efficiency comparisons, we must account for possible changes in the offer behaviour of the participants as well. To capture the change in the offer behaviour of participants, in response to a change in the market clearing mechanism, we appeal to the concept of Nash equilibria for describing a steady state behaviour of the participants under different mechanisms. The Nash equilibrium was introduced by Antoine Augustin Cournot in 1883 [9], and subsequently formalized by John Forbes Nash in 1951 [10]. It is defined as the state of the game, given by various players' strategies, where no player has incentive to change his/her strategy given that all other players' strategies remain fixed.

In the following example, we look at another stylized electricity market over a single node. We allow for demand to be price responsive, and we use two scenarios of demand, that similar to our previous examples could correspond to the introduction of an uncertain amount of generation due to wind. We are interested in examining the steady state offer behaviour of the generators when the market clears through a stochastic program as compared to when the market clears in two different settlements, one before and one after the realization of (demand) uncertainty. To make the model tractable, so we can compute equilibria, we restrict our generators to offer linear supply functions into the two settlement market. Here they submit a linear supply function offer stack once, and this same stack is used for any re-dispatch purposes (much like what currently happens in New Zealand). In the stochastic programming case, in addition to the linear supply function, we allow our generators to offer in a deviation cost parameter too. All offers are indicated at the start, and the stochastic program solves for dispatch knowing these parameters.

Let us represent the total amount of generation from generator i in scenario s by $y_{i,s}$ and deviation in the spot market by $x_{i,s}$. The quadratic cost function $\alpha y_{i,s} + \beta y_{i,s}^2$ determines generator i's cost of generation in scenario s, and $\delta x_{i,s}^2$ shows his cost of deviation, in this scenario.

Example 2. Consider a market with two symmetric thermal generators with marginal cost of generation equal to $30 + q$. Also, assume that the cost of deviation in the spot market is zero. Two equally likely states of demand uncertainty exist, which affect the y-intercept of demand function. Demand is elastic and $p_s = Y_s - D$ determines the relationship between spot price (p_s) and demand quantity (D). Also assume $Y_1 = 100$ and $Y_2 = 150$. The stochastic demand here can be thought of as a combination of fixed demand and variable generation from wind that would offset this demand. Figure 5 illustrates this market. Note that $G(\alpha, \beta, \delta)$ determines cost parameters introduced earlier.

If we compute the equilibrium quantities of this market under stochastic programming and the conventional two settlement mechanism, we end up with two different sets of offers. Table 3 shows that while stochastic programming leads to higher expected social welfare, it also increases expected total generation from thermal resources. This increase in the expected total generation from thermal resources then implies an increase in the CO_2 emissions. This is an unintended consequence of introduction of an intermittent renewable resource such as wind coupled with a market clearing mechanism (the stochastic program) that was designed to utilize wind more efficiently. This example serves to warn that change of behaviour of participants in response to a change in market clearing mechanism must be taken into account.

We have demonstrated this concept on a small example. The problems that arise as best response problems of individual generators for the above game, under the stochastic programming market clearing mechanism, are non-convex. Even though

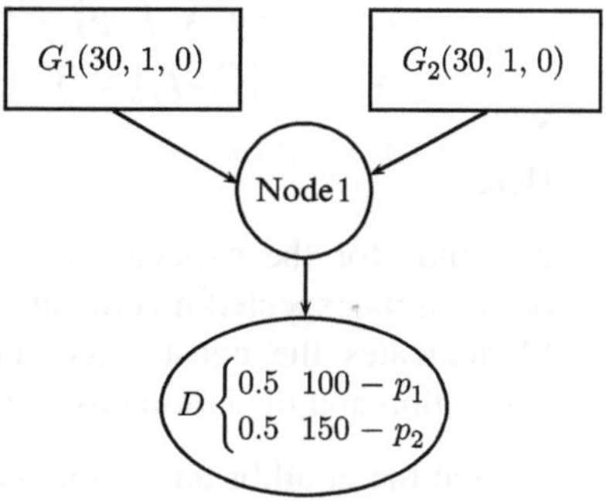

Fig. 5 Generators' cost parameters and demand scenarios for Example 2

Table 3 Social welfare and thermal generation in Example 2

		Social welfare	Thermal generation (each generator)
Stochastic programming	Scenario 1	1,627.34	24.7463
	Expected	3,210.59	31.6561
	Scenario 2	4,793.83	38.566
Two settlement model	Scenario 1	1,585.67	19.3475
	Expected	3,122.81	26.2574
	Scenario 2	4,659.94	33.1672

the size of these problems is relatively small, we have to solve many of these non-convex problems to global optimality before we reach an equilibrium. This process is time and computation intensive and at this stage prohibits solving large examples. Below we outline the full extent of the computational procedure that provides the equilibria of this example.

4.1 Computation of a Nash Equilibrium

For the above example, we have presented Nash equilibria under two different market clearing mechanisms. There are different ways of computing the Nash equilibrium of a game. One approach is to derive conditions for the best response of each player, given all other players' strategies, then find strategies where all of these conditions are satisfied simultaneously. The best response function of a participant determines that player's best strategy as a function of the other players' strategies (as fixed parameters).

For our example, for an electricity market with the stochastic programming clearing mechanism, player i solves the following optimization problem as his/her best response.

$$
\begin{aligned}
\text{BR}[i]\ :\quad & \max_{C_{i,\rho i}} \pi_i = f q_i + \sum_s \theta_s \left(p_s x_{i.s} - \alpha_i (q_i + x_{i.s}) - \frac{\beta_i}{2} (q_i + x_{i.s})^2 - \frac{\delta_i}{2} x_{i.s}^2 \right) \\
s.t \quad & \{\boldsymbol{q}, \boldsymbol{x}, \boldsymbol{f}, \boldsymbol{p}\} \text{ are optimal for SNSP} \\
& \{C_i, \rho_i\} \in \Psi_i
\end{aligned}
$$

Here

- π_i stands for the expected profit function of firm i which is the difference between its expected income and its expected cost of production.
- Ψ_i indicates the constraints imposed by the regulator on offered marginal generation and deviation cost functions of player i i.e. $\underline{C_i}$ and ρ_i.

To find the equilibrium of this game, by simultaneously solving all BR[i] ($\forall i$), we need to be able to derive a set of equations equivalent to each optimization problem BR[i]. Although it is tempting to replace BR[i] by its KKT conditions, note that BR[i] is not a convex optimization problem, due to the presence of some bilinear terms in the objective (such as $f\, q_i$) and quadratic terms in the constraints. Therefore, the KKT conditions are not equivalent to the optimization.

In order to find an equilibrium for the above game we use dynamic best response (fictitious play). The process we use is as follows.

1. The process starts from one (possibly random) strategy for each player.
2. While any player (e.g. j) has incentive to deviate from its current strategy,

   (a) Solve BR[j].
   (b) Update the current strategy vector with the solution of BR[j].

3. The resulting strategy vector is the equilibrium of the game, based on the definition of Nash equilibrium.

As we have already mentioned, the optimization problem BR[j] is not a convex optimization problem. This means that it is necessary to use global optimization to solve this problem. For solving this optimization problem, we have used the global solver of LINGO. The global solver of LINGO guarantees the optimality of its final solution using a branch and bound approach. Here a sequence of piecewise convex relaxations of the original (non-convex) problem are solved. The convex relaxations are derived using bounds on the variables. If the optimal solution of the relaxed problem is feasible for the original problem, it is also the optimal point of the original problem. If not, further enhancement is made through dividing up the domain of the objective function and creating more accurate, piecewise convex functions on each part of the domain. The process of branching continues until are branches end with an optimal point. Note that user defined tolerances on slitting procedure make this method a finite process. For more information about the mathematics behind this global solver see [11]. The tolerance that we have used as the minimum acceptable difference between best response strategies of firms in different turns to continue the process is of order of 10^{-10}.

5 Conclusion

In this article, we have discussed two market clearing mechanisms for settling electricity pool markets that involve uncertainty. The conventional approach is to rely on forecasts or expected values of random variables such as output of a wind farm, and plan dispatches. Only once the uncertainty is resolved does the system operator look for making changes to the plan intact and this may be very costly. In contrast, a smarter new mechanism has been introduced that takes into account the distribution of uncertainty in the future and makes plans with this distribution in mind. The latter mechanism is provably more efficient and desirable if the inputs (i.e. the generator offers and demands) are the same for both models. It should be kept in mind that changing a market clearing mechanism will likely change the offer behaviour of participants. When we closely examine this premise in the context of an example, we find that counter-intuitively the change over to the smarter stochastic programming market clearing can create undesirable effects. Such examples further verify that in markets where the exercise of market power is not ruled out, it is possible that a seeming improvement can lead to worsening of outcomes.

References

1. Pritchard G, Zakeri G, Philpott A (2010) A single-settlement, energy-only electric power market for unpredictable and intermittent participants. Oper Res 58:1210–1219
2. Bouffard F, Galiana FD, Conejo AJ (2005) Market clearing with stochastic security. IEEE Trans Power Syst 20:1818–1826
3. Wong S, Fuller JD (2007) Pricing energy and reserve using stochastic optimization in an alternative electricity market. IEEE Trans Power Syst 22:631–638
4. Schweppe F, Caramanis M, Tabors R, Bohn R (1988) Market operations in electric power systems. Kluwer, Boston
5. Oren SS, Spiller PT, Varaiya P, Wu F (1995) Nodal prices and transmission rights: a critical appraisal. Electr J 8(3):24–35
6. Downward A Carbon charges in electricity systems may increase emissions. Energy J (to appear)
7. Lipseyi RG, Lancaster K (1956) The general theory of second best. Rev Econ Stud 24:11–32
8. Birge JR, Louveaux F (1997) Introduction to stochastic programming. Springer, New York
9. Cournot A (1838) Researches into the mathematical principles of the theory of wealth (originally published in French, trans: Bacon N, 1897). MacMillan, New York
10. Nash JF (1950) Equilibrium points in n-person games. Proc Natl Acad Sci USA 36(1):48–49
11. Lin Y, Schrage L (2009) The global solver in the LINDO API. Optim Method Softw 24 (4):657–668

Stochastic Unit Commitment and Self-scheduling: A Review Considering CO_2 Emission Modeling

José Prina

Abstract Short term power system operation has traditionally been planned using optimization models that represent a cost minimization problem referred to as the unit commitment problem. Computational advances observed from the 1970s have made possible to solve increasingly complex and large scale formulations of this model, which usually are mixed-integer problems. More recently, the restructuring of many electricity markets in several countries has introduced competition among generators, and increased the interest in incorporating uncertainty into the unit commitment problem. Typical sources of uncertainty can be future demand, fuel prices, and unit availability. However, under competition, electricity spot price has also become an important uncertain variable, and has forced electric power generation companies to focus on operation strategies that may lead to the maximization of benefits. Short-term operation planning is not exempted from this shift and this has lead to the self-scheduling problem, where individual companies want to determine the unit commitment that would maximize their benefits. In this article, we review recent advances for the stochastic unit commitment and self-scheduling problems. After reviewing several recent modeling approaches we consider the relevance of environmental considerations, particularly CO_2 emissions, and review relevant propositions for modeling and incorporating emission constraints into the unit commitment problem.

Keywords Unit commitment • Self-scheduling • Stochastic unit commitment • CO_2 emissions • Stochastic programming

J. Prina (✉)
Department of Industrial and Systems Engineering, Pontificia Universidad Católica de Chile, Santiago, Chile
e-mail: jprina@uc.cl

Q.P. Zheng et al. (eds.), *Handbook of CO_2 in Power Systems*, Energy Systems,
DOI 10.1007/978-3-642-27431-2_14, © Springer-Verlag Berlin Heidelberg 2012

1 Introduction

Power system production planning and scheduling is a very complex task. In broad terms, it is possible to distinguish at least three types of planning problems, depending on the time horizon considered. For a time horizon of (roughly) 2–10 (or possibly more) years, we can talk of long-term planning. In this type of planning, operational decisions are relevant but are represented without a high level of detail, while capacity expansion decisions and long-term demand forecasting play an important role. The main objective in this case would be to ensure that the future operation of the system is capable of satisfying energy and power requirements while minimizing the total cost. For a 1-month to 2-year horizon, we can talk of medium-term planning. More detail in modeling should be included, such as unit availability. Monthly energy production plans can be obtained with this modeling, other than ensuring energy supply on a medium term with adequate reliability levels. Finally, short-term planning usually comprises a time horizon from 1 day (or even some hours) to 1 week. Generally, models contain a considerable degree of details, and hourly stages are used.

The classical unit commitment problem consists of finding an operation schedule which optimizes a given objective, commonly operation cost minimization. This schedule indicates which units need to be committed and when. Of course a relevant trade off is that committing more units increases security, but it is more expensive, so it is convenient not to commit more than necessary number of units [1].

A variety of models are possible, depending on the particular constraints taken into account. Typical constraints for thermal plants may include minimum operation level, maximum capacity level, ramp-up and ramp-down limits for each unit, minimum up and down times for each unit, spinning reserve, minimum off and on times. (If there are hydro units, reservoir levels and operational constraints have also to be taken into account.) Formulating this planning problem as a minimum cost problem results normally in a non-linear mixed-integer programming problem. Computational advances observed from the 1970s have made possible to solve increasingly complex and large scale formulations of this problem.

In recent decades, the restructuring of many electricity markets in several countries has introduced competition among generators, and increased the interest in incorporating uncertainty into the unit commitment problem. In addition, auction mechanisms have been used for short-term scheduling. Typical sources of uncertainty can be future demand, fuel prices, and unit availability. However, under competition, electricity spot price has also become an important uncertain variable, and has forced electric power generation companies to focus on operation strategies that may lead to maximization of benefits. Short term operation planning is not exempted from this trend and this increased interest in the self-scheduling problem, where individual companies want to determine the schedule for their units that would maximize their benefits.

In this article, we review recent modeling and algorithmic approaches for the stochastic unit commitment and self-scheduling problems. We present an overview of formulations and solution techniques for short-term unit commitment and self-scheduling models. Given that there is a very broad literature covering these, and related, problems, we restrict the scope to the thermal case.

This article is organized as follows. The next section describes in some details the unit commitment and self-scheduling problems, showing a general formulation. Section 3 briefly discusses typical solution methodologies. Section 4 focuses on emission constraint modeling, particularly CO_2 emissions. Section 5 reviews the stochastic version for the unit commitment and self-scheduling problems. Section 6 contains concluding remarks.

2 Basic Formulation of the Unit Commitment and Self-scheduling Problems

In this section, we present general concepts on unit commitment and self-scheduling problems as optimization models.

2.1 *Unit Commitment*

The traditional unit commitment problem consists essentially of deciding when a unit needs to be turn on and off, and the generation level when it is on. The planning horizon is generally 1 day to 1 week, with hourly time stages.

For a system with n thermal units, a basic formulation for the unit commitment problem as a cost minimization problem over T time periods (usually hours) would be as follows.

$$\min \sum_{t=1}^{T} \sum_{i=1}^{n} c_i(q_{i,t}) \tag{1}$$

$$\text{s.t.} \sum_{i=1}^{n} q_{i,t} \geq d_t, t = 1, ..., T \tag{2}$$

$$q_i^{min} \leq q_{i,t} \leq q_i^{max}, i = 1, ..., n; t = 1, ..., T \tag{3}$$

where $q_{i,t}$ is the output from unit i in hour t, c_i ( ) is the cost function for unit i and d_t is the demand (plus losses) in hour t, q_i^{min} and q_i^{max} are minimum and maximum operation levels for unit i. Here, costs include start-up costs and operation cost, resulting in a discontinuous function.

A variety of operational constraints can be included in the model, depending on the particular interests and needs. Among these are ramp-up and ramp-down limits for each unit, minimum up and down times for each unit, minimum operation level, maximum capacity for each unit, and minimum off and on times (in hours).

Given that constraints such as minimum operation level require the introduction of integer variables, the unit commitment problem can become very difficult to solve. In fact, the integrality constraints may result in a very large state space.

For more details about the traditional unit commitment problem see [1]. Now we deal with the self-scheduling problem.

2.2 *Self-scheduling*

The self-scheduling problem is similar to the unit commitment problem in the sense that both try to find appropriate short-term operation schedules. But, as it was mentioned above, the optimization is now viewed from the company's point of view. That is, the model tries to represent a decision making process where a power producer attempts to determine a feasible operation schedule that results in maximum profits. A major modeling challenge here is capturing the interaction between the company's power generation and the behavior of the rest of the system. Therefore, some types of representation of the rest of the system are needed. The simplest case is when the company is small, relative to the system, and can be reasonably assumed to be a price taker, i.e., its actions do not influence the market price. The case where the company's actions can modify the market price is more complex and requires modeling the market response, or even, in more complex cases, the strategic behavior of market participants could be modeled (using game theoretic tools).

The self-scheduling problem can also include conditions that are particular to a company, such as bilateral agreements or different attitudes towards risk.

3 Solution Methodologies

We briefly discuss typical solution methodologies that have been used to solve unit commitment and self-scheduling problems.

Priority list is possibly the most simple approach to unit commitment problem solving. The first step would be to compute the average cost for each plant. Then, units are sorted in increasing cost. A minimum cost schedule is obtained if units are dispatched from the least expensive, filling the load curve. Clearly, several technical aspects are not considered in this approach.

Dynamic programming (DP) is a very widely used methodology for intertemporal decision making problems. When the problem is deterministic, it is possible to use either forward or backward DP. For the unit commitment problem (and self-scheduling problem), the state of the system, depending on the constraints that are taken into account, would typically be the status of the units (in simple case), but it can also include the number of hours that has been on/off (when minimum up and down times are considered). A possible difficulty, as it is not uncommon with DP, is the dimensionality of the state space, that quite fast can reach levels that make the problem not solvable in practice.

Lagrangian relaxation, which is widely used for integer programming problems, is an alternative. The main step is to relax the original formulation by relaxing the constraints that link all the units. For instance, in the previous formulation (1, 2, 3), if the demand constraint is relaxed, it is possible to obtain a problem formulation that is separable on the generating units. In order to find a solution to the original problem, however, it is necessary to run an iterative scheme in order to determine the value of the multiplier.

It is possible to say that both DP and Lagrangian relaxation are the most common (maybe we can say "traditional") algorithmic techniques for solving the unit commitment problem. They are often combined, being DP the methodology chosen to solve the partial subproblems that originate when applying Lagrangian relaxation. These same methodologies can be applied to the solution of the self-scheduling problem.

In addition to the methodologies mentioned above, in Sheble et al. [2], exhaustive enumeration, integer and mixed-integer programming methodologies, branch-and-bound, linear programming, network flow models are also mentioned. Other possible approaches are heuristic and evolutionary methodologies. Among these, simulated annealing, genetic algorithms, and hybrid algorithms have been applied. This section does not intend to be a detailed review of solution methods. For the interested reader, a recent review of solution methodologies can be found in Padhy [3].

Next section specifically treats modeling approaches to incorporate CO_2 emission constraints into the unit commitment problem.

4 Modeling CO_2 Emission Constraints

Given the increasing relevance and impact of environmental considerations in the electric power generation industry, unit commitment models also have to take into account the impact of fossil fuel burning on the environment. Normally, coal, gas, and diesel combustion produce gases such as SO_2, NO_x and CO_2. In this section we review modeling alternatives that have been proposed to incorporate the emission effect into the unit commitment problem. In general, we keep the original notation employed by the authors.

A unit commitment model with emission considerations is proposed in Wang et al. [4]. The authors formulate and solve a unit commitment problem where emission constraints are included. The constraint is generic (for any kind of emission) and of one type, but can be adjusted to several emissions, being CO_2 one of the possibilities. The constraint can be written as

$$\sum_{t=1}^{N_t} \sum_{i=1}^{N} H_i(P_i(t))I_i(t) \leq EM,$$

where N_t is the total study time in hours, N is the number of units, $H_i()$ is the emission function of unit i, $P_i(t)$ is the power output of unit t in hour t, I_i (t) is a binary variable that is equal to 1 if unit I is on at hour t, and 0 otherwise, EM is an emission cap for the system emission output in the study horizon. The problem formulated corresponds to a short term generation scheduling problem and the solution method was an augmented Lagrangian relaxation algorithm (augmentation improves convexity of the objective function).

Gjengedal [5] addresses environmental objectives in short-term operation decision making, specifically the unit commitment. The paper presents a short-term unit commitment approach that incorporates environmental externalities into electric utility operations decision making. The objective is to determine least cost solutions for achieving daily or weekly emission targets. The problem formulation includes usual unit commitment technical constraints, and addresses variable emission during start-up, operation, and shut-down of generating units. The problem is solved using a Lagrangian relaxation-based algorithm.

Fossil fuel plants pollute the air during normal operation, start-up, ramping, and shut-down. The emission rate during start-up and up-ramping, and part-load operation deviate considerably from the levels of steady state full load operation. The model described in the article is a unit commitment model with emission constraints added to the formulation. Conceptually it can be summarized as follows.

$$\min \quad \text{total operation cost of system} \tag{4}$$

$$\text{s.t.} \quad \text{operational constraints} \tag{5}$$

$$\text{emission constraints} \tag{6}$$

More explicitly, the objective function is given by:

$$C_{tot} = \sum_{i=1}^{N} \sum_{t=1}^{T} (C_{it} + C_{up,i} + C_{d,i}) \tag{7}$$

where N is the number of units, T is the time horizon (in hours), C_{tot} is the total system operation cost, C_{it} is the operation cost of generator i in time interval t, $C_{up,i}$, is the start-up cost of unit i, and $C_{d,i}$ is the stop cost for unit i.

The fuel cost for unit i is built from the heat rate function, which is represented as a quadratic function:

$$C_{it} = \psi_i \ (k_{0i} + k_{1i}P_{it} + k_{2i}P_{it}^2)$$

where ψ_i is fuel price for unit i, k_{oi}, k_{1i}, and k_{2i} are polynomial heat rate coefficients for unit i, and $P_{i,t}$ is generation of unit i in time interval t.

The start-up cost of unit i is represented as

$$C_{up,i} = C_{0,i} + C_{1,i} \ (1 - e^{T_{d,i}/\tau_i})$$

where $C_{0,i}$ and $C_{1,i}$ are start cost coefficients, and $T_{d,i}$ is down time of unit i. The stop cost of unit i is considered constant, equal to $C_{d,i}$.

The emission function (valid both for SO_2 and CO_2) is given by

$$E_{jit} = e_{fij}(k_{0i} + k_{1i} \ P_{it} + k_{2i}P_{it}^2)$$

where E_{jit} corresponds to emissions of pollutant j from generator i in time interval t, e_{fij} is an emission factor (dependent on plant design parameters, fuel quality, and emissions type).

NO_x emissions are represented using a cubic equation having P_{it} as a variable.

Start-up emissions of unit i are given by the following exponential approximation.

$$E_{up,j,i} = e_{0,j,i} + e_{1,j,i}(1 - e^{T_{d,i}/\tau_i})$$

where $E_{up,j,i}$ is the emission of pollutant j during start-up of unit I,$e_{0,j,i}$ and $e_{1,j,i}$ are start-up coefficients (for pollutant j and unit i).

Emissions of pollutant j during shut-down of unit I are represented by a constant, $E_{d,j,i}$. Emission constraints in this case restrict entire system emission over the total scheduling period.

$$\sum_{i=1}^{N} \sum_{t=1}^{T} E_{jit} + \sum_{i=1}^{N} [E_{up,j,i} + E_{d,j,i}] \leq E_{j,tot}, \ j = 1, ..., J \tag{8}$$

where $E_{j.tot}$ is total emission of pollutant j from electric power production (kg).

System constraints are system load, and spinning reserve requirements. Additionally, other unit constraints are minimum and maximum generation limits, minimum up- and down-times, ramp rate limits, prespecified commitment and/or production status. The environmentally constrained unit commitment is a large nonlinear mixed-integer optimization problem.

The algorithm used to solve is based on Lagrangian relaxation. The emission constrained unit commitment is solved by a modified version of a two level primal-dual approach. Following steps similar to a Lagrangian relaxation, the algorithm was successfully applied on several cases. Results presented are for a 79-unit system.

In Catalao et al. [6], the authors take a different approach to restrict emissions during operation. Keeping in mind that the traditional unit commitment objective is to minimize fuel cost, and a minimum level of emissions would be also desirable, the authors incorporate emissions into the objective function. However, it is not possible to minimize fuel cost and emissions simultaneously. A multiobjective optimization is proposed to deal with this problem. This approach would allow to obtain "compromise solutions", also known as "non-inferior Pareto-optimal solutions".

The formulation would lead to the construction of trade-off curves between fuel cost and pollutant emissions criteria. Then, the decision-maker can select a point on the trade-off curve according to his/her judgement. The resulting problem is a dynamic mixed-integer non-linear mathematical programming problem.

The objective function can be written as

$$f(x,u,p) = \sum_{k \in K} \sum_{i \in I} C_{ik}(x_{ik}, u_{ik}, p_{ik}) \tag{9}$$

Here, k is the total number of hours in the scheduling time horizon, I is the total number of thermal units, C_{ik} is the total cost of unit i in hour k,x_{ik} is the unit state, u_{ik} is a discrete (binary) variable, p_{ik} is power production. Operation constraints in this model include hourly generation requirements, i.e. supply and demand balance, minimum power production requirement, and maximum emission by a group of units over the scheduling time horizon. A set of state equations describe the relation between $x_{i,k+1}$ and $x_{i,k}$.

The total cost of thermal unit i in hour k is given by:

$$C_{ik}(x_{ik}, u_{ik}, p_{ik}) = C_{ik}^{sc}(x_{ik}, u_{ik}) + C_{ik}^{fc}(u_{ik}, p_{ik})$$

The first term on the right hand side is the start up cost, which is considered constant; the second term is the fuel cost, which is a quadratic function, and is given by:

$$u_{ik}(\alpha_i + \beta_i p_{ik} + \gamma_i p_{ik}^2)$$

Where α_i,β_i, and γi, are coefficients.

The authors use the following expression for the pollutant emission by fossil-fueled thermal units

$$E_{ik}(u_{ik}, p_{ik}) = u_{ik}(a_i + b_i p_{ik} + c_i p_{ik}^2)$$

Where a_i, b_i, and c_i are the emission coefficients for thermal unit i. In order to obtain a compromise solution, the objective function consists of a convex combination of unit costs and emissions.

$$f(x,u,p) = w \sum_{k \in K} \sum_{i \in I} C_{ik} + (1-w)\lambda \sum_{kK} \sum_{i \in I} E_{ik}$$

Here, w is a weighting factor between 0 and 1, and λ is a scaling factor. The procedure consists of generating a set of solutions by varying the weighting factor over a range. This is combined with a so called ε-constraining method, which consists of constraining some terms of the objective function by allowable levels so that

$$\sum_{k\in K}\sum_{i\in I} C_{i\kappa}^{fc} \leq \epsilon_C^{req} \text{ and}$$

$$\sum_{k\in K}\sum_{i\in I} E_{ik} \leq \epsilon_E^{req}$$

Results from a case study with 11 thermal units and a time horizon of 168 h are presented. Higher emissions lead to lower cost and vice versa. The relation between these quantities is illustrated through trade-off curves that show percentage decrease in emission versus percentage increase in cost.

A previous work by the same authors, Mendes et al. [7], formulates a similar model but in this case the objective is to maximize the profits from energy trading, written as income minus operation cost. Since the electricity price is considered as known, this approach can be considered as a self-scheduling model for a price taker. The objective function is given by

$$f(x,u,p) = \sum_{k\in\kappa}\sum_{i\in I} f_{ik}(x_{ik},u_{ik},p_{ik})$$

and represents the sum of profits over a given time horizon, where profits in period k are given by

$$f_{ik}(x_{ik},u_{ik},p_{ik}) = \lambda_k p_{ik} - C_{ik}(x_{ik},u_{ik},p_{ik})$$

The resulting problem is a non-convex large-scale mixed-integer problem, which is solved using a Lagrangian relaxation approach combined with dynamic programming. A case study with 11 thermal units and a scheduling horizon of 168 h is presented. Results for the usual unit commitment (without emission constraints) are compared with the case with emission constraints. As a conclusion, it was observed that restricting emissions results in more units dispatched at a higher cost (but with less total emissions).

A different approach for incorporating carbon dioxide emission constraints in the solution of the short-term unit commitment problem is proposed in Simopoulos et al. [8]. In this case, the existence of an emission allowances market is assumed, which allows to restrict emissions by forcing the power companies to purchase additional emission allowances in case the total system emissions exceed a predefined limit. This is achieved by adding to the objective function the purchase cost of additional emission allowances.

Generally the unit commitment objective, which is to be minimized, corresponds to total operating cost during a given scheduling horizon, comprised by fuel cost and start-up cost. In Simopoulos et al. [8], emission costs are added to the objective.

As it is normally done, fuel cost is assumed to be represented adequately by a quadratic function. In addition to "typical" constraints (such as power balance, spinning reserve requirements, minimum capacity of units, minimum up and down time), a system emission constraint is included in the model.

If $ST_{i,t}$ represents the state of unit i at hour t (1, if unit is on; 0, otherwise) and $EMSS_{i,t}(P_{i,t})$ corresponds to the emissions produced by unit i at hour t when the generation level is $P_{i,t}$, the total emission constraint is given by

$$\sum_{t=1}^{T}\sum_{i=1}^{N} ST_{i,t}EMS_{i,t}(P_{i,t}) \leq EMCAP$$

where $EMCAP$ is a system emission cap, T is the number of time periods, and N is the number of units. The basic assumption is that the system emissions are constrained by a known emission cap that depends on the emission allowances that have been previously obtained by the system. The method proposed allows to violate the emission constraint, and in that case additional emission allowances can be purchased in order to overcome the shortfall in existing emission rights.

CO_2 emission produced by unit i at hour t is computed using the following equation:

$$EMS_{i,t}(P_{i,t}) = FUEL_{i,t}(P_{i,t})NCV_iEF_iOF_i$$

where $FUEL_{i,t}$ is the fuel consumed by unit i at hour t, NCV_i is the net calorific value of the fuel, EF_i is the emission factor of unit i, and OF_i is the oxidation factor of unit i (which is included because of incomplete oxidation of the carbon to CO_2 during the combustion process).

Moreover,

$$FUEL_{i,t}(P_{i,t}) = \frac{FC_{i,t}(P_{i,t})}{FSC_i}$$

where $FC_{i,t}(P_{i,t})$ is the fuel cost of unit i in hour $t(\$/h)$ and FSC_i is the fuel supply cost in $\$/ton$ or $\$/m^3$. Then,

$$EMS_{i,t}(P_{i,t}) = CF_i(A_iP_{i,t}^2 + B_iP_{i,t} + C_i)$$

where

$$FC_{i,t}(P_{i,t}) = A_iP_{i,t}^2 + B_i\,P_{i,t} + C_i$$

is the fuel cost (A_i, B_i, C_i are coefficients).

$$CF_i = \frac{NCV_i\, EF_i\, OF_i}{FSC_i} \tag{10}$$

where $FC_{i,t}(P_{i,t})$ is the emission conversion factor of unit i ($tCo_2/\$$).

The emission cost added to objective function is given by

$$EMC = (EMS - EMCAP)EMP.$$

This cost is added to the objective function whenever $EMS > EMCAP$, where EMP is the price of additional emissions allowances which must be purchased. This cost is 0 in other case.

The solution methodology is simulated annealing, a global optimization metaheuristic that in many cases is able to solve large combinatorial problems.

This methodology was tested successfully on a system with 26 units for a scheduling horizon of 24 h. Several cases were considered, fixing the percentage of emission reduction target and based on a base case study (without emission constraint) and varying the price of emission allowances. The results indicate that the price of the CO_2 emission allowances has a major effect on the final solution of the UC problem.

In the next section we review recent models for the stochastic unit commitment problem and the self-scheduling problem.

5 Recent Advances in Stochastic Unit Commitment and Self-scheduling

In this section, we review several recent works that propose formulations and solution methods for the stochastic unit commitment and self-scheduling problems. The stochastic version of the unit commitment and self-scheduling problems have attracted researchers's interest during the past decade, motivated, among other reasons, by the competition introduced in electricity markets.

We first consider the unit commitment problem and, subsequently, the self-scheduling problem.

5.1 *Stochastic Unit Commitment Modeling*

One of the earliest references that present a stochastic optimization approach to the unit commitment problem is Takriti et al. [9]. The authors develop a method for solving unit commitment problems having demand as a quantity subject to uncertainty. The motivation is that the inclusion of uncertainty improves the capacity of the model of representing actual observations, and should allow to achieve savings with respect to models that assume everything known with certainty.

The operation of n generating units over a horizon of T time periods (24-h horizon may be sufficient) is considered. The cost function $f_i(x^i_t)$ is assumed to be a convex quadratic function of x^i_t (level of unit i in stage t). In order to include start-up costs and shut-down costs, the function is modified to $f_i(x^i_t, u^i_{t-1}, u^i_t)$ which gives a non-continuous (and non-convex) function.

The general formulation is

$$\min_{x,u} \sum_{i=1}^{n} \sum_{t=1}^{T} f_i(x^i_t, u^i_{t-1}, u^i_t) \tag{11}$$

$$\text{s.t.} \quad \sum_{i=1}^{n} x^i_t \geq d_t \tag{12}$$

$$\text{other technical constraints} \tag{13}$$

where d_t is the total demand for electricity during period t.

In order to represent demand uncertainty, the authors use S demand scenarios. Each scenario has probability P_s, $s = 1, \ldots, S$. Then, the objective function becomes the expected value of the cost with respect to the scenario distribution. It is important to mention that non anticipativity constraints are common (and necessary) in multistage stochastic programming problems. The purpose is to make sure that decisions are made in every stage only with available information and that decisions over common portions of paths that reach different nodes are the same.

The resulting model is a large-scale mixed-integer quadratic programming problem.

In Takriti et al. [10], the authors present a stochastic unit commitment model that incorporates power trading and uncertainty in electricity demand and spot prices. It considers fuel constraints and prices (that may vary with electricity prices and demand). The resulting model is a mixed-integer program. The solution approach combines Lagrangian relaxation and Benders decomposition. Problems with 729 demand scenarios are solved to within 0.1% of the optimal solution in less than 10 min. One of the main observations is that it is possible to reach savings when spot market is considered and when a stochastic policy is used, instead of a deterministic one.

We show the general modeling presented, emphasizing the contrast between a deterministic formulation, that follows, and a stochastic version, shown below.

$$\min_{u.z} \quad \sum_{i=1}^{N} \sum_{t=1}^{T} \left(g_i(z^i_t) + c_i u^i_t + h_i(u^i_{t-1}, u^i_t)\right) \tag{14}$$

$$\text{s.t.} \quad \sum_{i=1}^{N} z^i_t \geq d_t, t = 1, \ldots, T \tag{15}$$

$$\sum_{i=1}^{N} Q_i u^i_t \geq d_t + r_t, \ t = 1, \ldots, T \tag{16}$$

$$q_i u_t^i \leq z_t^i \leq Q_i u_t^i,\ i = 1, ..., N; t = 1, ..., T \tag{17}$$

$$u_t^i - u_{t-1}^i \leq u_\tau^i, \tau = t+1, ..., \min\{t + L_i - 1, T\},\ t = 2, ..., T \tag{18}$$

$$u_{t-1}^i - u_t^i \leq 1 - u_\tau^i, \tau = t+1, ..., \min\{t + l_i - 1, T\},\ t = 2, ..., T \tag{19}$$

where d_t is the estimated load on period t ($t = 1, \ldots, T$), $u^i{}_{,t}$ is equal to 1 if unit i is on in period t, and 0 otherwise, $z^i{}_t$ is the amount of power generated by unit i during time period t, $g_i(z^i{}_t) + c_i\, u^i{}_{,t}$ is the cost of generating $z^i{}_t$ on period t with unit i. The operating range is from q_i to Q_i for unit i. r_t is a spinning reserve requirement. Constraints (18) and (19) refer to minimum up and down times. N can be normally in the range of 10–200, and T is usually 168 h (1 week).

As we mentioned, the stochastic model incorporates uncertain electricity spot-market prices and load uncertainty. Also, the cost function of generation units are allowed to change with fuel prices. Nodes in scenarios then have a (load, spot price) vector associated. The spot market is represented as two generating units (one representing buying and the other selling). An important assumption is that buying and selling activities do not affect spot prices.

(A summarized version of) the stochastic programming model is as follows.

$$\min_{u,z} \sum\nolimits_{s=1}^{S} p_s \sum\nolimits_{i=1}^{N+2} \sum\nolimits_{t=1}^{T} \left[g_t^{s,i}(z_t^{s,i}) + c_t^{s,i} u_t^{s,i} + h_t^{s,i}(u_{t-1}^{s,i}, u_t^{s,i})\right] \tag{20}$$

$$\text{s.t.} \qquad \sum\nolimits_{i=1}^{N+2} Z_t^{s,i} \geq d_t^s \tag{21}$$

$$u \in U, z \in Z(u) \tag{22}$$

Here, it is assumed that non anticipativity constraints are included in sets U_i and $Z_i(u^i)$.

Shiina and Birge [11] present a stochastic unit commitment model where uncertainty in demand is taken into account using demand scenarios.

Rebennack et al. [12] present a mid-term stochastic scheduling model for a hydro-thermal power system subject to CO_2 emission quotas. A CO_2 emission allowance is considered to be given over a fixed time horizon. Clearly, there is a tradeoff between using CO_2 quotas earlier, perhaps lowering operation costs on the short term, and deferring their use in order to prevent exceeding the quota, given that there is a fine for exceeding the emission allowance. In order to consider this factor, an emissions reservoir model is presented, where the CO_2 allowance usage is analog to the use of an amount of water initially stored in a reservoir. The objective is the minimization of expected operational costs over a time horizon. Operation costs consist of variable costs of thermal plants and fees paid if emission quotas are exceeded. The model assumes that for each thermal plant CO_2 is emitted at a constant rate (in *tons/MWh*) when the plant is on. The authors use this approach for incorporating emission quotas into a stochastic dual dynamic programming (SDDP) algorithm. A case study based on an actual electric power system is presented.

The next subsection focuses on self-scheduling models, where scheduling decisions are not centralized.

5.2 Self-scheduling Models

Since many markets have been restructured, introducing competition at the generation level, the focus of short term production planning has somewhat changed from minimization of costs to maximization of profits. Self-scheduling follows the idea of scheduling in the short term, but from the point of view of the generating company. That is, now the objective is the maximization of benefits.

Models that arise in this context many times only consider the power plants for a single company that is facing the market. Then, spot price modeling becomes crucial. One alternative is to simply assume that the company acts as a price taker, and then it is possible to build an exogenous model for the spot price behavior and use it in the optimization. Another option would be to assume an oligopolistic behavior, with the actions of the power producer influencing the price. In this case, the challenge is to represent the response of the other market participants. This can be attempted using some kinds of function, or trying to model the strategic behavior of other market participants using game theoretic tools.

A self-scheduling model with stochastic prices, multiple power plants, and allowing hourly adjustments is presented in Shresta et al. [13]. The authors assume that the price can only take three different values in each time stage in order to avoid a large state space.

Yamin and Shahidehpour [14] discuss self-scheduling of power plants considering forecasted market prices at different locations (buses) in the system.

Recently, Cerisola et al. [15] describe a unit commitment model, with a 1 week horizon, for a generation company that participate in an electricity spot market with two daily market mechanisms: a day-ahead market and an adjustment market. The objective is the maximization of profits of a power producer. A residual-demand function is used to model the influence of the decisions of the producer on the market price of electricity. This function corresponds to the amount of energy that the company can sell at different price levels.

Uncertainty is modeled using a scenario tree. Spot market scenarios are represented as residual-demand curves. The resulting model is a large-scale mixed-integer programming problem. In order to solve it, the authors use Lagrangian relaxation and two variants (proposed by the authors) of Benders decomposition.

Among the modeling features we can mention that the inverse residual-demand function (which is the clearing price) and the revenue function are approximated as piecewise-linear functions. Also, the hourly production cost of thermal units is approximated by a linear function. Minimum output, maximum capacity, and up and down ramp-rate limits are included as constraints. Minimum number of hours off line and on line are also taken into account. The authors illustrate the results of the model using a fictitious company.

The next section contains some concluding remarks.

6 Concluding Remarks

In this article we have attempted to summarize the most recent modeling approaches and solution methodologies for the stochastic unit commitment and self-scheduling problems. Special relevance has been given to CO_2 emission constraints in the context of the unit commitment problem.

We have seen that CO_2 emission modeling has been developed essentially for deterministic unit commitment models. Hence, CO_2 emission constraints in the context of stochastic unit commitment appears as an attractive research area. It appears that special attention is going to be needed in dealing with non average situations. For instance, if multiple scenarios are employed in a model, cases with extremely high level of emissions are likely to be problematic, even though the optimal operation schedule indicates that on average emission levels lie in an acceptable range.

As a counterpart to CO_2 emission modeling, other future modeling challenges in the context of the stochastic unit commitment will be the incorporation of renewable energy sources, which allow to decrease the emission levels but undoubtedly will bring another source of uncertainty that has to be taken into account.

References

1. Wood AJ, Wollenberg B (1996) Power generation, operation, and control, 2nd edn. Wiley, New York
2. Sheble GB et al (1994) Unit commitment literature synopsis. IEEE Trans Power Syst 11:128–135
3. Padhy N (2004) Unit commitment – a bibliographical survey. IEEE Trans Power Syst 19(2):1196–1205
4. Wang SJ, Shahidehpour SM, Kirschar DS, Mokhtari S, Irisarri GD (1995) Short-term generation scheduling with transmission and environmental constraints using an augmented Lagrangian relaxation. IEEE Trans Power Syst 10:1294–1301
5. Gjengedal T (1996) Emission constrained unit-commitment (ECUC). IEEE Trans Energy Convers 11(1):132–138
6. Catalao J, Mariano S, Mendes V, Ferreira L (2005) Unit commitment with environmental considerations: a practical approach. In: 15th PSCC, Liege, 22–26 Aug 2005
7. Mendes VMF, Mariano SJPS, Catalao JPS, Ferreira LAFM (2004) Emission constraints on short-term schedule of thermal units. In: 39th UPEC proceedings, Bristol, Sep 2004, pp 1068–1072

8. Simopoulos DN, Giannakopoulos YS, Kavatza SD, Vournas CD (2006) Effect of emission constraints on short-term unit commitment. In: Proceedings of IEEE MELECON 2006, 16–19 May, Benalmádena (Malaga), pp 973–977
9. Takriti S, Birge JR, Long E (1996) A stochastic model for the unit commitment problem. IEEE Trans Power Syst 11(3):1479–1508
10. Takriti S, Krasenbrink B, Wu LSY (2000) Incorporating fuel constraints and electricity spot prices into the stochastic unit commitment problem. Oper Res 48:268–280
11. Shiina T, Birge J (2004) Stochastic unit commitment problem. Int Trans Oper Res 11:19–32
12. Rebennack S, Flach B, Pereira MVF, Pardalos P (2011) Stochastic hydro-thermal scheduling under CO_2 emissions constraints (Accepted for publication in IEEE Trans Power Syst)
13. Shresta GM, Song K, Goel L (2004) An efficient stochastic self-scheduling technique for power producers in the deregulated power market. Electr Power Syst Res 71:91–98
14. Yamin HY, Shahidehpour S (2004) Self-scheduling and energy bidding in competitive electricity markets. Electr Power Syst Res 71:203–209
15. Cerisola S et al (2009) Stochastic power generation unit commitment in electricity markets: a novel formulation and a comparison of solution methods. Oper Res 57(1):32–46

CO_2 Capture: Integration and Overall System Optimization in Power Applications

Luis M. Romeo

Abstract It is generally accepted that CO_2 capture and storage technologies (CCS) will play an essential role in the reduction of greenhouse gases emission in a medium-large term. Despite the research efforts devoted to the development of more efficient capture processes, two of the main challenges of CCS are the efficiency penalty caused by the CO_2 separation, compression and conditioning, and the economic cost. Consequently, the minimizations of the energy requirements and/or the CO_2 avoided cost are the research priorities for the future implementation of CCS technology.

The objective of this chapter is to describe some examples of minimizing the CO_2 avoided cost in several applications of CCS. The first example illustrates a preliminary analysis for the selection of the appropriate option to overcome the energy requirement for regeneration in an amine scrubbing CCS application. The second case presents a problem for minimizing CCS cost depending on several operational variables in an emerging and promising option for CO_2 capture. The last example shows a formal optimization problem with a different objective function, minimizing the cost penalties associated to CO_2 compression. It is concluded that optimization will provided essential information to select the adequate process layout and the proper operational variables supported by the concepts of the Second Law Analysis of Thermodynamics.

Keywords CCS • CO_2 avoided cost • Process optimization • Amine scrubbing • Calcium looping • CO_2 intercooling compression

L.M. Romeo (✉)
Centro de Investigación de Recursos y Consumos Energéticos (CIRCE), Universidad de Zaragoza, Centro Politécnico Superior, Mariano Esquillor, 15, 50018 Zaragoza, Spain
e-mail: luismi@unizar.es

Q.P. Zheng et al. (eds.), *Handbook of CO_2 in Power Systems*, Energy Systems, DOI 10.1007/978-3-642-27431-2_15, © Springer-Verlag Berlin Heidelberg 2012

Nomenclature

CAPEX	Capital costs
CB	Carbonator
CCS	Carbon capture and storage
CFB	Circulating fluidized bed
Ci	Cost indexes
CL	Calciner
COE	Cost of electricity
CO_2kWh^{-1}	Specific CO_2 emissions
EC	Energy cost
EP	Electricity production
ER_i	Energy requirements
F	Annuity factor
GT	Gas turbine
HRSG	Heat recovery steam generator
IPCC	Intergovernmental Panel on Climate Change
K	Sorbent deactivation constant
MEA	Monoetalonamine
N	Number of sorbent cycles
NG	Natural gas
OF	Objective function
OPEX	Operation and maintenance costs
X_N	Average sorbent activity
X_r	Residual sorbent activity
Capture	Capture system process (with the power plant)
Max	Maximum value
Ref	Reference case (usually power plant without capture system)
sb	Steam bleeding

1 Introduction

Several relevant organizations have foreseen that CO_2 capture and storage technologies will play an essential role in the reduction of greenhouse gases emission in a medium-large term [1, 2]. Some studies [3] declare that they could reduce CO_2 emissions in industry sectors by over 50% by 2050. In spite of this interest, one of the main challenges of CCS is the cost and efficiency penalty reduction that is involved in the whole process. Minimizing the energy requirements and CO_2 avoided cost, as well as improvements in the efficiency of energy conversion processes, are key priorities for CCS future technology deployment [1].

To guarantee a suitable commercial development of this technology, the CO_2 avoided cost (including storage) should be, as a limiting value, the cost of the CO_2 emission trading. It is expected that this value would be in the range of

30–50 €/tCO_2 in a medium-term. Therefore, a target value for long-term is a cost reduction down to 25–30 €/tCO_2. Some calculations show that capture, compression and conditioning stand for 80–90% of this value and the 10–20% is for storage (excluding site selection and conditioning). So the capture, compression and conditioning cost should be reduced to 20–25 €/tCO_2.

Therefore in most cases, the objective function will be economic. Constraints of the problem will include: economic calculations, thermodynamic variables with mass and energy balances, and an adequate system layout. The layout defines the CAPEX and the thermodynamic efficiency of the whole system, therefore the OPEX. There are as many optimization results as potential system layouts. For this reason, as a preliminary step, optimization problems have to include Pinch and Second Law Analysis in order to work with an appropriate system layout and obtain valuable optimization results.

A detailed description of the application of exergy analysis and heat exchanger networks is out of the scope of the present chapter. Exergy analysis guides the determination of the most effective ways of improving complex processes through a better understanding of the irreversibilities (loss of work potential) for every equipment and for the whole system. A preliminary analysis of proposed system irreversibilities could leads to process modifications that improve the final optimization results. It will be assumed, in this chapter, that the researcher has a convenient system layout for analysis or this layout is fixed by a real process. When the optimization problem demands a system design, some guidelines have been proposed to guarantee a suitable process design based on the Second Law of Thermodynamics [4–6]. These guidelines could be complemented with the use of Pinch analysis that allows the design of optimum heat exchanger networks [7, 8]. Once the system is well-defined, the optimization problem might be solved.

2 Models Formulation

2.1 Objective Function

The definition of the objective function is one of the important issues for CCS applications. This function could focus on economy, energy or environment terms. Economic objectives include capital investment minimization, annual levelized net profit maximization, return of investment minimization or CO_2 capture costs minimization. The last one might include only the CO_2 capture system, or the overall system (incorporating the original power plant or combustion system). Energetic targets mainly include maximizing energy efficiency or minimizing the efficiency penalty. Finally, environmental target includes the maximization of the percentage of CO_2 captured or avoided and/or minimization of pollutants in the flue gases.

It is essential to define a valuable objective function (OF) that could integrate all these criteria and establish a basis for CO_2 capture system comparison. The Intergovernmental Panel on Climate Change (IPCC) has proposed Eq. 1, for the calculation of the CO_2 avoided cost [1]. Avoided cost refers to the decrease of CO_2 emissions due to capture, not the CO_2 captured that is always higher due to the additional energy used in this process. The evaluation of this cost includes the cost of electricity (COE), that incorporates CAPEX and OPEX for the CCS and the reference case (installation without capture).

$$Cost\ avoided\ \ tCO_2 = \frac{(COE)_{capture} - (COE)_{ref}}{(CO_2kWh^{-1})_{ref} - (CO_2kWh^{-1})_{capture}} \tag{1}$$

Equation 1 includes economic variables in the calculation of the cost of electricity (COE) both, in the original and the modified system (original system plus the CO_2 capture installation); energetic variables are included as OPEX within the COE definition; environmental quantities are taken into account when compares CO_2 emissions in the original and in the overall system including CO_2 capture.

In this equation, CAPEX has strong uncertainties because there are no commercial systems in operation that provide reliable economic data. The development of the commercial systems will improve the cost functions and therefore trustworthy results may be obtained. Other main cause of uncertainties is related to the fact that system layout could be altered. This connects with previous comments about the adequate definition of the system layout with the support of the Thermodynamics.

Then, the objective function use to be Eq. 2:

$$Minimize \frac{(COE)_{capture} - (COE)_{ref}}{(CO_2kWh^{-1})_{ref} - (CO_2kWh^{-1})_{capture}} \tag{2}$$

where:

$(COE)_{capture}$ is the cost of electricity in the capture system made up of OPEX and annual CAPEX

$(COE)_{ref}$ is the cost of electricity in the reference system (without capture) made up of OPEX and annual CAPEX

$(CO_2kWh^{-1})_{ref}$ are the CO_2 emissions per kWh produced in the reference system

$(CO_2kWh^{-1})_{capture}$ are the CO_2 emissions per kWh produced in the capture system

The independent variable is often included in the chemical reactions, mass and energy balances as an operational variable (see Sect. 4). When there is one independent variable in a non-linear model formulation, a frequent solving method is quadratic approximation. When two o more independent variables are selected, variable metric methods may behave better than direct search methods, although general optimization rules for non-linear problems could also be used. When the independent variable is the selection of one layout or another, the solving method change to a mixed-integer programming.

2.2 *Constraints: Second Law of Thermodynamics and Pinch Analysis*

Once the CCS system layout is defined, mains constrains are related with the thermal simulation and include chemical reactions, mass and energy balances of every equipment of the process and cost functions. The reference system has also to be calculated. In this case, thermal simulation and cost calculation are in the technology state-of-the-art and have lower uncertainties. These calculations establish a basis for comparison in Eq. 1. Again, care should be taken in the selection of the reference case, as low efficiency cases (high OPEX and specific CO_2 emissions) lead to lower CO_2 avoided cost and the minimization results could be misleading.

Finally, the problem to solve is summarized in Eq. 3.

$$Minimize \frac{(COE)_{capture} - (COE)_{ref}}{(CO_2kWh^{-1})_{ref} - (CO_2kWh^{-1})_{capture}}$$

$$Subject\ to: \begin{array}{l} COE = \left(\left(F^* \sum CAPEX_i\right) + OPEX\right)/EP \\ CAPEX_i = f(layout, equipment, size) \\ OPEX = Fixed\ cost + \sum (ER_i{}^* EC_i) \\ ER_i = f(capturesystem, CO_2captured) \\ EP = f(powerplant, CO_2captured, ER_i) \\ CO_2kWh^{-1} = f(powerplant, CO_2captured, EP) \\ CO_2captured = f(capturesystem, operational\ variables) \end{array} \quad (3)$$

3 Case Studies

3.1 *Integration of Amine Scrubbing for CO_2 Capture in Existing Power Plants*

3.1.1 System Description

CO_2 absorption by amine derived chemical solvents seems to be the most developed alternative from combustion flue gases at power stations. One of the important research objectives is the development of new solvents and blended amines that, mainly, reduce the energy requirements without adversely affecting chemical reaction rates, solvent stability and corrosion issues.

In these systems, flue gas from the desulphurization unit of a power plant, are introduced into an absorber packed column where CO_2 usually reacts with an amine aqueous solution. Recently, it has been proposed other absorbents based on blended amines [9], ammonia [10, 11], piperazine [12, 13]. Clean gases are directed to the stack. Amine solution is directed to regeneration column through a heat exchanger. Stripper demands high energy requirements to regenerate absorber releasing the CO_2, that is conditioning and pressurized to be transported. Energy consumption is the main drawback of the technology. Depending on the process and solvent used these requirements changes [9–19]. As a reference, this requirements are in the range of 2.0–4.0 GJ/tCO_2.

There could be several options to supply the energy required to capture CO_2 from flue gases with amine solvents. Figure 1 illustrates some possibilities: (1) using a natural gas auxiliary boiler to produce steam for the absorption process. It avoids any negative effect in original plant steam cycle efficiency and power output, but produces additional CO_2 from the boiler and increasing CAPEX and OPEX; (2) integrating the absorption process into the original power plant reducing the overall efficiency and the net power output; (3) using a gas turbine in partial repowering. It satisfies the required regeneration energy, increases the CO_2 emissions but also the power output and overall efficiency. The objective in this study is the selection of the layout that minimizes the CO_2 avoided cost and represent the initial analysis of a minimization problem. Once the layout that minimizes the CO_2 avoided cost was selected, the next step for this layout is to obtain the operational variables that minimizes the objective function. When there are not substantial layout modifications it is a mixed-integer programming problem. If there are significant modifications in the system it is easier to use a two-step optimization procedure than the formulation of an unique optimization problem with a lot of different constrains in the form of mass and energy balances.

3.1.2 Method and Assumptions

The objective function is the minimization of CO_2 capture cost, Eq. 4.

$$Minimize \rightarrow \frac{(COE)_{capture} - (COE)_{ref}}{(CO_2kWh^{-1})_{ref} - (CO_2kWh^{-1})_{capture}}$$

$$Subject\ to: \quad \begin{aligned} & COE = \left(\left(F * \sum CAPEX_i\right) + OPEX\right)/EP \\ & CAPEX_i = f(layout, equipment, size) \\ & OPEX = Fixed\,cost + \sum (ER_i * EC_i) \\ & ER_i = f(amine, CO_2captured) \\ & EP = f(powerplant, CO_2captured, ER_i) \\ & CO_2kWh^{-1} = f(powerplant, CO_2captured, EP) \\ & CO_2captured = f(amine, operational\ variables) \end{aligned} \tag{4}$$

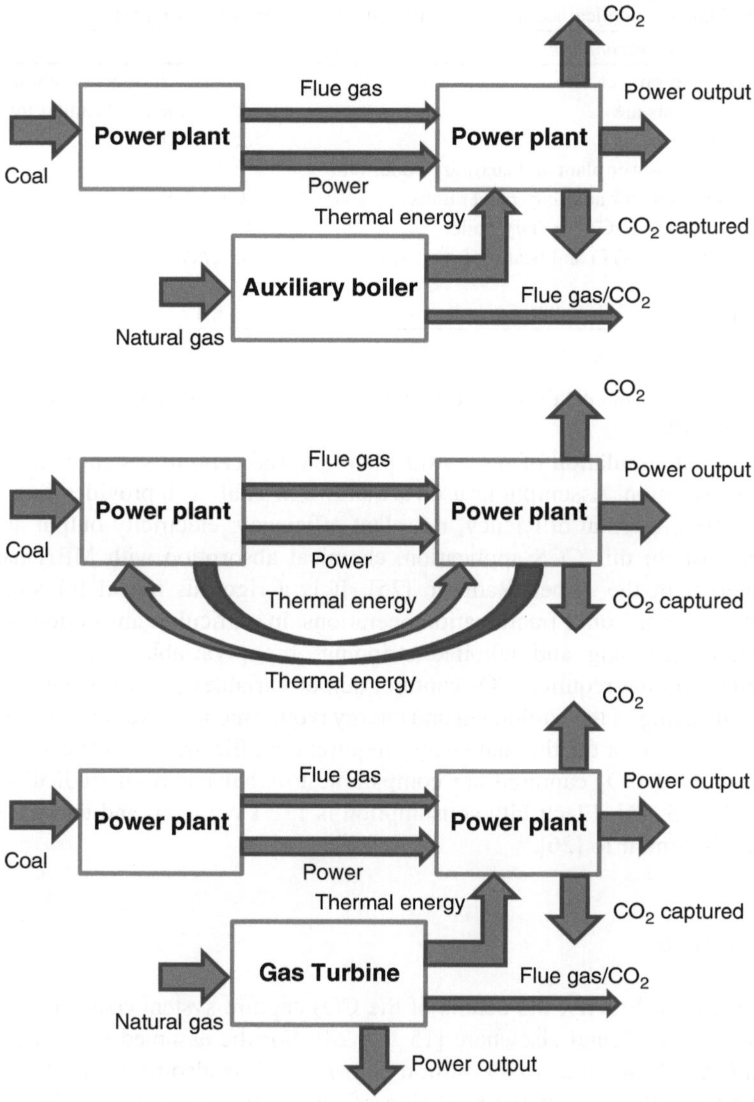

Fig. 1 Options to supply the energy required in amine scrubbing CO_2 capture

Reference case is a 350 MWe power plant with an efficiency of 36.93% (LHV). The combustion of coal supplied to each fired boiler produces 982.89 MWt at base load and yields approximately 630.0 kg/s of flue gas being 96.3 kg/s of CO_2 (9.76%v). In this case four separate absorption/regeneration column trains were necessary to capture around 60% and 65% of CO_2 produced.

The reference COE could be calculated, but it uses to be frequent to calculate the additional COE that is caused by the installation of the CO_2 capture plant. Despite

Table 1 Main economical assumptions for amine process optimization [27]

Main economical assumptions	
Interest rate and project life	5.0% and 20 years (7,500 h/year)
Maintenance costs are	2.2% of the fixed capital investment
Electricity price	5.29 ¢/kWh
Cost of coal for power plant and auxiliary boiler unit	2 €/GJ
Cost of natural gas for auxiliary power units	4 €/GJ
Cost of natural gas (NG) auxiliary boiler	75 €/kWh
Cost of gas turbine (GT) and heat recovery steam generator (HRSG)	265 €/kW
Cost of the make-up MEA, 981 €/t	981 €/t

equipment cost uncertainties, capital costs could be evaluated using different sources [20–24].

A thermal simulation of the power plant and the capture system is needed for OPEX calculation. Assumptions used are shown in Table 1. It provides data on coal consumption, thermal efficiency, net plant efficiency, electricity output and CO_2 capture cost. In this CCS application, chemical absorption with MEA has been simulated with the Aspen RadFrac [25]. It is a rigorous model for simulating multistage vapor-liquid fractionation operations, in particular: absorption, reboiled absorption, stripping and reboiled stripping. Input variables are the flue gas properties and the required CO_2 capture, output variables are the quantity of CO_2 separated, sizing of the equipment and energy requirements. A value of 3.1 GJ/tCO_2 has been chosen for the thermal energy requirement. Electricity and heat consumption per ton of CO_2 captured are comparable to what has been studied by other authors [9, 15, 26]. Electricity consumption is 112 kWh/tCO_2 and thermal energy required is similar to [26].

3.1.3 Results

For the sake of brevity, the details of the CO_2 capture system costs are not given here and can be found elsewhere [15, 27, 28]. For the assumed CO_2 installation, annual CAPEX amounts to 49 million euros/year. It is also necessary to take into account the influences of the reduction of power output, extra fuel for auxiliary equipment and the gas turbine and heat recovery steam generator, auxiliary boiler or plant modifications, depending on the option chosen to provide the energetic demand. These quantities are shown in Table 2.

For these options, the modifications of steam cycle for the stripper energy requirement are the optimum choice. However it has also an important power output reduction and a loss of electrical efficiency of 6.8 points. One of the main advantages is that no extra CO_2 emissions are generated and a minimization of the objective function is 25.3 euros per ton CO_2 avoided (4.8 million tCO_2 per year).

Gas turbine scheme shows intermediate annual costs due to the size of GT + HRSG. This option solves the energy demanded by CCS and increase the electricity

Table 2 Main results for amine process [27]

	CO_2 avoided cost (€/tCO_2)	Annual CAPEX (M€)	Annual OPEX (M€)	CO_2 avoided (Mt/year)	Overall (LHV) efficiency (%)
Gas natural boiler	60.58	54.5	162.1	3,58	26.18
Internal flows	25.25	49.0	72.6	4,82	30.11
Gas turbine	31.23	51.6	85.8	4,40	33.70

production. It has higher efficiency and power output than the others, however CO_2 avoided is slightly lower than previous configuration. As a consequence, the objective function increases up to 31.2 €/tCO_2 avoided. The same trend is observed for the extra NG boiler, but in this case there is no electricity production and shows the worst results. The equipment and operational cost of the auxiliary boiler option increase the total annual cost for this configuration. Moreover, the CO_2 emissions decrease the CO_2 avoided and increase the cost per tCO_2 avoided up to 60 €.

The second optimization step would be to define the operational variables that could influence the objective function minimization in the selected option. For absorption processes, it is well-known that the optimal option to provide the energy required for the amine solvent regeneration is a power plant steam turbine bleeding. This steam bleeding have to be saturated steam midway through the low pressure section of the turbine [15, 17, 26, 27, 29] with a pressure between 2.0 and 3.5 bar using the lowest quality steam available to fit with the reboiler requirements. However, most of the steam turbines do not have an extraction at this pressure range, as a consequence, perfect integration is only possible when steam cycle is designed taking into account a future amine scrubbing installation. In the other cases, a mixed-integer formulation with the available bleeding pressures as independent variable would solve the minimization problem. Next section shows an example of a problem formulation to obtain the operational variable that minimizes the objective function.

3.2 Calcium Looping and Oxyfuel Technologies for CO_2 Capture

3.2.1 System Description

A highly integrated process to capture CO_2 from an existing power plant based on calcium looping cycle with oxyfuel combustion-regeneration is used as an example to show the optimization of operational variables. For this CCS application, due to the novelty of the processes involved in CO_2 capture, it could be necessary to minimize the CO_2 avoided cost as function of several independent (operational) variables.

Figure 2 shows the general layout of the proposed CO_2 capture cycle. Combustion flue gases are led into a circulating fluidized bed (CFB) together with the

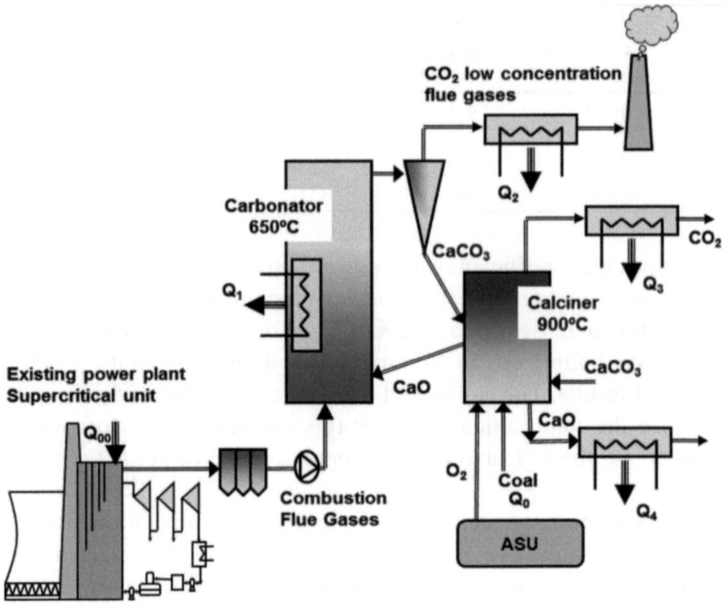

Fig. 2 Calcium looping with oxyfuel combustion scheme

sorbent, CaO, coming from an oxyfuel calciner. The sorbent carbonates in contact with CO_2 from combustion gases with an efficiency close to 90% [29–31]. This CFB used as a carbonator includes in-boiler heat transfer equipment in order to take advantage of the carbonation exothermic energy and control the temperature reaction. Flue gases from CFB carbonator are cleaned and solids enter an oxyfuel calciner for the regeneration of $CaCO_3$ at around 900–925°C. One critical aspect in this process deals with the conversion efficiency of carbonation reaction and its reduction after several cycles [32–34]. To compensate for sorbent activity losses, a make-up flow of fresh $CaCO_3$ is introduced in the calciner and a controlled solid purge of deactivated material is necessary. A high make-up flow increases the operational costs and improves the process performance, therefore a balance is required [35]. This will be one independent variable of the minimization problem. Another variable that influence the system layout and performance is the CaO/CO_2 ratio, high ratios increase the size of the system (CAPEX) but improve system performance (CO_2 captured) and reduce the influence of make-up flow. Minimizing CO_2 avoided cost with make-up and CaO/CO_2 ratio as independent variables is a process necessity and it will serve as an example of optimization.

3.2.2 Method and Assumptions

Using thermodynamics, the waste energy coming from CFB carbonator (at 650°C), clean flue gases (between 650°C and a final temperature of 150°C), carbon dioxide

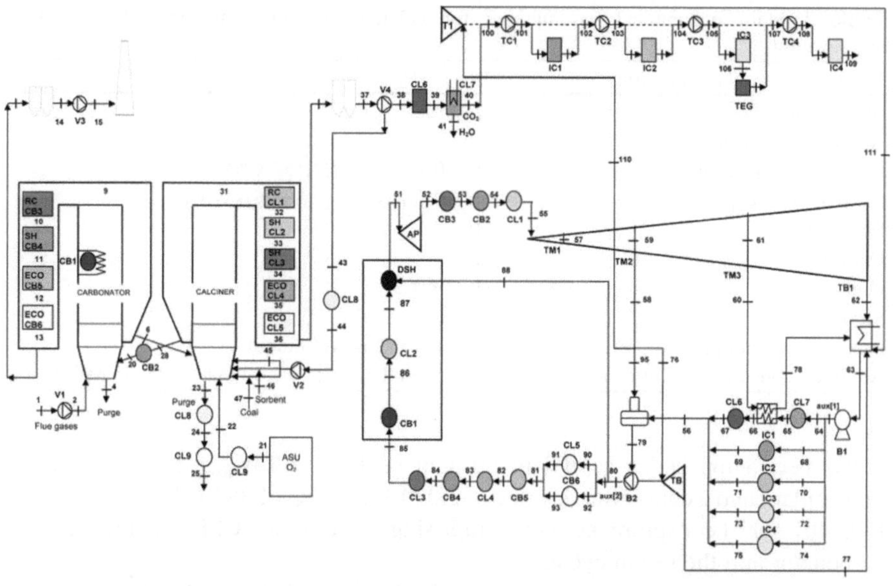

Fig. 3 Proposed calcium looping with oxyfuel combustion integration layout [37]

(between 900°C and 150°C), purge (between 650°C and 200°C) and CO_2 compression intercooling have to be used for water preheating and supercritical steam production. System layout is shown in Fig. 3, where CL and CB stands for calciner and carbonator and the energy transferred in heat exchangers (CL1 to CL9 and CB to CB6) go from boilers to steam cycle. Moreover, CO_2 intercooling energy is used as low pressure water preheaters together with the lower exergy content of CO_2 and flue gases. CO_2 and flue gases at high temperature are used as a reheating section in the steam cycle. Carbonator is used also as a boiler to produce supercritical (CB1, CL2) and reheat (CB2, CB3, CL1) steam.

Sorbent deactivation is modelled with equations at the literature based on the purged solids and average extent of carbonation reaction. Rodriguez et al. [36] proved the maximum conversion of a CaO particle, X_N, depends on the number of carbonation-calcination cycles (N) in accordance with Eq. 5. The variable k represents the deactivation, higher k means faster deactivation, and X_r is the residual sorbent activity. Values for $k = 0.52$ and $X_r = 0.075$, correspond to natural limestone.

$$X_N = \frac{1}{\frac{1}{(1-X_r)} + kN} + X_r \tag{5}$$

Reference plant and capture plant CAPEX have been taken from literature [37]. The extra auxiliary consumption in the capture system is mainly generated by the ASU (220 kWh/tO_2). Other assumptions for calculations are presented in Table 3,

Table 3 Main economical assumptions for calcium looping with oxyfuel combustion optimization

Main economical assumptions	
Interest rate	8.78%
Amortization period	25
Reference plant capital cost	1,100 €/kW_{gross} (1,158 €/kWnet)
Capture plant capital cost	2,070 €/kW_{gross} (2,383 €/kWnet)
Fixed costs	1%
O&M costs	2%
Fuel cost	1.43 €/GJ
Sorbent cost	12 €/t
Auxiliary consumption for reference plant	5% of gross power output
Auxiliary consumption for capture plant	5% of total gross power output + 28–85 MW (ASU)

it has been supposed a capital investment of 1,100 €/kW_{gross} for the reference power plant and a conservative value of 2,070 €/kW_{gross} (2,383 €/kW_{net}) according [39, 40] for the capture system comprising an oxyfuel CFB calciner, a CFB carbonator and the steam cycle.

Once again the objective function is the CO_2 avoided cost, Eq. 6. Constraints include mass and energy balances, the equation of extent of chemical reaction as a function of the independent variables (make-up and CaO/CO_2 ratio) and, finally, the capital and operational cost. For this optimization example with two independent variables and the variable metric method is chosen for minimization.

$$Minimize \rightarrow \frac{(COE)_{capture} - (COE)_{ref}}{(CO_2kWh^{-1})_{ref} - (CO_2kWh^{-1})_{capture}}$$

$$\text{Subject to :}\quad \begin{aligned} &COE = \left(\left(F * \sum CAPEX_i\right) + OPEX\right)/EP \\ &CAPEX_i = f(layout, equipment, size) \\ &size = f(CO_2captured) \\ &OPEX = Fixed\,cost + \sum (ER_i{}^* EC_i) \\ &ER_i = f(X_N, CO_2captured) \\ &X_N = \frac{1}{\frac{1}{(1-X_r)} + kN} + X_r \\ &N = f(make - up, Ca/CO_2) \\ &EP = f(powerplant, ER_i, CO_2captured) \\ &CO_2kWh^{-1} = f(powerplant, CO_2captured, EP) \\ &CO_2captured = f(k, X_r, make - up, Ca/CO_2) \end{aligned} \qquad (6)$$

3.2.3 Results

The COE for the reference case is 31.9 €/MWh. The minimization of the avoided cost show the system dependency of the sorbent characteristics: deactivation (k) and residual activity (X_r), Tables 4 and 5. For each pair of data about the sorbent, minimization find the optimum values of make-up flow and Ca/CO_2 in the system to obtain the most economical system design and operational performance.

If sorbent deactivates quickly (high k values), Table 4, the average activity decreases and Ca/CO_2 ratio should increase to achieve enough CO_2 separation efficiency (around 90%). Sorbent make-up flow remains in low values (1–2% of total mass of sorbent) and approximately constant for all cases. Avoided cost increases with deactivation, k, due mainly to an increase in system size and CAPEX. Larger Ca/CO_2 ratios forces to increase the size of the overall system for solid circulation (increasing gross power) and use the energy released in carbonation stage.

The same effects could be observed with decreasing the residual activity of the sorbent (X_r), Table 5. In this case the effect on the objective function is more important, due to the system has to be small and it affects directly to CO_2 capture capital cost.

A sensitivity analysis as a function of main cost assumptions might complement the conclusions. Capital cost of base case are 1,100 €/kW$_{gross}$ for power plant and 2,070 €/kWh for additional capacity in the capture system. As it has shown in Table 7 for residual activity of 0.08, the avoided cost is 14.5 €/tCO_2 with a purge of 1.8% and a Ca/CO_2 ratio of 5.8. The size of the overall system is around 950 MW. If reference capital cost are increased a 15%, Table 6, results show a reduction of avoided cost to 13.1 €/tCO_2 due to reference COE increases and it reduces the CAPEX difference between CO_2 capture and reference. Evidently, increasing 15% the capture system capital cost, up to 2,380 €/kW$_{gross}$, that is the variable with large uncertainty, implies an increase of 3 €/tCO_2 in the avoided cost. Overall system has

Table 4 Minimization results depending on sorbent deactivation constant (k)

k	X_r	CO_2 avoided cost (€/t)	Make-up flow	Ca/CO_2	System gross power	System net power	Overall COE
0.3	0.08	13.76	0.015	5.00	938.1	813.8	44.34
0.4	0.08	14.21	0.019	5.11	921.4	798.4	44.75
0.5	0.08	14.49	0.017	5.85	952.7	825.0	45.00
0.6	0.08	14.70	0.016	6.47	979.4	847.7	45.20

Table 5 Minimization results depending on sorbent residual activity (X_r)

k	X_r	CO_2 avoided cost (€/t)	Make-up flow	Ca/CO_2	System gross power	System net power	Overall COE
0.5	0.04	15.59	0.019	7.24	987.1	852.4	45.99
0.5	0.08	14.50	0.018	5.81	949.6	822.3	45.01
0.5	0.16	13.22	0.013	4.13	909.7	790.2	43.85
0.5	0.24	12.55	0.013	3.19	892.5	776.6	43.24

Table 6 Sensitivity analysis depending on cost functions

Original cost	Capture cost	CO_2 avoided cost (€/t)	Make-up flow	Ca/CO_2	System gross power	System net power	Overall COE
Ref	Ref	14.50	0.018	5.81	949.6	822.3	45.01
+15%	Ref	13.12	0.015	6.10	970.5	840.5	46.5
Ref	+15%	17.51	0.022	5.37	921.8	797.8	47.73
+15%	+15%	16.21	0.020	5.56	933.4	808.1	49.30
+25%	Ref	19.46	0.024	5.15	909.4	800.6	52.15

to be small to limit the total capital cost (decreasing also the Ca/CO_2 ratio). Despite this numbers, the CO_2 capture cost remains very attractive. If, both, reference and capital cost increase 15%, the objective function and the system size have intermediate results. Finally, an increase of 25% in CO_2 capture equipment cost (2,590 €/kW$_{gross}$) makes the avoided cost increases up to 19.46 €/tCO_2 and system size reduces in 40 MWe. Summarizing, even in the worse situation analysed, the CO_2 avoided cost remains below 20 €/tCO_2, make-up flow stays in low values (2%) and Ca/CO_2 should be maintained high (5–6%).

3.3 *Integration of CO_2 Compression into the Steam Cycle of a Power Plant*

3.3.1 System Description

CO_2 conditioning is an essential process in the development of carbon capture and storage technologies. Moisture elimination and gas compression up to 100–120 bar are the main objectives of this stage [41–43]. Around 25% of the total capital requirements are caused by the compression equipment. Energy requirements for gas conditioning are typically 90–120 kWh/tCO_2 [41], which means around an electrical consumption of 100 kWh/MWh produced and COE increases 10%. These values highlight the necessity for a detailed analysis of minimizing both, the energetic requirements of the compression and the additional capital requirement for the capture system. One of the strategies to reduce power needs, compressor size and capital cost is compression intercooling (split compression in several stages with intermediate CO_2 intercooling).

Figure 4 illustrates a basic configuration for a CO_2 intercooling compression system. Compression process is divided in four stages. Waste energy (Q_1 to Q_5) is a useful energy and it could be used in the low-pressure part of the steam cycle for water preheating. It reduces the steam turbine bleedings mass flow, increases the gross power output and the compressor specific consumption.

There are three variables that influence the compressor power consumption and cooling necessities: compressor efficiency, pressure ratio (i.e. compression stages) and CO_2 compressor inlet temperature. The compressor efficiency would depend on

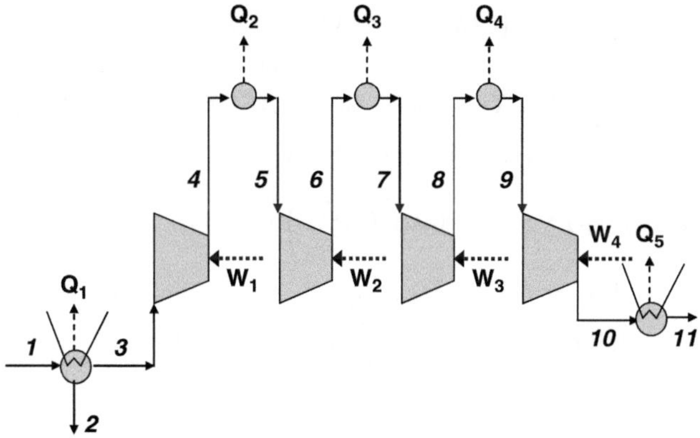

Fig. 4 Simplified intercooling compression layout

the manufacturer and it is supposed constant. Pressure ratio depends on the system layout (3–4–5 compressor stages). Then, several optimization problems should be solved to select the optimized results for each layout. Approximately the effect of inlet temperature reduction from 50°C to 30°C is a decrease of 10% in power requirements [44]. Cooling equipment cost depends on the cooling duty and temperature difference between cooling media and compressed CO_2. Reducing CO_2 temperature implies high CAPEX of heat exchangers but low power requirements. It will be the independent variable of the minimization problem.

3.3.2 Method and Assumptions

In this case, the objective function is slightly different from previous ones. In this example the CO_2 quantity is constant (denominator of Eq. 1) therefore the additional COE associated with CO_2 compression will be the function to minimize. This function includes CAPEX and OPEX (net compression power consumption). Annual capital cost are included with the CAPEX and the annuity factor (F, supposed 0.1), and OPEX are covered by a reduction in the net power production compared with reference case. For reference purposes a COE of 40 €/MWh has been considered. It has been supposed a 350 MW(gross) power plant with specific emissions of 741 kg/MWh and 80% capacity factor. CO_2 mass flow to compression is assumed 72 kg/s, plus a water content of $10\%_{\mathrm{w}}$.

$$Minimize \rightarrow \Delta COE = \frac{(COE_{ref} \times MWh_{ref}) + F \times \sum CAPEX}{MWh} - COE_{ref}$$

$$
\begin{aligned}
&\Delta COE = \frac{(COE_{ref} \times MWh_{ref}) + F \times \sum CAPEX}{MWh} - COE_{ref} \\
&\sum CAPEX_i = CAPEX_W + CAPEX_{TEG} + CAPEX_{HE} \\
&CAPEX_{W\_i} = 517.5^{*}2.11 * W_i^{0.82*} ci \\
\text{Subject to:}\quad &CAPEX_{HE\_i} = 3650^{*}\left(\frac{Q_i}{\Delta T_{lm}}\right)^{0.7948*} ci \\
&\Delta T_{lm} = f(T_{input}, T_{cooling}) \\
&MWh = MWh_{ref} - \sum (W_i{}^{*}h) + W_{st}{}^{*}h \\
&W_i = f(r_p, \eta_i, T_{input}) \\
&Q_i = f(r_p, \eta_i, T_{input}) \\
&\sum Q_i \leq Q_{max} \\
&Q_{max} = f(m_{sbj}, p_{sbj}, t_{sbj})
\end{aligned}
\tag{7}
$$

CAPEX is the sum of main components capital cost: Compressor, dehydration unit (supposed unchanged in this calculations) and heat exchangers. Every of these costs have to be evaluated with literature and/or manufacturer equations. For compressors cost estimation, a recent proposal (assuming a centrifugal, reciprocating compressor) depends on the power requirements (W_i), compressor type and cost indexes (ci) [45]. The function capital cost for heat exchangers is taken from [46], where Q_i is the energy transferred in each heat exchanger in kW, ΔT_{lm} the logarithmic mean temperature difference in K.

For a specific integration of this scheme into a power plant an additional constraint is needed, i.e. the total energy transferred to the steam cycle (ΣQ_i) must not exceed the maximum specified that depends on steam bleeding conditions (mass flow, pressure and temperature). In parallel with solving the optimization problem, it is necessary to know the possibility of integrating the energy from intercooling into a steam cycle, eliminating two or three low-pressure heaters and increasing steam turbine power output. In this problem, pinch analysis forces to split the water mass flow in several streams for an efficient use of the waste energy. A final heat exchanger configuration is showed in Fig. 5. Steam turbine bleedings are reduced and steam power output increases so, the net power consumption (turbines minus CO_2 compressor increases).

3.3.3 Results

For reference case the energy from CO_2 intercooling is removed in the cooling system and compressor inlet temperature is assumed 30°C. Assuming four compression stages with an isentropic efficiency of 80% the compression total cost is 43.4 M€, approximately 90% due to compressors and 10% of heat exchangers and gas treatment (dehydration). Compressor power necessities are 26.5 MW and the

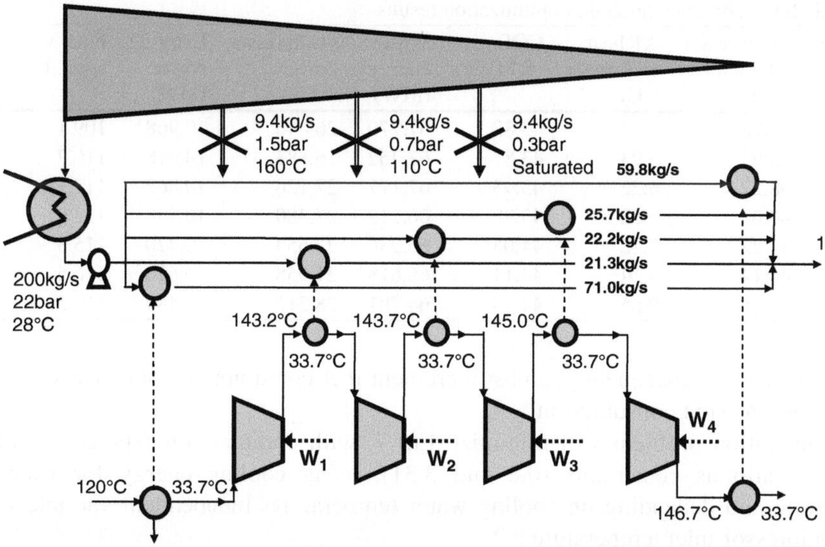

Fig. 5 Proposed intercooling compression waste energy use layout [44]

Table 7 Dependency of the COE with compression inlet temperature

Cooling temp. (°C)	Compressor inlet temp. (°C)	COE (€/MWh)	Waste energy (MW)	Compressor power (MW)	Extra ST power (MW)	Final water temp. (°C)
28	50	44.04	64,662	28,359	10,465	105.4
28	46	43.96	65,649	27,944	10,625	106.6
28	42	43.90	66,469	27,525	10,757	107.6
28	38	43.85	67,151	27,103	10,868	108.4
28	34	43.82	67,733	26,676	10,962	109.1
28	32	43.83	67,996	26,461	11,004	109.4
28	30	43.88	68,243	26,244	11,044	109.7

COE is 45.19 €/MWh. Therefore, the incremental COE due to CO_2 compression is around 13%, 5.19 €/MWh. Table 7 illustrates the variation of the COE with compressor inlet temperatures (heat exchanger outlet temperatures). Reducing compressor inlet temperature, COE values also decrease but the tendency changes around 34°C. Lower temperatures means an increase in HE cost (due to the reduction of ΔT_{lm}) and the reduction in steam bleeding do not compensate these cost increase.

Evidently, water cooling coming from condenser has an important impact on optimization results. In cold locations is possible to obtain lower condenser pressures and water temperatures (increasing original power plant efficiency). This reduction of temperatures enables lower temperatures at intercooling heat exchanger outlet and compressor inlet. Nevertheless, for hot locations, this temperature augmentation causes an important reduction of temperature in heat

Table 8 Intercooling compression optimization results

Cooling temp. (°C)	Compressor inlet temp. (°C)	ΔT heat exchanger (°C)	COE (€/MWh)	Waste energy (MW)	Compressor power (MW)	Extra ST power (MW)	Final water temp. (°C)
28	33.72	5.72	43.82	67,770	26,646	10,968	109.1
30	35.93	5.93	43.87	67,462	26,883	10,918	110.7
32	38.22	6.22	43.93	67,117	27,126	10,862	112.3
34	40.62	6.62	43.98	66,718	27,380	10,798	113.9
36	43.22	7.22	44.05	66,236	27,653	10,720	115.3
38	46.14	8.14	44.11	65,618	27,958	10,620	116.5
40	49.55	9.55	44.19	64,781	28,312	10,484	117.5

exchangers and, consequently, a cost increment that could not be compensated by lower compression consumption.

Optimization problem will minimize Eq. 7, with compression efficiency and pressure ratio as constraints (0.8 and 3.31), using cooling energy for water preheating and depending on cooling water temperature. Independent variable is the compressor inlet temperature.

Table 8 shows the minimization results for different cooling temperatures. Optimum intercooling outlet temperature is 33.7°C, with a COE of 43.82 €/MWh, that represents a reduction of 1.37 €/MWh (26% lower that reference case). In this case, 67.8 MW from intercooling are used for water preheating. This amount reduces steam bleeding, Fig. 5, and steam turbine gross power could increase up to 11 MW, reducing compressor electric consumption from 26.65 to 15.56 MW. Total capital requirements increases up to 45.5 M€ (2.1 M€ an increase of 4.8%) from which compressor cost represent 83%.

Table 4 also illustrates as this temperature influences optimization results. Depending on these values, the intercooling system should be designed. Temperature difference from intercooling outlet and water inlet is small for low water temperature around 5.7°C. Increasing water temperature to 40°C, this temperature difference increases up to 9.55°C, and the COE increases from 43.82 to 44.19 €/MW. Figure 2 shows a possible integration scheme of intercooling compression and low-pressure section of a steam cycle. The reduction in each steam extraction mass flow is 8.7 kg/s for the 1.5 bar pressure bleeding, Fig. 5, and 9.0 kg/s for the 0.7 and 0.3 bar pressure bleedings, and therefore the extra power in steam turbine is 10.4 MWe.

4 Conclusions

One of the main drawbacks of CO_2 capture systems is the cost associated with the separation and CO_2 storage. This cost is expressed in an accepted equation that includes a comparison of COE and specific CO_2 emissions between a reference installation and an installation with CO_2 capture. CAPEX and OPEX have significant influence in the total cost. CAPEX is influenced by the system layout and OPEX by the energy penalty caused by the CO_2 separation. Thus, cost reduction is a driving force to increase the interest for this greenhouse gases mitigation.

The use of formal optimization problems do assist to find strategies that minimize the cost function. In this chapter, several examples have been presented to come up with different objective functions and methodologies. Firstly, minimizing CO_2 avoided cost (€/tCO_2) by selecting one of the proposed options to come thought the stripper thermal energy requirements in amine based CO_2 post-combustion capture. Secondly, minimizing the same objective function depending on several operational variables in a complex system integrated by sorbent postcombustion plus oxyfuel calciner with a new designed power plant. Finally, a formal COE minimization to design the characteristic of heat exchangers for CO_2 compression intercooling with waste energy utilization.

References

1. Metz B, Davidson O, de Coninck H, Loos M, Meyer L (eds) (2005) Special report on carbon dioxide capture and storage. Intergovernmental panel on climate change. Cambridge University Press, Cambridge/New York
2. European Technology Platform for Zero Emission Fossil Fuel Power Plants (2007) EU demonstration programme for CO_2 capture and storage (CCS). ZEP's proposal. European Technology Platform for zero emission fossil fuel power plants (ZEP)
3. Stangeland A (2007) A model for the CO_2 capture potential. Int J Greenhouse Gas Control 1:418–429
4. Sama DA (1195) The use of the second law of thermodynamics in process design. J Energy Resour Technol 117:179–185
5. Sama DA, Qian S, Gaggioli R (1989) A common-sense second law approach for improving process efficiencies. In: Proceedings of the international symposium on thermodynamics analysis and improvement of energy systems, Beijing, pp 520–531
6. Leites IL, Sama DA, Lior N (2003) The theory and practice of energy saving in the chemical industry: some methods for reducing thermodynamic irreversibility in chemical technology processes. Energy 28:55–97
7. Linnhoff B, Townsend DW, Boland D, Hewitt GF, Thomas BEA, Guy AR, Marsland RH (1982) A user guide on process integration for the efficient use of energy. IChemE, Rugby
8. Hall SG, Linnhoff B (1994) Targeting for furnace systems using pinch analysis. Ind Eng Chem Res 33:3187–3195
9. Aroonwilas A, Veawab A (2007) Integration of CO_2 capture unit using single- and blended-amines into supercritical coal-fired power plants: implications for emission and energy management. Int J Greenhouse Gas Control 1:143–150
10. Darde V, Thomsen K, van Well WJM, Stenby EH (2010) Chilled ammonia process for CO_2 capture. Int J Greenhouse Gas Control 4:131–136
11. Puxty G, Rowland R, Attalla M (2010) Comparison of the rate of CO_2 absorption into aqueous ammonia and monoethanolamine. Chem Eng Sci 65:915–922
12. Freeman SA, Dugas R, Van Wagener DH, Nguyen T, Rochelle GT (2010) Carbon dioxide capture with concentrated, aqueous piperazine. Int J Greenhouse Gas Control 4:119–124
13. Cullinane JT, Rochelle GT (2005) Thermodynamics of aqueous potassium carbonate, piperazine, and carbon dioxide. Fluid Phase Equilibria 227:197–213
14. Chapel D, Ernst J, Mariz C (1999) Recovery of CO_2 from flue gases: commercial trends. Presented at the Canadian Society of Chemical Engineers annual meeting, Saskatoon
15. Desideri U, Paolucci A (1999) Performance modelling of a carbon dioxide removal system for power plants. Energy Convers Manage 40:1899–1915

16. Mimura T, Shimojo S, Suda T, Iijima M, Mitsuoka S (1995) Research and development on energy saving technology for flue gas carbon dioxide recovery and steam system in power plant. Energy Convers Manage 36:397–400
17. Mimura T, Simayoshi H, Suda T, Iijima M, Mituoka S (1997) Development of energy saving technology for flue gas carbon dioxide recovery in power plant by chemical absorption method and steam system. Energy Convers Manage 38:S57–S62
18. IEA Greenhouse Gas R&D Programme (2004) Improvement in power generation with post-combustion capture of CO_2. Report no. PH4/33
19. Abu-Zahra MRM, Schneiders LHJ, Niederer JPM, Feron PHM, Versteeg GF (2007) CO_2 capture from power plants. Part I. A parametric study of the technical performance based on monoethalonamine. Int J Greenhouse Gas Control 1:37–46
20. Peters MS, Timmerhaus K, West RE (2003) Plant design and economics for chemical engineers. McGraw-Hill, New York
21. Chemical Engineering (2010) Updating the CE plant cost index. Chem Eng 60:52
22. Rao A, Rubin E, Berkenpas M (2004) An integrating modelling framework for carbon management technology. US Department of Energy. http://www.cmu.edu/epp/iecm/documentation/tech_04.pdf
23. Fischer K, Beitler C, Rueter C, Searcy K, Rochelle G, Jessim M (2005) Integrating MEA regeneration with CO_2 compression and peaking to reduce capture costs. US Department of Energy. http://www.trimeric.com/report%20660905.pdf
24. Stoll HG, Smith RW, Tomlinson LO (1994) Performance and economic considerations of repowering steam power plants. GE Industrial & Power Systems Report GER-3644D
25. Aspen Plus 12.1 (2003) User guide, Aspen Technology, Inc., Cambridge, MA
26. Bozzuto CR, Nsakala N, Liljedahl GN, Palkes M, Marion JL (2001) Engineering feasibility of CO_2 capture on an existing US coal-fired power plant, Final report, vol 1. ALSTOM Power Inc.
27. Romeo LM, Bolea I, Escosa JM (2008) Integration of power plant and amine scrubbing to reduce CO_2 capture costs. Appl Thermal Eng 28:1039–1046
28. Singh D, Croiset E, Douglas PL, Douglas MA (2003) Techno-economic study of CO_2 capture from an existing coal-fired power plant: MEA scrubbing vs. O_2/CO_2 recycle combustion. Energy Convers Manage 44:3073–3091
29. Alie C (2004) Simulation and optimisation of a coal-fired power plant with integrated CO_2 capture using MEA scrubbing. Master thesis, University of Waterloo
30. Abanades JC (2002) The maximum capture efficiency of CO_2 using a carbonation/calcination cycle of $CaO/CaCO_3$. Chem Eng J 90:303–306
31. Abanades JC, Alvarez D (2003) Conversion limits in the reaction of CO_2 with lime. Energy Fuel 17:308
32. Lee DK (2004) An apparent kinetic model for the carbonation of calcium oxide by carbon dioxide. Chem Eng J 10:71–77
33. Shimizu T, Hirama T, Hosoda H, Kitano K, Inagaki M, Tejima K (1999) A twin fluid-bed reactor for removal of CO_2 from combustion processes. Trans IChemE 77-A:62
34. Wang C, Jic L, Tan Y, Anthony EJ (2008) Carbonation of fly ash in oxy-fuel CFB combustion. Fuel 87:1108–1114
35. Romeo LM, Lara Y, Lisbona P, Escosa JM (2009) Optimizing make-up flow in a CO_2 capture system using CaO. Chem Eng J 147:252–258
36. Rodriguez N, Alonso M, Grasa GS, Abanades JC (2007) Heat requirements in a calciner of $CaCO_3$ integrated in a CO_2 capture system using CaO. Chem Eng J 138:148–154
37. Lisbona P, Martínez A, Lara Y, Romeo LM (2010) Integration of Carbonate CO_2 Capture Cycle and Coal-Fired Power Plants. A Comparative Study for Different Sorbents. Energy Fuels 24:728–736
38. Rubin ES, Chen C, Rao AB (2007) Cost and performance of fossil fuel power plants with CO_2 capture and storage. Energy Policy 35:4444–4454

39. Abanades JC, Grasa G, Alonso M, Rodriguez N, Anthony EJ, Romeo LM (2007) Cost structure of a postcombustion CO_2 capture system using CaO. Environ Sci Technol 41:5523–5527
40. Romeo LM, Abanades JC, Escosa JM, Pano J, Giménez A, Sanchez-Biezma A, Ballesteros JC (2008) Oxyfuel carbonation/calcination cycle for low cost CO_2 capture in existing power plants. Energy Convers Manage 49:2809–2814
41. Aspelund A, Jordal K (2007) Gas conditioning – the interface between CO_2 capture and transport. Int J Greenhouse Gas Control 1:343–354
42. Heggum G, Weydahl T, Roald W, Mølnvik M, Austegard A (2005) CO_2 conditioning and transportation. In: Thomas DC, Benson SM (eds) Carbon dioxide capture for storage in deep geologic formations, vol 2. Elsevier, Burlington
43. Aspelund A, Mølnvik MJ, De Koeijer G (2006) Ship transport of CO_2 – technical solutions and analysis of costs, energy utilization, energy efficiency and CO_2 emissions. Chem Eng Res Des (Official Journal of the European Federation of Chemical Engineering: Part A) 84-A9:847–855
44. Romeo LM, Bolea I, Lara Y, Escosa JM (2009) Optimization of intercooling compression in CO_2 capture systems. Appl Thermal Eng 29:1744–1751
45. Duc NH, Chauvy F, Herri JM (2007) CO_2 capture by hydrate crystallization – a potential solution for gas emission of steelmaking industry. Energy Convers Manage 48:1313–1322
46. Olsommer B (1998) Méthode d'optimisation thermoéconomique appliqué aux centrales d'incinération d'ordures à cogénération avec appoint énergétique. Ph.D. dissertation, Swiss Federal Institute of Technology of Lausanne, Lausanne

Modeling the Costs of Carbon Capture

Erin Baker, Gregory Nemet, and Peter Rasmussen

Abstract This paper explores the fundamental concepts required to model intertemporal carbon capture costs. A technical overview of post-combustion, pre-combustion, and alternative combustion carbon capture technologies is followed by a discussion of cost measures that allow for side-by-side comparison of one technology to the next. Carbon capture cost measures include capital cost per unit output power, levelized electricity cost (LEC) and CO_2 avoidance cost. The relationships between these cost measures are mathematically explored and cost data for several carbon capture technologies obtained from a survey of the literature are provided. We then discuss the most significant drivers of intertemporal carbon capture cost reduction, which include learning by doing (LBD), production returns to scale and research and development (R&D).

Keywords Carbon capture • Cost modeling • Research and Development (R&D) • Technical change

1 Introduction

In order to meet the challenge presented by climate change, low carbon energy technologies must supplant high carbon, fossil-based energy technologies. Uncontrolled carbon dioxide emissions from traditional fossil fuels such as coal and

---

E. Baker (✉) • P. Rasmussen
Mechanical and Industrial Engineering Department, University of Massachusetts Amherst, Amherst, MA 01003-1027, USA
e-mail: edbaker@ecs.umass.edu

G. Nemet
La Follette School of Public Affairs, University of Wisconsin – Madison, Madison, WI 53706-1211, USA
e-mail: nemet@wisc.edu

Q.P. Zheng et al. (eds.), *Handbook of CO_2 in Power Systems*, Energy Systems,
DOI 10.1007/978-3-642-27431-2_16, © Springer-Verlag Berlin Heidelberg 2012

natural gas constitute about 60% of the world's carbon dioxide emissions from fossil fuels and generate over 66% of the world's electricity [1, 2]. In the absence of climate policy, coal combustion is likely to increase in conjunction with global energy demands, especially in populous, coal-rich, developing countries such as China and India. This is true even under a wide range of climate change damage and policy scenarios [3]. Accordingly, coal plants are likely to continue to play an integral role in providing the world's power but must do so emitting significantly less carbon dioxide than they do now. Incremental improvements to operating efficiencies alone are unlikely to be sufficient to meet carbon stabilization goals. In fact, it is suggested that emissions need to be reduced by 80–90% in order to meet such goals. Given the continued reliance on combustible fuels along with anticipated climate change-induced carbon dioxide emissions reductions, carbon capture and storage (CCS) is likely to factor into any low carbon energy technology portfolio [3]. There are a variety of CCS technologies in various stages of research, development, and demonstration (RD&D), although none have been demonstrated on a commercial level yet.

Carbon capture and storage describes two consecutive processes that control coal fired power plant carbon dioxide emissions – the capture and compression of CO_2 followed by the transport and storage of it. This chapter focuses on carbon capture, which constitutes over 80% of CCS costs [3]. The goal of this chapter is to highlight significant technical and economic aspects of carbon capture addressed by the literature and identify knowledge gaps in the field, with a focus on cost modeling.

The rest of the chapter is organized as follows. In Sect. 2 we provide an overview of the key capture technologies. In Sect. 3 we review the current literature on cost estimates for capture technologies. In Sect. 4 we discuss the possibilities for technological improvement in CCS. We distinguish between technical change that results from increases in production, such as learning-by-doing and returns to scale, and technical change that results from focused investment in research and development (R&D). We conclude in Sect. 5.

2 Overview of the Technologies

CCS technologies involve removing carbon during the production of energy so that it can be stored rather than released into the atmosphere. There are three main categories of CCS corresponding to three points in the process: pre-combustion carbon capture, alternative combustion (sometimes called oxy-fueling), and post-combustion removal [3]. In this section we discuss each of these in turn. Initial investment costs for each technology will be discussed in Sect. 3.

2.1 Pre-Combustion

Pre-combustion carbon capture technologies, which are used in conjunction with integrated gasification combined cycle (IGCC), separate carbon dioxide from the process stream prior to combustion. IGCC is a technology that is a modification of natural gas-fired combined cycle (NGCC). IGCC gasifies less expensive, otherwise "dirty" fuels such coal and refinery residuals (e.g. petroleum coke) before the gasified stream is combusted in a gas turbine. Waste heat is partially recovered in an integrated steam turbine to maximize efficiency. CO_2, along with criteria pollutants, are removed prior to combustion from the gasified stream. At the beginning of the pre-combustion process, slurry or dry-fed coal is fed along with an oxygen-rich stream to a gasifier where the two streams react to form syngas, a mixture of hydrogen and carbon monoxide:

$$C + H_2O \rightarrow H_2 + CO \tag{1}$$

There are three broad categories of gasifiers: moving-bed gasifiers (Sasol-Lurgi dry ash), fluidized-bed gasifiers (GE and Shell), and entrained-flow gasifiers (i.e. ConocoPhillips E-Gas) [3, 4]. In a conventional IGCC system the next step is sulphur removal. For a plant with capture, the syngas stream is then reacted with high-pressure steam, a process known as the water-gas shift (WGS) reaction, which converts CO into CO_2:

$$CO + H_2O \rightleftharpoons CO_2 + H_2 \tag{2}$$

Research is being conducted on reducing the energy penalty incurred from the water-gas shift reaction through the use of improved WGS catalysts and advanced configurations of shift reactors in combination with various gasifier types (i.e. dry-feed vs. slurry) [5].

The hydrogen and carbon dioxide stream is then purified to remove sulfur dioxide before the carbon dioxide is separated from hydrogen. The CO_2 separation takes place via a physical or chemical absorption process. For example, EPRI has produced a design using Selexol, a physical solvent consisting of a proprietary mixture of glycols [6]. Research is being performed on other physical solvents, such as chilled methanol, and on chemical solvents such as methyldiethanolamine (MDEA) [7]. The captured carbon dioxide is compressed to a supercritical fluid so that it can be transported for storage. The hydrogen product stream can be combusted to produce electric power or stored as chemical potential energy for later use.

It has been noted that *integrated* gasification and carbon dioxide separators could significantly increase IGCC energy efficiency [8].

While few IGCC plants have been built at this point, they have multiple advantages over conventional Pulverized Coal (PC) plants, including higher efficiencies, lower pollutant emissions, excellent performance even when using less expensive, high ash coals [4], and the ability to use biomass fuels. This last feature could result in lower carbon emissions since biomass is part of the carbon cycle. In addition, power plants equipped with pre-combustion technologies have hydrogen and electricity coproduction capabilities [4, 9], which allows for variable power output [9]. Hydrogen and electricity coproduction performance is highest using entrained-flow gasifiers [4]. Finally, IGCC plants are essentially "capture ready".

The main challenges of pre-combustion CCS are to make it energy efficient and reliable. The water-gas shift reaction consumes the largest amount of additional energy, followed by CO_2 compression and CO_2 recovery [3].

2.2 *Alternative Combustion*

Technologies that remove CO_2 during combustion are more varied and involve changing the environment in which combustion occurs so that carbon is not released. The most commonly discussed of these are versions of oxy-fueling. The critical aspect of oxy-fueling technologies is the use of an air separation unit (ASU) which produces an oxygen rich stream that is fed to boiler or superheater. Oxygen-rich combustion produces CO_2-rich flue gas that does not require any CO_2 separation, only purification of the CO_2 stream prior to compression. In addition, the oxygen-rich combustion reaction occurs at higher temperature, resulting in higher combustion efficiency than air-fed counterparts. On the other hand, the ASU creates an energy penalty. The oxy-fueling chemical reactions for coal and gas are identical to the post-combustion reaction equations given below, except for the absence of nitrogen.

Chemical looping, a promising oxy-fueling technology, uses a metal oxygen carrier that removes oxygen from the air and transports the oxygen to be reacted with a hydrocarbon. One ongoing project that uses chemical looping to capture carbon dioxide is being demonstrated on a 10 MWe pilot plant [10]. This technology faces the challenge of finding particles that are both effective and durable at high enough temperatures.

2.3 *Post-Combustion*

Unlike pre-combustion CCS, post-combustion technology builds on existing, industry-standard PC combustion technologies. Post-combustion is an "end-of-the-pipe" pollution control which scrubs exhaust gas of CO_2 emissions. Air and fuel are fed to and combusted in a boiler or superheater. The PC combustion chemical reaction is (assuming complete combustion):

$$C_xH_y + \left(x + \frac{y}{4}\right)O_2 \rightarrow xCO_2 + \left(\frac{y}{2}\right)H_2O \tag{3}$$

The resulting heat of combustion power is transferred via superheated steam to a steam turbine or generator to produce electric power. For coal-fired processes, combustion by-products consist of ash and wet solids bottoms and flue gas. The flue gas contains nitrogen in addition to a variety of air pollutants and particulates, including CO_2. Upon gravity-fed removal of the bottoms, the flue gas is treated to remove carbon monoxide (CO), particulate matter (PM), nitrogen oxides (NO_X) sulfur dioxide (SO_2), and mercury. A separation process captures CO_2 which is then compressed for transportation and storage. Post-combustion separation can be achieved via absorption (using amines), adsorption, membranes, or cryogenic processes, all of which are currently being researched. Amine-based CO_2 separation occurs via the following reaction:

$$2RNH_2 + CO_2 + H_2O \rightleftharpoons (RHN_3)_2CO_3 \qquad (4)$$

where R represents a generalized organic group, in this case the rest of the amine compound involved in the chemical reaction. The double arrows indicate that the CO_2 separation reaction is reversible.

PC combustion includes, in increasing order of efficiency, the following technologies: subcritical pulverized coal combustion (PC), supercritical pulverized coal combustion (SC), and ultra-supercritical pulverized coal combustion (USC). Overall plant efficiency increases as the boiler temperature and pressure increases from subcritical to supercritical to ultra-supercritical [3].

Post-combustion capture can also be used with natural gas combined cycle (NGCC). NGCC is similar to IGCC except that natural gas is used instead of a solid coal or biomass feed. The combustion reaction for natural gas (represented here as pure methane) is:

$$CH_4 + 2O_2 \rightarrow CO_2 + 2H_2O \qquad (5)$$

An advantage of post-combustion capture technologies is that the majority of existing plants are PC and can be retrofitted with bolt-on carbon dioxide separation units and CO_2 compressors. However, retrofitting existing plants results in a significant drop in plant efficiency that is more pronounced for subcritical PC plants than for SC and USC plants [3].

2.4 A Brief Comparison Between Each Technology Branch

These three technology branches are at different levels of maturity, and are potentially useful in different cases. Table 1 was modified from Terry F. Wal [11] and compares each technology to the other. Post-combustion is the most technologically mature of the three technologies, and is most suitable for bolt-on retrofits of existing, uncontrolled plants. In addition, given post-combustion CO_2

Table 1 Comparison between post-combustion, pre-combustion, and oxy-fueling technologies

Technology	Retrofit capability	Partial capture capability	Requires O_2 supply	H_2 production capability
Post-combustion	Yes	Yes	No	No
Pre-combustion	No	Unlikely	Yes	Yes
Oxy-fueling	Sometimes	No	Yes	No

capture is an end-of-the-pipe process and it does not affect upstream combustion, post-combustion could most suitably be applied for partial CO_2 capture, which could be desirable given incremental policy changes mandating CO_2 controls. Finally, post-combustion also does not require an oxygen supply, unlike pre-combustion and oxy-fueling. Post-combustion of course has drawbacks compared to pre-combustion and oxy-fueling: process integration for post-combustion lags behind the other two technologies, an attribute which has a direct impact on plant efficiency.

Pre-combustion can only be used in conjunction with IGCC or NGCC plants, thus limiting its use. But, pre-combustion has the potential for the production and storage of hydrogen (H_2), a capability that allows for better plant operational flexibility, especially given future increases in renewable energy sources that provide variable power over time (e.g. solar and wind).

Oxy-fueling technologies are the least mature, but present opportunities for great efficiencies if development proceeds.

2.5 *International Perspective*

There are numerous overviews on the various strengths and weaknesses of carbon capture technologies in western countries [3, 12], but there has also been a recent increase in the number of papers that discuss advanced, low-cost carbon capture technologies in China. For example, both the popular press and the literature report that Chinese companies are beginning to export advanced coal technologies to the United States, especially more efficient, less expensive gasifiers that are being produced in China [13].

3 Cost Estimation of Capture Technologies

In this section we will discuss the ways that the cost of carbon capture is measured, and present a range of estimates. There are three main ways that the costs of CCS technology options are compared:

- Capital cost per unit of output power ($/kW);
- Levelized electricity cost (LEC) ($/kWh); and
- Unit CO_2 avoidance cost ($/t CO_2 avoided).

3.1 Capital Costs

The capital cost per unit of output power refers to the cost of the additional capital required to capture carbon. It is measured in terms of the net power output of the host plant. For example, PC and NGCC post-combustion capture plants require CO_2 capture and CO_2 compression process units; pre-combustion plants require a WGS reactor, CO_2 capture, and CO_2 compression process units; oxy-fuel capture plants require ASU, CO_2 distillation, and CO_2 compression process units [14]. It should be noted, however, that the failure to consider downstream, integrated carbon capture and storage processes neglects economic benefits that can be attained by holistic changes [3, 14]. This conclusion serves as a reminder that carbon capture and carbon storage are interdependent processes. The IEA gives estimates of additional capital cost ranging from $600 to $1,700/kW [15]. MIT estimates the additional capital cost for post combustion to be in the range of about $700–$1,000/kW, pre-combustion and oxy-fueling about $450–$600/kW [3].

3.2 Levelized Electricity Cost

Measuring additional capital cost is useful, particularly as capital costs are often a deciding factor in which new technologies get adopted. Moreover, it does not require an assumption about discount rates. However, it does not capture the full cost of CCS technologies, as it misses operating costs, and importantly, the *parasitic energy loss* that is associated with most of the technologies. In order to account for both capital costs and energy loss, it is convenient to use the Levelized Electricity Cost (LEC). Additionally, this provides a metric that allows easy comparison across a wide variety of energy generation technologies that have very different cost structures. Here we start by describing measures for the energy loss, and then discuss how this can be used to calculate the LEC.

3.2.1 Energy Penalty

Capture technology requires energy to work, thus the total energy output per unit of fuel will be reduced by any CCS system. Some of the systems have quite high energy losses. This loss is measured in a number of ways in the literature. Rubin et al. [16] define the fractional reduction in plant output per unit of energy input as

$$FR = 1 - \left(\frac{\eta_{capture}}{\eta_{base}}\right) \tag{6}$$

where $\eta_{capture}$ and η_{base} measure net efficiency of a plant with and without capture, respectively. This can also be interpreted as the fraction of energy produced

Table 2 Summary of pre-combustion capture costs surveyed in the literature

Citation[a]	Technology[b]	Energy Penalty	LEC_{base} c/kWh	LEC_{cap} c/kWh	LEC_{add} c/kWh	C_a \$/tonne
MIT	IGCC	0.23	5.6	7.2	1.5	21
Rubin	IGCC	0.23	5.1	6.9	1.7	27
IPCC	IGCC	0.19	5.6	7.4	1.8	27
SFA	GEQ	0.19	6.7	8.8	2.1	**30**
EPRI	GEQ	–	6.2	8.7	2.6	31
EPRI	GERQ	–	6.8	9.3	2.5	34
EPRI	CoP	–	6.2	9.2	3.0	43
EPRI	Shell	–	6.7	9.6	2.9	55
IEA	IGCC	0.15	4.2	8.4	4.2	57
IEA	Biomass IGCC	0.50	7.1	12.4	5.3	**82**
Al-Juaied	IGCC	–	8.0	18.2	10.2	148
		Mean	*6.2*	*9.6*	*3.4*	*51*

[a]The key to the citations is as follows: Al-Juiaed [8]; EPRI from EPRI Report 1013355 as reported in [8]; SFA from SFA: Pacific study as reported in [8]; MIT [3]; IEA [15]; IPCC [12]; Rubin et al. [16]

[b]The key to the technologies is as follows: *CFB* circulating fluidized bed technology, *PC* subcritical pulverized coal technology, *SC* supercritical pulverized coal technology, *USC* ultra-supercritical pulverized coal technology

which is devoted to capture rather than production of electricity. Another metric is the energy penalty *EP* – the increase in plant energy input per unit of output – defined as

$$EP = \left(\frac{\eta_{base}}{\eta_{capture}}\right) - 1 = \frac{FR}{1 - FR} \tag{7}$$

The estimated energy penalty varies considerably across the technologies. We have reported some estimates in Tables 2, 3 and 4.

3.2.2 Calculating the LEC

The LEC is the total cost of generating electricity amortized over the total amount of electricity produced by the plant. For fossil fuel plants the LEC is made up of the levelized cost of capital, the fuel cost, and Operating and Maintenance (O&M) costs. The levelized capital cost amortizes the initial capital cost over the total energy output. Specifically, we amortize the capital cost of a kW of power, K_p, at a discount rate δ over a lifetime L, and divide the result by a factor to convert power to the energy produced in a year: the capacity factor, CF multiplied by the number of hours in a year (h), 8,760. Let the amortization factor be represented by $f(\delta, L) \equiv \frac{\delta}{(1-(1+\delta)^{-L})}$. Then

Table 3 Summary of post-combustion capture costs surveyed in the literature

Citation[a]	Technology[b]	Energy Penalty	LEC_{base} c/kWh	LEC_{cap} c/kWh	LEC_{add} c/kWh	C_a \$/tonne
MIT	CFB	0.36	5.1	8.5	3.4	44
MIT	SC	0.23	5.3	8.4	3.2	44
MIT	USC	0.27	5.2	8.1	2.9	45
MIT	PC	0.37	5.3	9.0	3.6	45
SFA	SC	0.27	3.7	9.8	6.2	47
Rubin	SC	–	4.9	8.1	3.1	49
IPCC	SC	0.31	5.5	8.7	3.2	49
EPRI	SC	–	5.7	9.9	4.2	59
IEA	SC	0.24	4.0	8.1	4.1	**69**
		Mean	*5.0*	*8.7*	*3.8*	*50*

[a]The key to the citations is as follows: EPRI from EPRI Report 1013355 as reported in [8]; SFA from SFA: Pacific study as reported in [8]; SFA from a study by SFA: Pacific as reported in [8]; MIT [3]; IEA [15]; IPCC [12]; Rubin et al. [16]
[b]The key to the technologies is as follows: *CFB* circulating fluidized bed technology, *PC* subcritical pulverized coal technology, *SC* supercritical pulverized coal technology, *USC* ultra-supercritical pulverized coal technology

Table 4 Summary of oxy-fueling combustion capture costs surveyed in the literature

Category	Citation[a]	Technology[b]	Energy Penalty	LEC_{base} c/kWh	LEC_{cap} c/kWh	LEC_{add} c/kWh	C_a \$/tonne
Post-combustion and oxy-fueling	IPCC	NGCC	0.16	4.4	6.4	2.0	63
	Rubin	NGCC	–	4.1	5.9	1.9	63
	EPRI	NGCC	–	6.0	8.4	2.4	**71**
Natural gas	SFA	NGCC	0.13	6.5	8.9	2.3	78
	IEA	NGCC	0.09	4.8	7.6	2.8	**93**
	IEA	NGCC OXY	0.28	4.3	6.7	2.4	**68**
			Mean	*5.0*	*7.3*	*2.3*	*73*
Oxy-fueling	MIT	SC	–	5.6	7.7	2.0	33
	SFA	SC	–	6.7	10.2	3.4	49
	Rubin	SC	–	4.9	8.7	3.7	**59**
	IEA	SC	0.24	4.4	8.5	4.1	**59**
			Mean	*5.4*	*8.7*	*3.3*	*50*

[a]The key to the citations is as follows: EPRI from EPRI Report 1013355 as reported in [8]; SFA from SFA: Pacific study as reported in [8]; SFA from a study by SFA: Pacific as reported in [8]; MIT [3]; IEA [15]; IPCC [12]; Rubin et al. [14, 16]
[b]The key to the technologies is as follows: *NGCC* natural gas combined cycle technology, *NGCC OXY* oxy-fueled natural gas combined cycle technology, *SC* supercritical technology

$$LEC_{CAP} = \frac{K_p}{CF \cdot h} \cdot f(\delta, L) \tag{8}$$

The levelized fuel cost, LEC_{fuel} will depend on the cost of the fuel, C_{fuel} and the efficiency of the plant η, as follows:

$$LEC_{fuel} = \frac{0.003412 \cdot C_{fuel}}{\eta} \tag{9}$$

where C_{fuel} is the fuel price in dollars per MMBtu in terms of High Heating Value (HHV), the constant 0.003412 converts from MMBtu to kWh, and η is the efficiency of the plant. O&M costs can also be expressed as a levelized cost, $LEC_{O\&M}$. The total LEC is the sum of the levelized capital, fuel, and O&M costs:

$$LEC = LEC_{CAP} + LEC_{fuel} + LEC_{O\&M} \tag{10}$$

Both baseline (i.e. non-capture) levelized cost, $LEC_{\mathbf{base}}$ and capture levelized cost, $LEC_{\mathbf{capture}}$ can be calculated using Eq. 10. Then, we can calculate the additional LEC from CCS, LEC_{add} as the difference between $LEC_{capture}$ and LEC_{base}:

$$LEC_{add} = LEC_{capture} - LEC_{base} \tag{11}$$

In Tables 2, 3 and 4, which each provide a range of levelized baseline, capture and additional costs, we calculate LEC_{add} using Eq. 11. We can also calculate $LEC_{\mathbf{add}}$ directly from the additional capital cost, K_{add}, the energy penalty EP, the baseline efficiency of the plant η_{ref}, the fuel price C_{fuel}, and any additional O&M cost $LEC_{O\&M,add}$, as follows:

$$LEC_{add} = \frac{K_{add}}{CF * h} f(\delta, L) + \frac{0.003412 \cdot C_{fuel} \cdot EP}{\eta_{ref}} + LEC_{O\&M,add} \tag{12}$$

This assumes there is no reduction in the capacity factor with CCS. We are most interested in the total levelized capture costs $LEC_{capture}$ (since this is what will drive the market share of fossil generation with CCS versus other low- or no-carbon options), and the additional LEC due to CCS, LEC_{add}. We have reported some estimates in Tables 2, 3 and 4.

3.3 CO_2 Avoidance Cost

Finally, the third category of cost metrics is the CO_2 avoidance cost. This is useful, since it gives an idea of how high a carbon price would have to be before CCS is adopted. The CO_2 avoidance cost can be calculated from the LEC as follows:

$$C_a = \frac{LEC_{capture} - LEC_{base}}{Q_{base} - Q_{capture}} \tag{13}$$

Where $LEC_{capture}$ and LEC_{base} refer to the LEC of plants with and without capture, respectively, and Q_{base} and $Q_{capture}$ refer to the carbon emissions per unit of electricity (i.e. kWh or MWh depending on how the LEC is measured) for plants without and with capture, respectively. Tables 2, 3 and 4 show a range of estimates for CO_2 avoidance cost.

3.4 Summary

Tables 2, 3 and 4 summarize CCS capture costs for pre-combustion, coal post-combustion, and natural gas and oxy-fueling methods as reported in the literature. Cost estimates within each category are presented in order of lowest to highest CO_2 avoidance cost. We present the energy penalty (*EP*), the LEC for no capture (LEC_{base}), the additional LEC (LEC_{add}), the total LEC with capture ($LEC_{capture}$), and the CO_2 avoidance cost (C_a), as well as averages across all references. Some papers did not explicitly report the CO_2 avoidance cost, so we have calculated it. These values are emphasized in bold sans serif type in the table. If emissions rates were not explicitly given we assumed pre-capture rates of 1,700 and 836 lb/MWh for coal and gas respectively, and a capture rate of 90%. All monetary amounts are presented in 2009 dollars.

Note that the additional LEC and the CO_2 avoidance cost are not perfectly correlated across the different studies. For example, in Table 3 the MIT CFB and SC technologies have higher additional LEC than MIT USC, but lower CO_2 avoidance cost. This is due to differences in the baseline rate of emissions for the technologies, as well as to actual capture rate. The USC technology has a lower baseline emissions rate, therefore fewer emissions are captured with the technology. We see a similar issue with the natural gas technologies in Table 4. They have the lowest additional LECs, but higher cost of avoidance, since gas has lower emissions in the baseline.

Pre-combustion and post-combustion mean CO_2 avoidance costs are approximately the same, but pre-combustion has much more variance between the estimates, largely due to two outliers – Al-Juaied and IEA IGCC with biomass. Without those outliers the mean CO_2 avoidance cost is only \$36/t (close to the median of \$34), much lower than the means for the other technologies, and in fact lower than any individual estimate of post-combustion. The least robust are the estimates for oxy-fueling, with the fewest papers that have attempted this, and a lower amount of data reported in these papers.

A very important issue to consider is that the additional LEC and the CO_2 avoidance cost are in each case comparing the given technology with and without CCS. To get a better estimate, we would want to compare the actual technology that would be used in the absence of climate change policy with the given technology with CCS. For example, we may want to use the lowest baseline LEC for each coal study. In this case the pre-combustion additional LEC increases to 5.4/kWh and the CO_2 avoidance cost is \$80.2/t.[1] This gives an estimate of reducing CO_2 below

[1] The mean LEC_{add} corresponding to the lowest levelized baseline cost LEC_{base} was calculated by (1) selecting the lowest LEC_{base} in 2009 USD, (2) subtracting the lowest LEC_{base} from each estimate's $LEC_{capture}$ to generate a new LEC_{add} for each estimate, and (3) averaging across the new LEC_{add} values to calculate the mean LEC_{add} corresponding to the lowest LEC_{base}. The mean CO_2 avoidance cost C_a corresponding to the lowest LEC_{base} can be calculated by simply multiplying the original mean C_a by the ratio of the new LEC_{add} over the original LEC_{add}.

the otherwise lowest cost technology. Which cost is appropriate depends on assumptions about the technology that will be otherwise most competitive. If IGCC turns out to be competitive because of other pollution emissions, for example, then the costs given in the table above are appropriate. But, if IGCC is being used primarily to reduce CO_2 emissions, then the higher cost is more appropriate. Note that while the CO_2 avoidance cost is greater, the total LEC with capture is much lower for post-combustion than for pre-combustion.

4 Technical Change and CCS

In this section we discuss how technical change can be modeled for CCS. We consider two large categories of technical change. The first is technical change that is related to the quantity of production. These production-related changes include learning-by-doing (LBD) and scale effects. Knowledge acquired from production and economies of scale drives the resulting cost reductions. Ultimately they are spurred by increases in demand. The second category is non-incremental technical change that results from investments in research and development (R&D). Understanding the potential improvements from both of these avenues is very important for policy design.

4.1 *Production Effects*

Increased production has been noted to decrease production costs through two channels – LBD and returns to scale. Experience effects have been noted empirically in manufacturing since Wright [17] and in services since Bryan and Harter. First, it was qualitatively observed that workers became more efficient as they produce more units. Second, it was observed that unit costs decrease with accumulated production experience. Arrow [18] went on to formalize the concept of LBD in his seminal paper, focusing mainly on labor costs. In Sect. 4.1.1 we will discuss how the concepts of LBD can be applied to CCS. Several studies, however, have criticized the attribution of cost reductions to LBD (see for example Dutton and Thomas [19]). Argote and Epple [20] proposed four alternative hypotheses for the observed technical improvements: economies of scale, knowledge spillovers, organizational forgetting, and employee turnover. In Sect. 4.1.2 we will focus on the first of these, returns to scale.[2] In particular, it has been argued that capital costs respond more to returns to scale, while O&M costs respond more to LBD [22]. Finally, in Sect. 4.1.3 we discuss some applications of production effects and

[2] Knowledge spillovers are especially important in firm-level analyses of technical change. The latter two categories are typically addressed by depreciating the value of experience over time [21].

provide an illustrative example of combining LBD and returns to scale to estimate future costs.

Although there are gains that result from increasing the size of individual CCS plants, we ignore this increasing unit scale effect by assuming that: (1) optimal scale is similar to that of current coal plants, (2) this optimal scale is achieved quickly, and (3) that it is uniform so that every future CCS plant operates at this optimal scale. Other work modeling the cost of CCS generally assumes plants are built at steady scale and that the main returns to unit scale are in increasing the size of the transport infrastructure [23].

4.1.1 LBD

LBD has most commonly been represented as a power function, for its simplicity and generally very good fit to observations. Measures of fit for energy technologies are often well above 0.90 [24]. Let total O&M costs at time t be represented by M_t and O&M costs for the n individual components (see the top of Sect. 1 for a discussion of the different units) be represented by $m_{i,t}$. Then the learning curve can be represented as follows:

$$M_t = \sum_{i=1}^{n} m_{i,t-1} \cdot \left(\frac{E_t}{E_{t-1}} \right)^{-b_i} \tag{14}$$

where experience (E) in operating CCS plants is the sum of the CCS electricity output since $t = 0$. Studies of learning rates for O&M costs have been calculated for an array of similar technologies [25–27]. Estimates of O&M learning rates for CCS in particular also exist [14, 28, 29]. Based on these studies, Table 5 provides the results of a survey of learning rates for technologies relevant to carbon capture and sequestration. The median value from these studies, 0.11, is slightly below the value from a survey of a broad set of learning rates in the energy sector [30], suggesting that CCS may improve at a slower rate than smaller scale technologies which may involve many more, albeit smaller, units.

4.1.2 Returns to Scale

Increases in demand for CCS will lead to increases in manufacturing capacity for the individual components, which enable the manufacturing firms to reduce costs through economies of scale. Specifically, as demand for power plants with CCS increases, manufacturing of the components – gasifiers, air separation units, etc. – will increase so that production will grow to a more efficient scale. Manufacturers of these components take advantage of the resulting opportunities for cost reductions through, for example, spreading fixed costs, investing in automated processes, and developing specialized equipment. These changes reduce the cost

Table 5 Estimates of learning rates for technologies related to carbon capture

Analogous technology	Capital cost	Operations and maintenance
Flue gas desulfurization (FGD)	0.110	0.220
Hydrogen steam methane reforming	0.270	0.270
Coal IGCC	0.000	0.000
	0.100	0.050
	0.100	0.060
	0.110	0.220
	0.120	0.220
	0.140	0.120
Liquefied natural gas (LNG)	0.140	0.120
	0.000	0.000
	0.100	0.060
	0.100	0.060
	0.110	0.220
Oxygen production	0.100	0.050
Pulverized coal	0.000	0.000
	0.050	0.180
	0.060	0.150
	0.110	0.220
	0.120	0.220
Selective catalytic reduction (SCR)	0.120	0.130
	0.168	0.269
Summary statistics		
Mean	0.101	0.135
Median	0.110	0.130
Standard deviation	0.060	0.092
n	21	21
Maximum	0.270	0.270
Minimum	0.000	0.000

of components for power plants. The general relationship is that costs of components fall as demand for CCS increases. Specifically, the total capital costs of CCS M_t will respond to scale in the following way, where $m_{i,t}$ is the capital cost of technology component i at time t, n is the total number of components, k_t and k_{t-1} are the levels of demand for CCS plants in period t and the previous period $t-1$, respectively; and a_i is the component-specific scaling factor for component i.

$$M_t = \sum_{i=1}^{n} m_{i,t-1} \cdot \left(\frac{k_t}{k_{t-1}}\right)^{a_i} \tag{15}$$

This gives the total cost, unlike in the learning curve above, which gives the per-unit cost. Total cost increases as scale increases, but less than proportionally. To get the per-unit cost, we divide M_t by k_t.

Since CCS components are not being manufactured at any large scale currently, we suggest using empirical estimates from other industries about the effects on unit

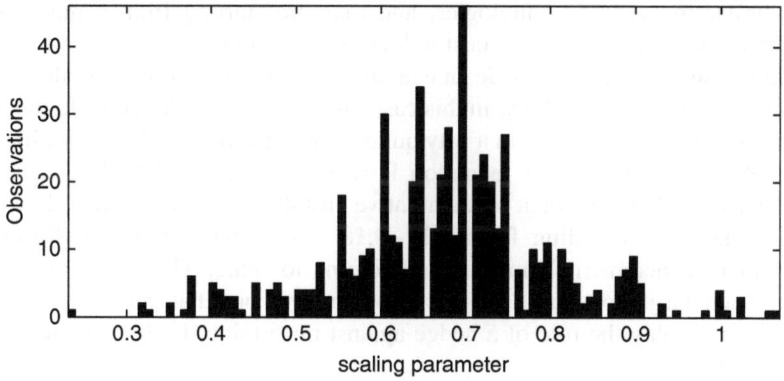

Fig. 1 Histogram of scaling factors relevant to CCS (n = 570)

cost of increases in manufacturing scale. One way to address this is to use estimates of cost reductions that result from up-scaling in the chemical and related industries, which probably includes the largest set of empirical estimates on scale [31]. Moreover, the processes in the chemical industry are rather similar to those in a CCS plant. We use estimates of scaling parameters from many studies surveyed by Remer and Chai [32]. Figure 1 shows scaling factors for 570 manufacturing processes cited in their study. These include chemical processes, oil refineries, power plants, and pollution controls. For the 570 scaling factors, we calculated the following descriptive statistics: mean = 0.68; median = 0.68; standard deviation = 0.13: minimum = 0.13: maximum = 1.07.

4.1.3 Production Effects on Future Costs

The production effects on future costs depends on the future demand for CCS, which in turn depends on costs, as well as carbon policies, the broader economy, and the characteristics of competing technologies. Thus, estimating the size of production effects on future costs requires a sophisticated dynamic model. Here we discuss two papers – one on applying learning curves, the other on returns to scale. We go on to provide a very simple illustration of how to combine these two effects in order to estimate the impact that increases in demand can have on the cost of CCS.

Rubin et al. [14] provides an example of applying learning curves to CCS. In this paper they first develop a set of experience curves for seven technologies relevant to CCS. They then decompose CCS plant designs into sub-systems, and apply a learning rate from the related technologies to each of the sub-systems based on judgement of similarity. These learning rates ranged from 0.05 to 0.27, with a median of 0.12. From this, they estimated the overall learning rate for CCS plants, and found that this was only between 0.03 and 0.05, much lower than most of the component learning rates. The reason for this is that many of the subsystems are very similar to current technologies, and thus are starting from a much higher experience base, leading to less cost reduction over time.

Nemet and Baker [33] provide an example of applying returns to scale to a pre-commercial energy technology, in this case purely organic solar cells. In this paper they modeled returns to scale in a way quite analogous to LBD. The key difference is that the scale of the cost reduction is related to new installed capacity in a particular period, rather than to cumulative installed capacity. Thus, they apply Eq. 15, but used a scaling factor of −0.18. They considered both R&D with uncertain non-incremental impacts and returns to scale. They found that R&D had a much larger potential for technical change, but that subsidies to induce demand could play the role of a hedge against uncertainty in the outcomes of the R&D program.

Since CCS will probably have both LBD and returns to scale, the ideal model would incorporate both. Here we present a simplistic example of exogenous technical change through both LBD and returns to scale. It is exogenous because we are taking the demand for CCS as given. In reality, production effects have positive feedbacks – the lower the price, the higher the demand, the lower the price in the next period.

We will apply LBD to O&M and returns to scale to capital costs. In general, the dynamics of O&M costs respond to different influences than do the capital cost components. There is much more evidence that O&M costs change as a function of learning-by-doing than there is for capital costs, which tend to respond to economies of manufacturing scale [34]. As in many other industrial facilities, plant operators find ways to improve the operation of their plants as they gain experience. For example, we might expect improvements in capacity factor because less maintenance is needed as operators make better decisions.

We have taken the demand from the CCTP scenario called "CCS Stabilization" [35]. The demand for energy generated using coal with CCS is approximately 10 EJ in 2035, 100 EJ in 2050, and 250 EJ in 2100. We assume that the demand increases linearly between these points, and model 5 year periods. We translate EJ into kWh, and then translate that into installed capacity measured in kW. We then calculate the demand for new capacity, as the difference between installed capacity in successive periods. Table 6 shows the values we used up to 2100. It can be seen from this table that the demand for new installations increases dramatically in 2040.

For illustration, we will use the mean CO_2 avoidance cost of \$51/t from Table 2 above. This is equivalent to an additional LEC of 3.54 cents/kWh. We assume that this LEC can be apportioned into capital costs, O&M costs, and fuel costs in the ratio 69:22:09. We have taken this ratio from David and Herzog [36], Table 2, using the values for IGCC 2000. We apply Eq. 14 to the O&M costs, using a learning ratio of 0.13 and total installed capacity as representing the total experience. We apply returns to scale to the total capital cost of the new capacity, using a scaling factor of 0.68, and then divide total capital cost by the new capacity, to get a per-unit cost. We assume that the energy penalty will not be affected by learning or scale, and so the fuel cost remains the same throughout the simulation. Figure 2 shows the results of this exercise, with the CO_2 avoidance cost measured on the left axis and

Table 6 Summary of simulation parameters and results

Year	2020	2030	2040	2050	2060	2070	2080	2090	2100
Demand (EJ)	1.7	6.7	40.0	100.0	130.0	160.0	190.0	220.0	250.0
Demand (TWh)	463	1,852	11,111	27,778	36,111	44,444	52,778	61,111	69,444
New CCS capacity (GW)	67	133	1,267	1,333	1,867	1,933	2,467	2,533	3,067
Cumulative CCS capacity (GW)	67	267	1,667	4,267	6,867	10,667	14,468	19,468	24,468
CO_2 avoidance cost ($/t)	51.3	42.4	25.9	24.8	23.0	22.5	21.4	21.1	20.3

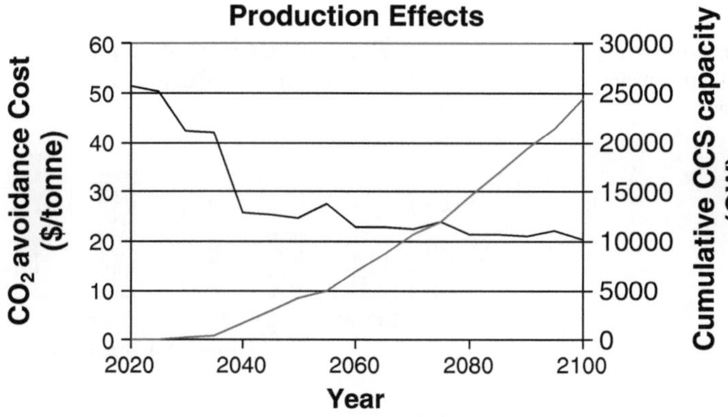

Fig. 2 Cost of CO_2 avoidance and total installed capacity

cumulative installed capacity measured on the right axis. The CO_2 avoidance cost moves downward through time, but not smoothly. The large decreases are due to large increases in the scale of production. In particular, as we point out above, there is a very large increase in scale between 2035 and 2040 in order to meet the large demand in 2050. We also note that the cost *increases* in some periods. This is due to the decrease in scale when demand starts to increase at a slower pace. In particular, given the data above, the pace of new plants decreases after 2050, thus, there is a bump up in costs. The learning effects are much smoother, but not as pronounced. The LEC due to capital costs decreases from 2.45 cents in 2020 to 0.72 cents in 2050; while the LEC due to O&M decreases from 0.78 to 0.36 during the same period.

The results are quite sensitive to assumptions about the quantity of the initial stock. In Fig. 2, we have assumed that the initial CO_2 avoidance cost is for demand equal to 1.7 EJ, as shown in Table 6. To investigate the importance of this assumption, we vary this initial stock between 0.01 and 10. Figure 3 shows how the reduction in the cost is related to the assumption about the initial stock. This is the point that Rubin et al. [14] make in their paper, and indicates that care must be

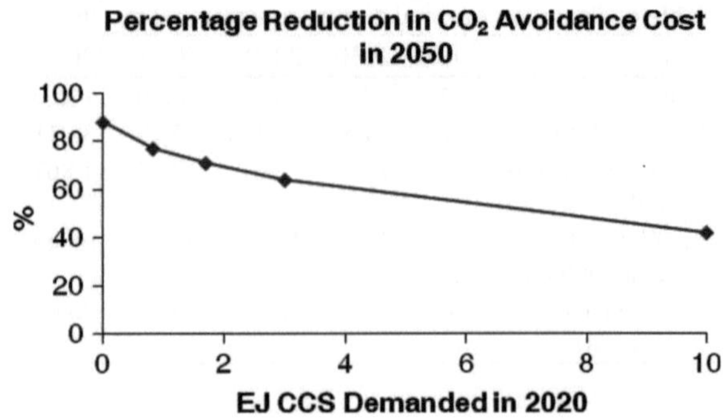

Fig. 3 Impact of the initial capacity stock on cost reductions

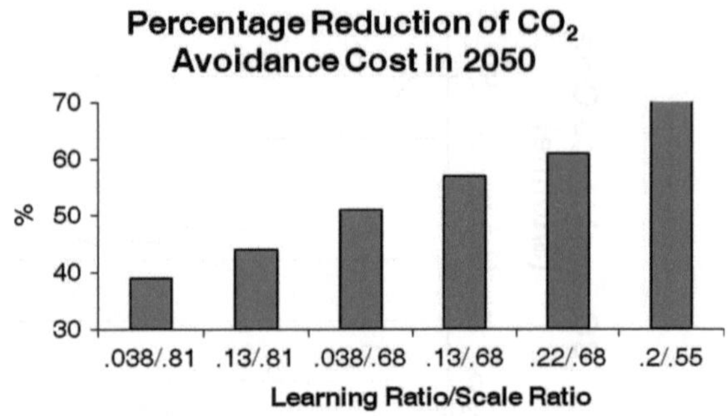

Fig. 4 Impact of learning ratio and scale parameter on cost reductions

taken to correctly assess the initial experience and level of production when modeling cost reductions through production effects.

The calculations are somewhat less sensitive to the values of the learning ratio and scale factor, but still impacted. Figure 4 shows the percentage reduction in cost under different combinations of the learning ratio and scale factor. In the worst case, the learning ratio is lower by one standard deviation and the scale factor is higher by one standard deviation; in the best case these are reversed. A change in the scale factor by one standard deviation has a larger impact than a change in the learning ratio.

These calculations are meant to be illustrative only, showing the impact of various assumptions on the potential for cost reduction through production impacts. Carefully combining these two avenues of production impacts is open to future work.

4.2 *Research and Development*

A second avenue for technical change is through R&D, sometimes called learning through searching. This refers to a focused investment into finding new technological solutions to a problem. It is possible to have non-incremental technical change through this avenue. The relationship between R&D investments and technical

Table 7 Funding trajectories from Baker et al. [40]

Technology	Years	Low	Medium	High
Pre-combustion	10	5	20	50
Oxy-fueling[a]	10	0.5/5	1/10	5/10
Post-combustion	15	5	15/30	50

[a]With chemical looping

Table 8 Technological endpoints from Baker et al. [40]

Technology	Energy Penalty	LEC_{base} c/kWh	LEC_{add} c/kWh	LEC_{cap} c/kWh	C_a $/tonne	$LEC_{capture}$ % Increase
Pre-combustion	0.11	5.0	*0.8*	5.8	13.4	16%
Post-combustion	0.30	4.4	1.7	6.1	25.0	39%
Oxy-fueling[a]	–	4.4	0.6	5.0	8.9	14%

[a]With chemical looping

change is inherently uncertain. For more mature technologies, historical data and historical comparisons, e.g., Moore's law from the semiconductor industry, may exist and be useful. With highly innovative technologies, however, history provides only sketchy guidance. In such cases, common to R&D management, decision analytic techniques are often used to obtain the necessarily subjective judgment of experts who are most familiar with the specific technologies [37–39]. Here we review three papers that report on the results of expert elicitations into CCS technology.

Baker et al. [40] elicited probabilities of technical success conditional on US funding trajectories, for three categories of CCS: pre-combustion, chemical looping, and post-combustion. They defined technological endpoints in terms of a number of factors, and assessed the probability of achieving these endpoints conditional on US government funding (see Table 7 for funding trajectories). Table 8 summarizes the technological endpoints using the metrics discussed in Sect. 3 above. The numbers in bold are based directly on data from the paper. All other data used for calculations are from David and Herzog [36]. Baker et al. [40] found (1) that the collection of more information appears to be of high value, especially on chemical looping (an oxy-fueling technique), but on the other two categories as well; and (2) a significant amount of disagreement between experts, even over the most mature technology, which was most pronounced in regards to cost estimates. The goal of the study was to determine the impact on the entire marginal abatement cost curve, and so the elicitations were specifically NOT conditioned on a carbon tax. Thus the study did not get at the possible impact of demand-pull instruments.

An NAS study [41] focused on implementation of CCS technologies by 2025. They compare a no-DOE funding scenario with the DOE program as currently projected to be funded. The DOE projected spending a mean of about $220 M a year over a 20 year period starting in 2001, on the CCS program. This program, however, includes six research streams: CO_2 capture; carbon storage; monitoring, mitigation and verification; breakthrough concepts; non-CO_2 greenhouse gases; and infrastructure development. The first research stream is most similar to the

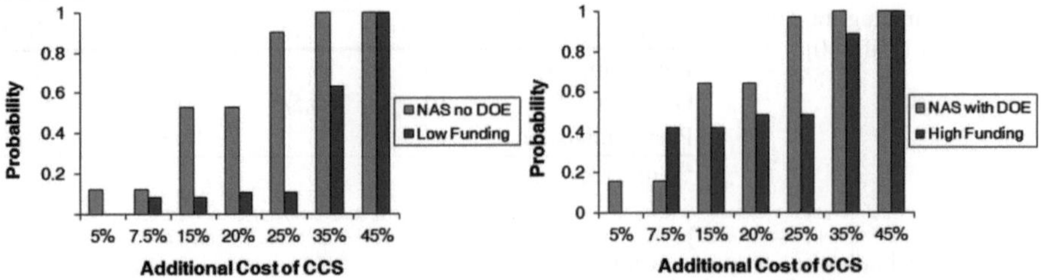

Fig. 5 The cumulative probability over the additional COE needed to achieve CCS

technologies assessed in Baker et al. [40] and discussed in this review. The research stream on "breakthrough technologies" also considers CO_2 capture technologies and so is relevant.

The NAS study [41], rather than defining endpoints in terms of specific technology, considered the overall success in terms of increase in the LEC. Four possible choices of increases in LEC were given to the expert panel: less than 10%; between 10% and 20%; between 20% and 30%; or greater than 30%. The mean increase in LEC from the panel assessments was about 15–18% with the DOE program and about 16–20% without it, depending on the level of carbon tax [41].[3,4]

Figure 5 compares the mean probabilities in the NAS study [41] with Baker et al. [40].[5] The left panel compares the NAS assessments assuming no DOE funding, with the Baker assessments assuming a low funding trajectory for each technology, a total of about \$15 M per year. The right panel compares the NAS assessments assuming DOE funding (\$220 M over six categories), with the Baker assessment assuming a high funding trajectory for each technology, a total of about \$110 M per year. The mid-points of each range are used for the NAS figures; that is, "less than 10%" is represented by 5%, etc. The figures show the probability of achieving the additional cost to LEC on the horizontal axis, or better. So, for example, in the left panel with no or low funding, the NAS has a probability of achieving an additional cost of 15% or less of about 55%; Baker et al. has about 7.5%. Note that the estimates without breakthrough technical change given in Tables 2, 3 and 4 above report an additional LEC ranging from 27% to over 100%. With the means for each technology ranging between 50% and 80%. The NAS panel, even in the absence of DOE funding find that additional LEC will be *no more* than 35% whereas only

[3] The carbon tax is assumed to impact the private sector incentive to innovate; but does not impact the increase in LEC directly.

[4] These estimates of LEC only include additional cost of capture and compression, but exclude the cost of geologic storage. They are presented only for comparison purposes. The actual LEC may vary by fuel type, sequestration method and distance, and changes in carbon price and demand for fuel. The estimates presented here assume zero carbon price and a baseline coal price of \$1.7/GJ of primary energy. The detailed assumptions and calculations are available in Baker et al. [40].

[5] Reprinted from Baker et al. [40].

seven of the studies find an increase in LEC of 40% or less. NAS finds a 50–50 chance that the additional LEC will be no more than 15%. This corresponds to a CO_2 avoidance cost of about \$10/t CO_2 avoided.[6] Thus the NAS panel appears quite optimistic next to Tables 2, 3 and 4.

Rao et al. [42] elicit technological parameters and then extrapolate the cost impacts of those parameters using a model. They consider one sub-technology of post-combustion – amine-based capture – focus on the near term (2015), and condition only on the current US funding trajectory ("with modest increases"). The results suggest that the CO_2 avoidance cost and the additional LEC will be reduced by about 18% on average, from baseline values of \$47/t and 3.8 cents/kWh to \$39 and 3.1 cents. The most optimistic outcomes, with probabilities of well under 5%, were \$30 and 2.5 cents, respectively. If we assume a baseline LEC of 6.8 cents/kWh, then the mean and most optimistic estimates are equivalent to increases in the LEC of 46% and 37%, respectively. If we compare this with the two studies above, this implies either that this technology is much less promising than the other CCS technologies considered in the studies above, or that this group of experts is much more pessimistic. These values imply a small improvement over the values given in Table 2, 3 and 4.

5 Conclusions

In this paper we have briefly reviewed the three categories of carbon capture technologies – pre-combustion, oxy-fueling methods, and post-combustion. We have defined and discussed key metrics for evaluating the costs of these technologies, namely, capital cost, energy penalty, the levelized cost of electricity, and the CO_2 avoidance cost. In Sect. 3 we provide a summary of the cost estimates from the literature and tie together these various metrics. We then go on to discuss the avenues of technical change for CCS, distinguishing between production effects and R&D. We provide an overview of two key production effects, LBD and returns to scale. In Sect. 4.1.3 we provide an illustrative example of combining these two effects. In Sect. 4.1.2 we reviewed three papers that provide estimates for the efficacy of R&D into carbon capture technologies.

We find that there is a wide range in the current estimates of the cost of the technology, both between different groups of researchers, and between different technologies for given groups of researchers. Pre-combustion generally has a lower CO_2 avoidance cost, but a higher total LEC. The MIT group consistently has the lowest estimates, and the IEA consistently has the highest. One challenge is to pin down exactly what assumptions about the state of the technology have been made for these baseline estimates so that they can be used appropriately with LBD and returns to scale models.

[6] Assuming 0.018 t of carbon per kWh.

We saw that production effects can appear to lead to dramatic cost reductions, potentially reducing the costs by a factor of 4 or 5. However, great care must be taken when attempting to model cost reductions through these avenues. First, it seems likely that both LBD and scale will play a role, but it is less clear how these different effects should be applied, and to which components and aspects of cost. Second, it is crucial to very carefully align initial costs with an associated initial production level and experience level. In particular, it can sometimes be difficult to determine how much experience is assumed in the cost estimates provided in Tables 2, 3 and 4. Moreover, the components of CCS vary considerably in the amount of experience that has been accrued in their manufacture, with some being very similar to mature technologies. Finally, there is always a question about how far costs can really decrease given physical limitations.

In this paper we have focused on *modeling* future costs, but if these models are to be used to analyze polices, then further care must be taken. In particular, LBD does not necessarily justify policy intervention on its own; only if there are spillovers is policy justified. Similarly, returns to scale have existed in many industries and has not always proven a problem. However, if the firms likely to produce CCS components have a history of being particularly myopic; or are likely to have trouble raising the required capital (perhaps because the demand for CCS will be entirely policy-driven), then policies aimed at increasing the scale may be justified. Of course, in the absence of a direct climate policy, such as a carbon tax or cap and trade, it may be argued that directed technology policy is a second-best response.

In reviewing the three R&D studies, we see that as the study becomes more technologically detailed, the predictions become less optimistic. Moving from Rao et al. [42], to Baker et al. [40], to NAS [41], we move from a very detailed study of a particular technology, to a broad study of three technological categories, to a general characterization of all carbon capture technology. The best guesses (means or medians) of these three studies for the percent increase in LEC due to capture given minimal government funding are about 46%, 30%, and 15% respectively. On the one hand, the chance of success must increase when considering a wider range of technologies, and so this progression is logical. On the other hand, it is possible that when the technological detail is suppressed, some of the barriers to achieving success become less obvious. It would be ideal to combine a set of technologically detailed elicitations with a broad overview of the potential for change across all technologies.

Finally, in order to craft the most efficient and effective policies, there is a need to model a combination of production effects and non-incremental technical change resulting from R&D. An understanding of how these effects are likely to interact will help provide support to the design of an optimal, coherent portfolio of policies to address climate change in the presence of technical change.

Acknowledgement The research leading to these results was completed while Baker was visiting the Precourt Energy Efficiency Center at Stanford University; and was partially supported by funding from the European Research Council under the European Community's Seventh Framework Programme (FP7/2007–2013)/ERC grant agreement n 240895 – project ICARUS "Innovation for Climate Change Mitigation: a Study of energy R&D, its Uncertain Effectiveness and Spillovers".

References

1. Energy Information Administration (2009) International energy outlook. Technical report, U.S. Department of Energy
2. Energy Information Administration (2006) Independent statistics and analysis: international electricity generation. U.S. Department of Energy
3. Katzer J, Moniz EJ, Deutch J, Ansolabehere S, Beer J (2007) The future of coal: options for a carbon-constrained world. Technical report, Massachusetts Institute of Technology, Cambridge, MA
4. Cormos C, Starr F, Tzimas E (2010) Use of lower grade coals in IGCC plants with carbon capture for the co-production of hydrogen and electricity. Int J Hydrogen Energy 35:556–567
5. Carbo MC (2008/2009) Staged water-gas shift configuration: key to efficiency penalty reduction during pre combustion decarbonisation in IGCC. Energy Procedia 1(1):16–20
6. Esber GS III (2006) Carbon dioxide capture technology for the coal-powered electricity industry: a systematic prioritization of research needs. Master's thesis, Massachusetts Institute of Technology
7. Heintz YJ, Sehabigue L, Morsi BI, Jones KL, Pennline HW (2008) Novel physical solvents for CO_2 capture from fuel gas streams at elevated pressures and temperatures. Energy Fuel 22:3824–3837
8. Al Juaied MA, Whitmore A (2009) Realistic costs of carbon capture discussion paper. Technical report, Belfer Center for Science and International Affairs, Cambridge, MA
9. Haines MR, Davison JE (2009) Designing carbon capture power plants to assist in meeting peak power demand. Energy Procedia 1:1457–1464
10. Mattisson T, Lyngfelt A, Leion H (2009) Designing carbon capture power plants to assist in meeting peak power demand. Int J Greenhouse Gas Control 3:11–19
11. Terry F. Wal (2007) Combustion processes for carbon capture. In: Proceedings of the Combustion Institute, doi:10.1016/j.proci.2006.08.123
12. Metz B, Davidson O, de Coninck HC, Loos M, Meyer LA (eds) (2005) IPCC special report on carbon dioxide capture and storage. Cambridge University Press, Cambridge
13. Gallagher KS (2009) Key opportunities for U.S. China cooperation on coal and CCS. Technical report, Brookings Institution
14. Rubin ES, Yeh S, Antes M, Berkenpas M, Davison J (2007) Use of experience curves to estimate the future cost of power plants with CO_2 capture. Int J Greenhouse Gas Control 1(2):188–197
15. Kerr T (ed) (2008) CO_2 capture and storage: a key carbon abatement option. OECD/IEA, Paris. ISBN 978-92-64-041400
16. Rubin ES, Chen C, Rao AB (2007) Cost and performance of fossil fuel power plants with CO_2 capture and storage. Energy Policy 35:4444–4454
17. Wright TP (1936) Factors affecting the costs of airplanes. J Aeronaut Sci 3:122–128
18. Arrow KJ (1962) The economic implications of learning-by-doing. Rev Econ Stud 29:155–173
19. Dutton JM, Thomas A (1984) Treating progress functions as a managerial opportunity. Acad Manage Rev 9(2):235–247
20. Argote L, Epple D (1990) Learning curves in manufacturing. Science 247(4945):920–924
21. Thompson P (2007) How much did the liberty shipbuilders forget? Manag Sci 53(6):908–918
22. Arrow K (1962) Economic welfare and the allocation of resources from invention. In: Nelson R (ed) The rate and direction of innovative activity. Princeton University Press, Princeton
23. Bielicki J (2008) Returns to scale in carbon capture and storage infrastructure and deployment. Technical report, Energy Technology Innovation Policy research group, Belfer Center for Science and International Affairs
24. McDonald A, Schrattenholzer L (2001) Learning rates for energy technologies. Energy Policy 29:255–261

25. Taylor MR, Rubin ES, Hounshell DA (2005) Control of SO_2 emissions from power plants: a case of induced technological innovation in the U.S. Technol Forecast Soc Change 72(6):697–718
26. Taylor MR, Nemet GF (2006) The interaction between policy and innovation in clean energy technologies. In: USAEE/IAEE/ASSA meetings: current issues in energy economics and energy modeling, Boston
27. Nemet G (2007) Policy and innovation in low-carbon energy technologies. Ph.D. dissertation, University of California
28. Yeh S, Rubin ES (2007) A centurial history of technological change and learning curves for pulverized coal-fired utility boilers. Energy 32(10):1996–2005
29. van den Broek M, Hoefnagels R, Rubin E, Turkenburg W, Faaij A (2009) Effects of technological learning on future cost and performance of power plants with CO_2 capture. Prog Energy Combust Sci 35(6):457–480. ISI document delivery no.: 523KV
30. Nemet GF (2009) Interim monitoring of cost dynamics for publicly supported energy technologies. Energy Policy 37(3):825–835
31. Sinclair G, Klepper S, Cohen W (2000) Sources of cost reduction in a large specialty chemicals producer. Manag Sci 46(1):28–45
32. Remer DS, Chai LH (1990) Design cost factors for scaling-up engineering equipment. Chem Eng Prog 86(8):77–82
33. Nemet GF, Baker E (2009) Demand subsidies versus R&D: comparing the uncertain impacts of policy on a pre-commercial low-carbon energy technology. Energy J 30:49–80
34. Rosenberg N (1982) Inside the black box: technology and economics. Cambridge University Press, Cambridge, MA
35. Clarke L, Kyle P, Wise MA, Calvin K, Edmonds JA, Kim SH, Placet M, Smith S (2008) CO_2 emissions mitigation and technological advance: an updated analysis of advanced technology scenarios. Technical report PNNL-18075, Pacific Northwest National Laboratory
36. David J, Herzog H (2001) The cost of carbon capture. MIT. http://www.netl.doe.gov/publications/proceedings/01/carbon_seq_wksp/David-Herzog.pdf. Accessed on March 21, 2012
37. Clemen RT, Kwit RC (2001) The value of decision analysis at Eastman Kodak Company, 1990–1999. Interfaces 31:74–92
38. Peerenboom JP, Buehring WA, Joseph TW (1989) Selecting a portfolio of environmental programs for a synthetic fuels facility. Oper Res 37:689–699
39. Sharpe P, Keelin T (1998) How SmithKline Beecham makes better resource-allocation decisions. Harv Bus Rev 76:45–57
40. Baker E, Chon H, Keisler J (2009) Carbon capture and storage: combining expert elicitations with economic analysis to inform climate policy. Clim Chang 96(3):379
41. National Research Council (2007) Prospective evaluation of applied energy research and development at DOE (phase two). The National Academies Press, Washington DC. http://www.nap.edu/catalog/11806.html
42. Rao AB, Rubin ES, Keith DW, Morgan MG (2006) Evaluation of potential cost reductions from improved amine-based CO_2 capture systems. Energy Policy 34:3765–3772

Operation System Optimization

Haslenda Hashim, Shuhaimi Mahadzir, Woon Kok Sin, and Mahmoud Ahmed

Abstract This chapter presents a new advancement in energy minimization and CO_2 mitigation research. It describes the development of a combined technique to simultaneously synthesize energy recovery network and offers fuel switching options to satisfy CO_2 emission reduction target using graphical and mathematical programming approaches. The approach is illustrated using petroleum refinery and palm oil refining process as the case study. The application of this technique yields significant CO_2 emission reduction with short payback period with or without clean development mechanism (CDM).

Keywords Fuel switching • Emission reduction • Palm oil refinery • Petroleum refinery • Graphical technique • Mathematical model • Process integration

---

H. Hashim (✉)
Faculty of Chemical Engineering, Universiti Teknologi Malaysia, Johor, Malaysia
e-mail: haslenda@cheme.utm.my

S. Mahadzir
Department of Chemical Engineering, Universiti Teknologi PETRONAS, Bandar Seri Iskandar, Tronoh, Perak, Malaysia
e-mail: shuham@petronas.com.my

W.K. Sin
Civil and Environmental Engineering Department Hong Kong, University of Science and Technology
e-mail: woonkoksin@gmail.com

M. Ahmed
Department of Chemical Engineering, Universiti Teknologi PETRONAS, Bandar Seri Iskandar, Tronoh, Perak, Malaysia
e-mail: mahmod_gu@yahoo.com

Q.P. Zheng et al. (eds.), *Handbook of CO_2 in Power Systems*, Energy Systems,
DOI 10.1007/978-3-642-27431-2_17, © Springer-Verlag Berlin Heidelberg 2012

1 Introduction

The emission of carbon dioxide (CO_2) from the burning of fossil fuels has been identified as the major contributor to global warming and climate change. The challenge for the fossil fuel industry is to find cost-effective solutions that will improve energy savings as well as reduce the release of CO_2 into the atmosphere. There are currently many technologies available to reduce CO_2 emission significantly, with carbon capture being the leading technology. Among promising technologies to capture CO_2 are chemical solvent absorption, physical absorption, physical adsorption in solid and liquid, cryogenic separation, membrane separation, O_2/CO_2 combustion process and biological capture process. Meanwhile, as fuel switching process (transferring from high carbon intensive fuels to low carbon intensive fuels or renewable energies) poses a gigantic impact on the environment, the study of fuel switching process is getting more attention from the researchers. Beside fuel switching, process integration is also an important activity to improve energy savings and hence reduce emissions from chemical process plants. In this chapter, a combined process integration and fuel switching strategy using graphical and mathematical programming approaches are presented. These cost-effective techniques offer a potential to improve the overall energy utilization in process industry and satisfy targeted carbon emission reduction. Application of these techniques on petroleum refinery and palm oil refinery showed significant energy recovery potentials and their ability to satisfy specific target for CO_2 emission reduction.

2 Combined Process Integration and Fuel Switching Strategy: Graphical Technique

2.1 Case Study 1: Crude Oil Distillation Unit Pre-heat Train

Case study 1 demonstrates the application of the new graphical method, starting by fuel switching and then retrofit of heat exchanger network, HEN, for a preheat train of crude oil distillation unit [1]. As illustrated in Fig. 1, the crude oil is fed to the distillation tower from storage at ambient temperature. It is preheated in two sections by heat exchange with the hot fractions from the distillation column product streams. The first section runs from storage to a desalter unit. The second section runs from the desalter to the crude tower. The final process heating is provided by a furnace prior to the crude entering the fractionation tower. Any improvement in the heat recovery within the HEN will reduce the amount of the external heat required by the process, leading to lower fuel requirement and hence to emissions reduction.

2.2 Development of Graphical Technique

Emission reduction of CO_2 by fuel switching is limited to alternative fuels availability and the ability of the boiler combustion system to accept different fuel type. In addition, there is no available practical and economical flue gas treatment to

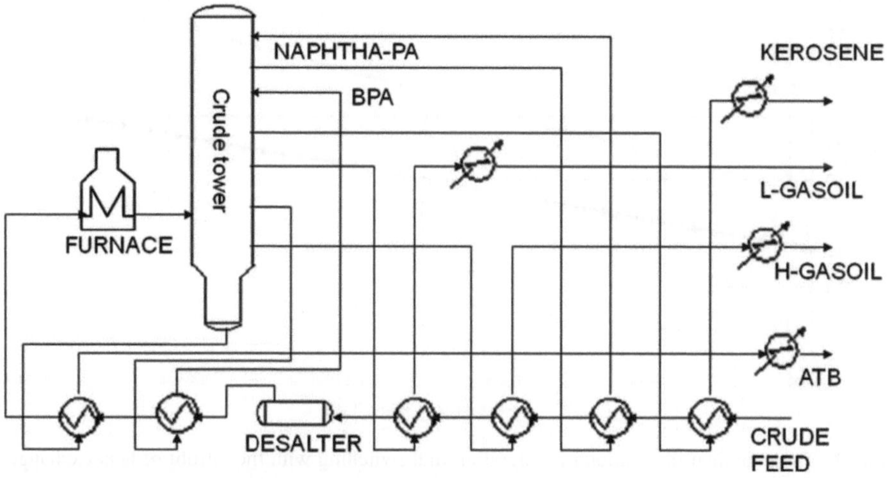

Fig. 1 Flowsheet for the crude oil pre-heat train

remove post combustion CO_2 on-site [2, 3]. Therefore, a strategy to integrate fuel switching with retrofit of HEN is more useful to achieve a more practical emissions reduction target. For any given heat transfer process, the heat duty can be translated into fuel heat duty and hence to emissions flowrate using Eqs. 1, 2 and 3.

$$Q_{FUEL} = \frac{Q_{PROCESS}}{\eta_F} \tag{1}$$

$$\eta_F = \frac{T_{TFT} - T_{STACK}}{T_{TFT} - T_o} \tag{2}$$

$$M_{pol} = \frac{Q_{fuel}}{NHV} * \beta * \phi \tag{3}$$

where

Q_{fuel} = heat duty from fuel (kW)
Q_{PROC} = process heat duty (kW)
η_F = furnace efficiency
T_{TFT} = theoretical flame temperature of the furnace flue gases (°C)
T_{STACK} = stack temperature (°C)
T_o = ambient temperature (°C)
M_{pol} = mass flow rate of pollutant
NHV = net heating value of the fuel
β = mass percentage of the pollutant in non-oxidized form
φ = ratio molar masses of the oxidized form to the non-oxidized form of the pollutant

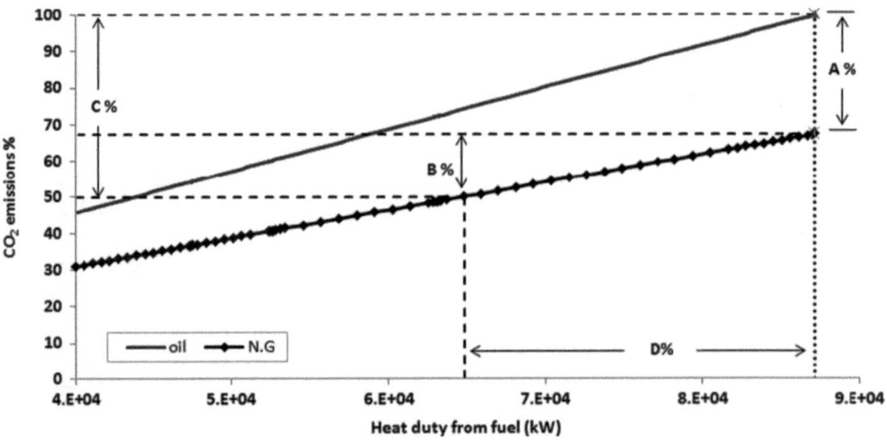

Fig. 2 The graphical illustration of integrating fuel switching with the retrofit of heat exchanger network

The graphical technique of combined fuel switching with retrofit of HEN is demonstrated below. In Fig. 2, for target emissions reduction of C%, switching from fuel oil to natural gas fuel can achieve a limited emissions reduction amounting to A%. Suppose there is a set target for emission reduction, which is at B%. In order to reach the target emission reduction, B% of the existing emissions has to be reduced through retrofit of the HEN. The intersection of the lowest dotted horizontal line with the operating line of natural gas fuel shows the new point of the fuel consumption. Therefore savings in natural gas fuel consumption amounting to D% through the retrofit of the HEN need to be achieved. This amount of savings is required in order to meet the target B% emissions reduction from the existing emissions. The amount of B% emissions reduction as a result of HEN retrofit can be saved by investing on new heat exchanger surface area. The target added area due to retrofit of HEN can be predicted by retrofit of HEN through Pinch Design Method [4–9].

The total capital investment of fuel switching with retrofit of HEN for emission reduction is made up of two parts. Firstly, the investment required to upgrade the burner for combustion of a new fuel. Secondly, an investment is also required to upgrade heat exchangers that require larger heat transfer area for the retrofit exercise. However, if replacing the fuel oil burners with natural gas burners is the only modification that will occur in the existing furnace to accommodate the new fuel, then the investment of changing burners will be smaller compared to that of the retrofit of HEN. Thus, if the total investment of combining fuel switching with the retrofit of the HEN is approximated to the investment of retrofitting the HEN, then the relationship between emissions reduction (%), investment and energy savings (%) can be set as shown in Fig. 3.

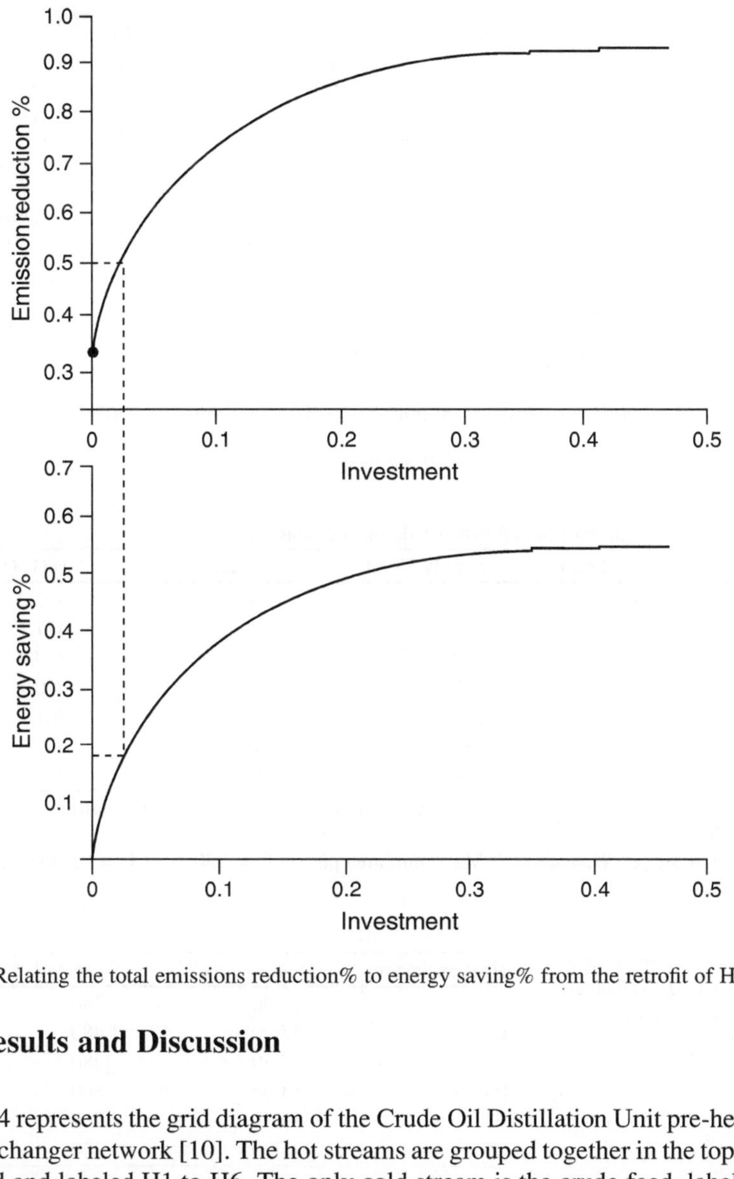

Fig. 3 Relating the total emissions reduction% to energy saving% from the retrofit of HEN

3 Results and Discussion

Figure 4 represents the grid diagram of the Crude Oil Distillation Unit pre-heat train heat exchanger network [10]. The hot streams are grouped together in the top part of the grid and labeled H1 to H6. The only cold stream is the crude feed, labeled C1. There are six process-to-process heat exchangers (E1–E6), five coolers (E7–E11) and a furnace (E12) in the crude pre-heat train HEN. The temperatures are shown at the two ends of the stream while the heat loads are shown below the heat exchangers. For example, stream H6 has an initial temperature of 290°C and a final temperature of 190°C. Stream H6 is cooled down by cooler E5 by exchanging the heat with stream C1 at heat load of 38,480 kW. Table 1 show the streams and cost data and Table 2 shows the existing process-to-process heat exchange area for

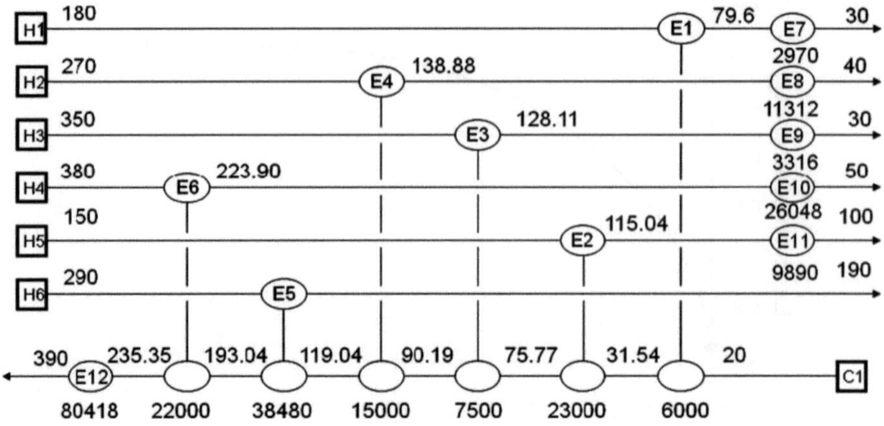

Fig. 4 Existing network for the crude oil pre-heat train

Table 1 Streams and cost data for the crude oil pre-heat train

Stream	Flow (kg/s)	Supply Ts (°C)	Target Tt (°C)	HTC (W/m^2.°C)
H1	23	180	30	492.2
H2	44	270	40	477.8
H3	13	350	30	439.8
H4	56	380	50	470.7
H5	253	150	100	561.5
H6	148	290	190	432.6
C1	200	20	390	343.0

Note: Exchanger capital cost (£) = 8,600 + 0.83 × 670 × (area). Hot utility cost (£/kW. year) = 70
Cold utility cost (£/kW.year) = 7. Maximum area per shell = 580 m^2. CP = 2,600 (J/kg °C) for all streams

Table 2 Existing process-to-process heat exchanger area of the crude oil pre-heat train

Heat exchanger	Heat load (kW)	Existing HT area (m^2)
E1	6,000	280
E2	23,000	1,480
E3	750	280
E4	15,000	800
E5	38,480	2,760
E6	22,000	1,360

the HEN of the case study. The two types of selected fuels and their properties are further shown in Table 3. For base case, the existing fuel in the combustion device is assumed to be fuel oil at 25°C. The combustion air is also assumed to be fed at the same temperature.

For the current heat requirement of the process of 80,418 kW, which is the amount of heat provided by furnace E12, the mass flowrate of a pollutant is calculated depending on the type of fuel used. The reductions in SO_x and CO_2

Table 3 Fuel properties

Compositions	Oil	Natural gas
Carbon (wt%)	87.26	76.0
Hydrogen (wt%)	10.94	22.8
Oxygen (wt%)	0.64	–
Nitrogen (wt%)	0.28	1.1
Sulfur (wt%)	0.84	0.1
Ash (wt%)	0.04	–
NHV (kJ. kg^{-1})	39,830	51,550

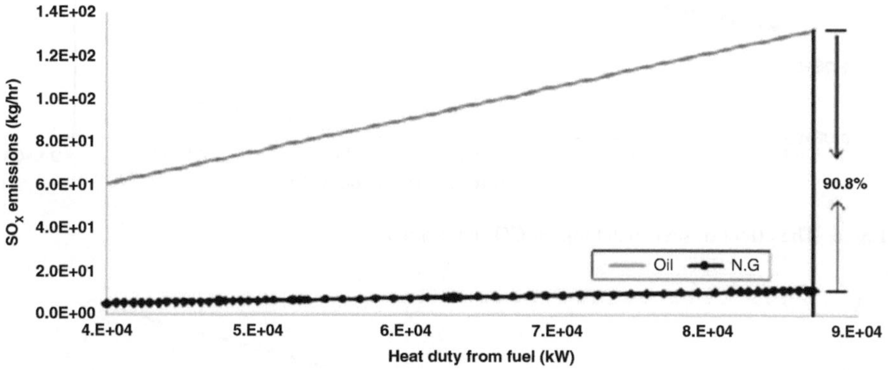

Fig. 5 The effect of fuel switching on SO_X emissions

emissions due to fuel switching from oil to natural gas are illustrated in Figs. 5 and 6, respectively. Figure 5 shows that the reduction in SO_x emissions can reach 90.8% because the sulfur content of natural gas fuel is very low compared to that of fuel oil. From Fig. 6, the reduction in CO_2 is 32.7%. This is because the natural gas fuel has lower carbon-to-hydrogen ratio whilst maintaining high NHV.

Fuel switching from fuel oil to natural gas can provide satisfactory emissions reduction for fuel NO, SO_x and particulates. The emissions reduction of CO_2 due to fuel switching is achieved by almost 32.7%. Furthermore, there is no available practical flue gas treatment to remove CO_2 emissions. Consequently a combined strategy of fuel switching with retrofit of HEN is implemented to get more emissions reduction.

Figure 7 shows the application of the graphical method of combined fuel switching and HEN retrofit to cut down the emissions of CO_2. For target emissions reduction of 50% (C), fuel switching from fuel oil to natural gas has achieved only 32.7% emissions reduction (A). So, in order to reach the target emission reduction, 17.3% of the existing emissions (B) have to be reduced through the retrofit of the HEN. The intersection of the lowest dotted horizontal line with the operating line of natural gas fuel shows the new point of the fuel consumption. Therefore 26% saving in fuel natural gas consumption (D) through the retrofit of the HEN have to be achieved to reach the 17.3% emissions reduction from the existing emissions.

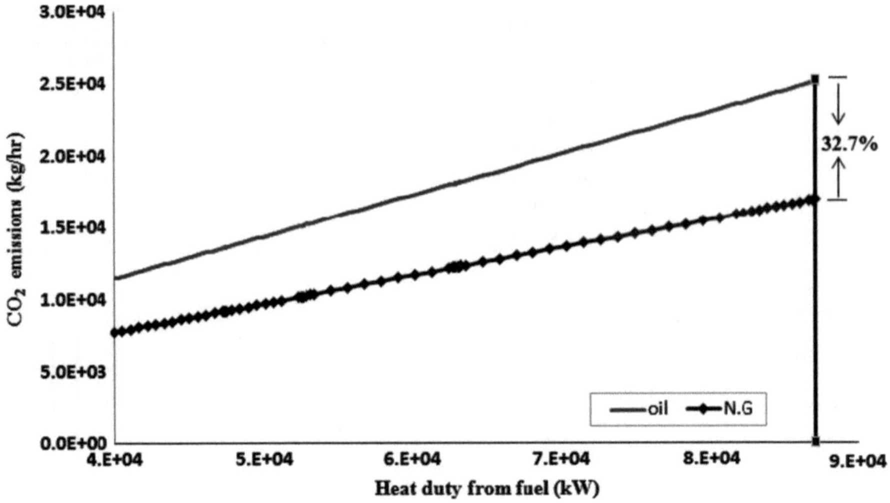

Fig. 6 The effect of fuel switching on CO_2 emissions

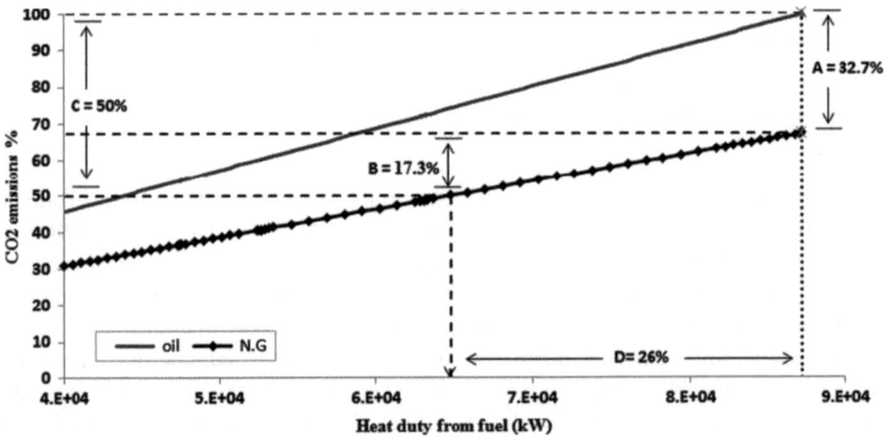

Fig. 7 The graphical illustration of integrating fuel switching with the retrofit of the HEN of the case study

If replacing the fuel oil burners with natural gas fuel burners is the only modification that will occur in the existing combustion device to accommodate the new fuel, then the investment of changing burners will be smaller compared to that of the retrofit of the HEN. If the total investment of combining fuel switching with retrofit of the HEN is approximated to the investment of retrofitting the HEN, then the relationship between emissions reduction (%), investment and energy savings (%) can be built as shown in Fig. 8.

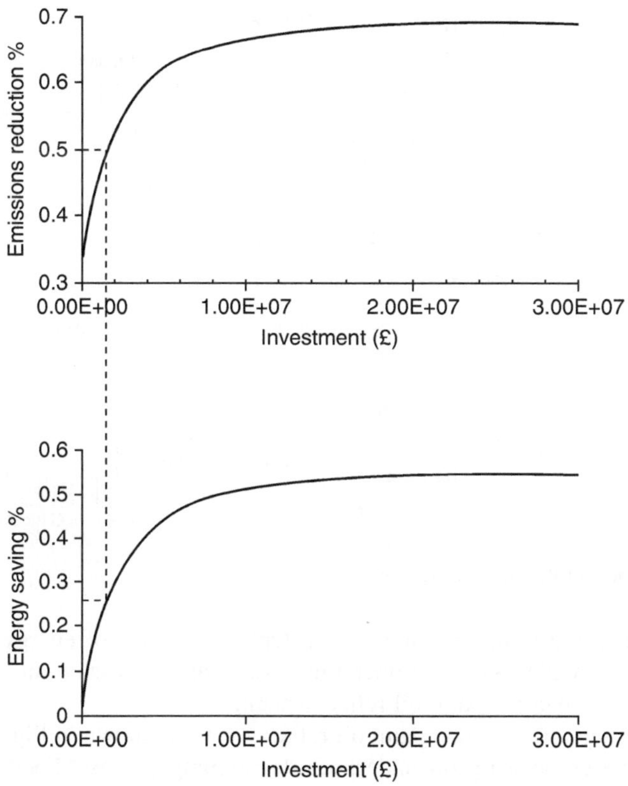

Fig. 8 Relating the total emissions reduction % to energy saving % from the retrofit of the HEN

Figure 8 indicates that the maximum emissions reduction from this is 69%. This is due to the fact that the thermodynamic minimum hot utility consumption of the process (i.e. $Q_{process}$ at $\Delta T_{min} = 0°C$) is greater than 0.

However, the major contribution from this technique is that, it can determine the maximum Greenhouse Gas emissions (GHG) reduction which can be achieved for any given chemical process plant. Furthermore, it is able to determine the amount of energy savings which should be achieved through retrofit of heat exchanger network (HEN) for any target emissions reduction.

4 Combined Process Integration and Fuel Switching Strategy: Mathematical Modeling Techniques

4.1 Case Study 2: Palm Oil Refinery

Palm oil refining involves gigantic energy during its operation. According to Zainuddin et al., 20–30% of the production cost in a crude palm oil physical refining

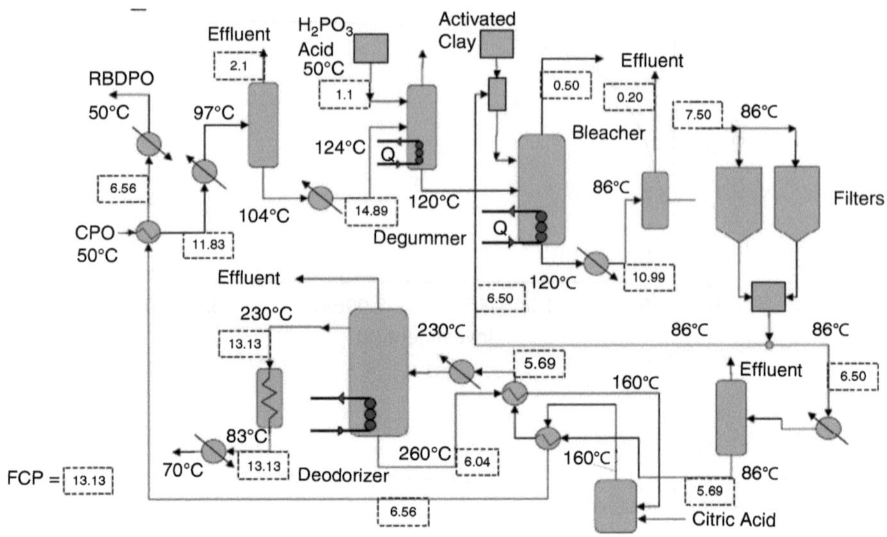

Fig. 9 Flowsheet of the crude oil refinery

plant is due to the usage of energy [11]. Optimum usage of energy is therefore imperative in order to save the operating cost, reduce emission and increase the profitability of the crude palm oil refinery plant.

The flowsheet of the crude palm oil refinery plant is shown in Fig. 9.

Firstly, the crude palm oil undergoes degumming process. Food grade phosphoric acid, which acts as a solvent, is added into the process. This is done at a temperature in excess of 100°C in order to get rid the oil from gums (phosphatide compounds). Then, it undergoes bleaching process at which the bleaching earth is used to remove color pigments from the degummed oil. The process is implemented under vacuum condition at 120°C. After that, the slurry product from the bleaching section travels through a series of Niagara Filter in order to remove the excess bleaching earth and impurities. After the filtration, the oil undergoes deodorization process at which free fatty acids, aldehyde and ketone compounds is removed from the bleached oil by stream stripping. The deodorizer column operates under high vacuum and temperature in the range of 240–270°C. The product is called as Refined, Bleached and Doedorised Palm Oil (RBDPO) [12]. A set of hot streams that must be cooled and a set of cold streams that must be heated for a palm oil refinery plant are given in Table 4.

In this case study, auxiliary heating and cooling are available from hot steam and cooling water. Diesel is the typical fuel used to generate the steam for hot utility purpose in a palm oil refinery plant. However, due to the fact that diesel fuel produce high amount of carbon dioxide, which is one of the greenhouse gases, study on switching fuels to less carbon intensive fuel e.g. natural gas, biomass and biogas to generate heat and power in a refinery plant if of great interest. The data for fuel switching is provided is Table 5.

Table 4 Process stream data for the palm oil refinery plant

Stream no.	Type	Stream name	Supply temperature, T_S (°C)	Target temperature, T_T (°C)	Heat capacity flow rate (kW/°C)
C1	Cold	Crude palm oil feed, CPO	50	97	11.83
C2	Cold	Degaser outlet, DEGAS	104	124	14.89
H1	Hot	Bleached palm oil, BPO	120	86	10.99
C3	Cold	Dried oil, DRIED	86	230	5.69
H2	Hot	Deodorizer outlet, DEO	260	160	6.04
H3	Hot	Free fatty acids, FFA	83.3	70	13.13
H4	Hot	Refined, bleached and deodorized palm oil, RBDPO	160	50	6.56

Table 5 Data for fuel switching analysis

Parameter	Diesel	Natural gas	Biomass (empty fruit bunch)	Palm oil mill effluent (POME)
Heating value (kJ/kg)	45,900	42,500	18,795	55,400
Capital cost of boiler (RM/MW)	0	617,000	15,900,000	1,530,000
Capital cost of steam turbine (RM/kW)	0	0	760	760
Efficiency of boiler	0.56	0.83	0.61	0.4
Fuel cost (RM/kg)	2.94	2.5289	0	0
Operating and maintenance cost (RM/MWh)	34.63	95.33	150	150
Efficiency of steam turbine	0	0	0.629	0.5
Heat to power ratio	1.0	1.0	0.8	0.7
Carbon dioxide emission (kg/kg fuel)	3.2	2.30	0	0

4.2 Mathematical Model Development

The objective of this study is to simultaneously synthesize maximum energy recovery and satisfy CO_2 emission reduction via heat integration and fuel switching process. The total cost includes: retrofit or capital cost for switching fuel from diesel to natural gas or biomass, operational cost mainly for fuel price, operating and maintenance cost (O&M), carbon emission reduction (CER) revenue and electricity saving cost due to electricity generation from combined heat and power (CHP) system. This can be written as,

$$\begin{aligned} Min\ cost = & \underbrace{\sum\nolimits_i (W_i \times CA_i \times HV_i \times HTPR_i)}_{Capital\ Cost} + \underbrace{\sum\nolimits_i (W_i \times FUELC_i)}_{Fuel\ Price} + \\ & \underbrace{\sum\nolimits_i (W_i \times OAM_i \times HV_i \times HTPR_i)}_{O\&M\ Cost} - \underbrace{\sum\nolimits_i (W_i \times CER_i \times HV_i)}_{CER\ Revenue} \\ & - \underbrace{\sum\nolimits_i (W_i \times HV_i \times PTHR_i \times ETARIFF)}_{Electricity\ Saving\ Cost} \end{aligned} \quad (1)$$

Equation 1 represents the objective function at which the total cost is minimized. $\boldsymbol{i}$ is the only set which represents the alternative fuels used in the fuel options, includes, biomass (EFB), POME and natural gas with subsets BMS, MTHGAS and NG respectively. $\boldsymbol{W_i}$ (kg/year) is represents the flow rate of each fuels switching option. It is a variable of the model at which it will affect the capital and operational cost for fuel switching process. $\boldsymbol{CA_i}$ (RM/kJ) is the capital cost of boiler of each alternative fuel. The more fuel is consumed, the larger size of the boiler is needed to generate more energy, hence the higher the capital cost incurred. $\boldsymbol{HV_i}$ (kJ/kg) is the calorific value or heating value of respective alternative fuels. $\boldsymbol{HTPR_i}$ is represents a heat to power ratio for co-generation system used biomass (EFB) and methane gas fuel to generate heat and electricity. In this case, the heat to power ratio for EFB and POME are 0.8 and 0.7 respectively. Meanwhile, $\boldsymbol{FUELC_i}$ (RM/kg) is the fuel cost for each alternative fuels. In this case, the fuel cost for EFB and POME is considered zero due to the reason that it is assumed to be taken from nearby palm milling plant. $\boldsymbol{OAM_i}$ (RM/kg) is the operating and maintenance cost for each alternative fuels. $\boldsymbol{CER_i}$ (RM/kg) is the certified emission reduction cost at which the usage of green renewable energies will help to gain profit to the system. In this case study, it is expected that a reduction of 1 t of carbon dioxide will provide a profit of RM 72 to the plant owner. As electricity generation is considered in this case study, the saving cost of electricity generated from renewable energies using combined heat and power system, CHP system is considered in order to obtain minimum operational cost. $\boldsymbol{PTHR_i}$ is the power to heat ratio used for EFB and POME in generating the electricity to the plant. The respective values are 0.2 and 0.3. Meanwhile, $\boldsymbol{ETARIFF}$ is a scalar which represents the electricity tariff in this case study. The given electricity tariff is RM 0.28 per kWh.

Equation 2 indicates the total heat required from different alternative fuels for generating the minimum hot utility target, Q_s. $\boldsymbol{EFF_i}$ is the efficiency of boiler for each alternative fuels. Meanwhile, $\boldsymbol{f}$ is the scalar factor which represents the demand factor. In this case, the value for the demand factor is 1. Should there is an increase in demand in future, the scalar factor will change accordingly in order to fulfill the heat demand requirement.

$$\sum_i (W_i \times HV_i \times EFF_i \times HTPR_i) = Q_s \times f \tag{2}$$

For above equation, Qs is obtained using heat balance equation for each interval as shown in Eq. 2. In this case, global (shifted) temperature interval is used as which all supply temperatures and targeted temperatures are increased (cold stream) or reduced (hot stream) by half of the minimum temperature approach, $\boldsymbol{\Delta T_{min}}$ (10°C) respectively and arranged from highest value to lowest value in order to build up the temperature interval as shown in Tables 6 and 7.

The $\boldsymbol{K}$ equations are heat balances around each temperature interval $\boldsymbol{k}$. Heat residual entering interval $\boldsymbol{k}$, $\boldsymbol{R_{k-1}}$, and heat contents of hot stream $\boldsymbol{i}$ in interval $\boldsymbol{k}$, $\boldsymbol{\Delta H_k}$, balance to heat residual exiting interval $\boldsymbol{k}$, $\boldsymbol{R_k}$.

$$R_{k-1} - -R_k + \Delta H_k = 0, k = 1, \ldots, K \tag{3}$$

Table 6 Actual temperature and shifted temperature

Stream no.	Actual T		Shifted T	
	T_S (°C)	T_T (°C)	T_S (°C)	T_T (°C)
C1	50	97	55	102
C2	104	124	109	129
C3	86	230	91	235
H1	120	86	115	81
H2	260	160	255	155
H3	83.3	70	78.3	65
H4	160	50	155	45

Table 7 Set up temperature interval

Stream no.	T_S/T_T	Remove duplicates (if any)	Set up intervals (T_{int})
C1	55	55	255
	92	102	235
C2	109	109	155
	119	129	129
C3	91	91	115
	225	235	109
H1	125	115	102
	81	81	91
H2	265	255	81
	155	155	78.3
H3	88.3	78.3	65
	65	65	55
H4	160	155 – remove duplicate	45
	45	45	–

If $R_k = 0$ at the optimal solution, the temperature at interval $\boldsymbol{k}$ will be corresponding to a pinch point and Qs is the minimum heat utility required for the process. Equations 4, 5 and 6 indicate the constraints which show that the hot utility/steam consumption, $\boldsymbol{Q_S}$, cold utility, $\boldsymbol{Q_{CW}}$, as well as residual heat, $\boldsymbol{R_k}$, is a positive value at any interval.

$$Q_S \geq 0 \tag{4}$$

$$Q_{CW} \geq 0 \tag{5}$$

$$R_k \geq 0, k = 1, \ldots, K \tag{6}$$

Equation 7 is a constraint to indicate the electricity generation from CHP plant satisfies the annual electricity requirement for the plant. $\boldsymbol{ECHP_i}$, represents the efficiency of turbine used in generating the electricity. The flow rate of alternative energy, $\boldsymbol{W_i}$, multiplies with the heating value, $\boldsymbol{HV_i}$, power to heat ratio, $\boldsymbol{PTHR_i}$, efficiency of the turbine, $\boldsymbol{ECHP_i}$, and efficiency of the boiler, $\boldsymbol{EFF_i}$, to give the electricity consumption which is in unit of MWh.

$$\sum_i (W_i \times HV_i \times PTHR_i \times ECHP_i \times EFF_i) \leq ELECONSM \tag{7}$$

Meanwhile, Eq. 8 shows that the appropriate fuel will be selected according to carbon dioxide emission reduction target. $\boldsymbol{CDE_i}$ is the carbon dioxide emission in kg per year. Meanwhile, ***RPER*** represents the reduction of percentage of carbon dioxide emission in the plant. ***TCO2*** is a scalar at which it represents the current carbon dioxide emission in the plant per year.

$$\sum_i (W_i \times CDE_i) \leq (1 - RPER) \times TCO2 \tag{8}$$

Equations 9, 10, 11 and 12 show the constraints for total availability of the alternative fuels for plant generation respectively. In this case study, the total availability for biomass (EFB), ***SBMS*** methane gas (POME), ***SMTHGAS***, diesel, ***SDIESEL*** and natural gas, ***SNG*** is 1,000,000 kg/year, 20,000 kg/year, 500,000 kg/year and 350,000 kg/year respectively. Therefore, the total flowrate of fuel required for biomass, $\boldsymbol{W_{BMS}}$, methane gas, $\boldsymbol{W_{MTHGAS}}$, diesel, $\boldsymbol{W_{DIESEL}}$, and natural gas, $\boldsymbol{W_{NG}}$, must be less than availability.

$$W_{BMS} \leq SBMS \tag{9}$$

$$W_{MTHGAS} \leq SMTHGAS \tag{10}$$

$$W_{DIESEL} \leq SDIESEL \tag{11}$$

$$W_{NG} \leq SNG \tag{12}$$

Equations 13 and 14 indicate the non-zero constraints for electricity consumption, ***ELECONSM***, and total carbon dioxide emission, ***TCO2***, respectively.

$$ELECONSM \geq 0 \tag{13}$$

$$TCO2 \geq 0 \tag{14}$$

4.3 Results and Discussion

Figure 10 illustrates the effect of different combination cases for heat integration and fuel switching process in terms of total operational cost.

From the figure, it can be seen that base case requires highest amount of total operational cost while the integration of diesel, heat integration and fuel switching process with clean development mechanism (CDM) requires lowest total operational cost. Compared to base case scenario, the introduction of different combination cases reduces the total operating cost as much as 90.5%, 18.1%, 16.7%, 93.9% and 92.5% respectively. This is mainly due to the fact that the implementation of heat integration will significantly reduce the amount of fuel usage, hence reducing

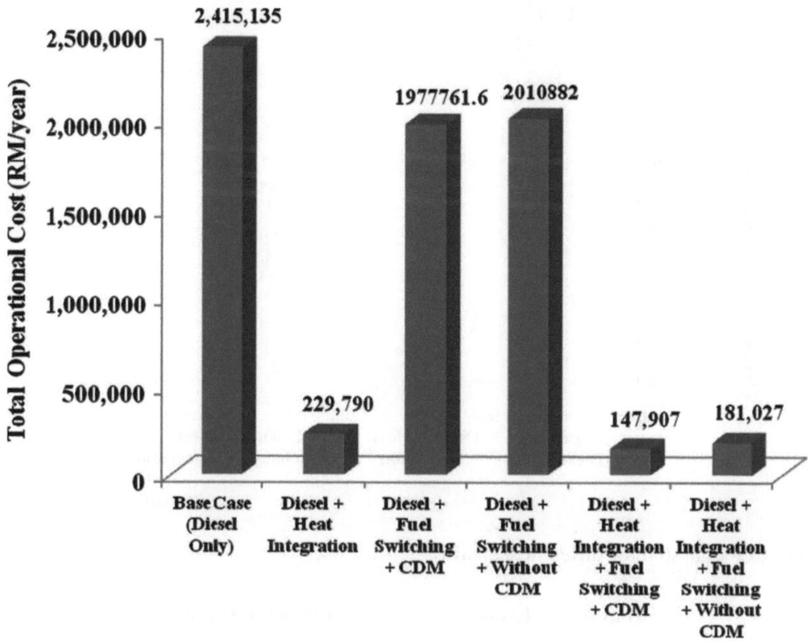

Fig. 10 Total operational cost for different scenarios

the purchase cost and operating cost for respective fuels. As a conclusion, the integration of diesel, heat integration and fuel switching with CDM yields the best result among all in terms of total operational cost saving.

Figure 11 shows the effect of various combination cases towards the CO_2 emission. From the figure, it can be seen that the introduction of heat integration and fuel switching process will reduce the emission of CO_2 considerably compared to base case scenario. The introduction of heat integration, fuel switching process and combination of heat integration and fuel switching will reduce 92.3%, 42.3% and 96.9% of CO_2 emission respectively. The introduction of heat integration will tremendously reduce the usage of diesel fuel. A reduction of usage of diesel fuel signifies a reduction in CO_2 emission to environment. Meanwhile, through the application of methane gas release from POME and biomass (EFB), it will significantly reduce the release of CO_2 as they are considered as green energy and will not release CO_2 to atmosphere while burning. The effect will be more obvious while heat integration and fuel switching are implemented together. Therefore, the combination of heat integration and fuel switching shows the highest percentage reduction among all.

Figure 12 shows the percentage of CO_2 reduction with energy consumption from conventional energy (diesel) and renewable energy due to fuel switching. Base case applied for this graph is after the implementation of heat integration. It can be seen that the percentage of fuel switching to renewable fuels is increased when the percentage of CO_2 reduction is increased. The trend for conventional energy (diesel

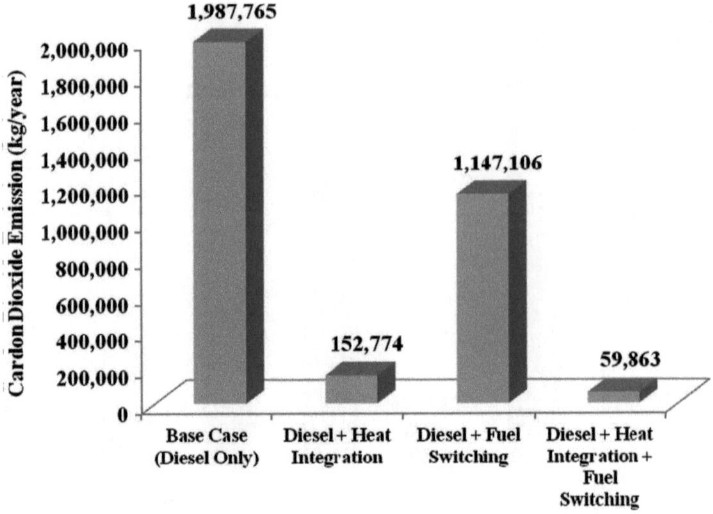

Fig. 11 CO_2 emission for different scenarios

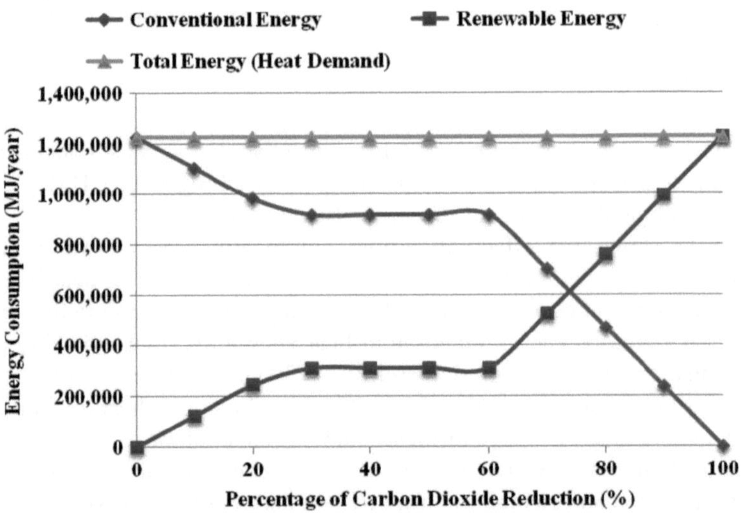

Fig. 12 Percentage of CO_2 reduction versus energy consumption after the implementation of heat integration

consumption) is in reverse order compared to renewable energy. From the figure, it shows that the usage of renewable fuels starts to outweigh the conventional fuels when 74.5% of CO_2 reduction is intended to be achieved in the plant. As CER revenue keeps increasing for this time being, this will considerably give a handsome profit through the implementation of renewable energy in the future.

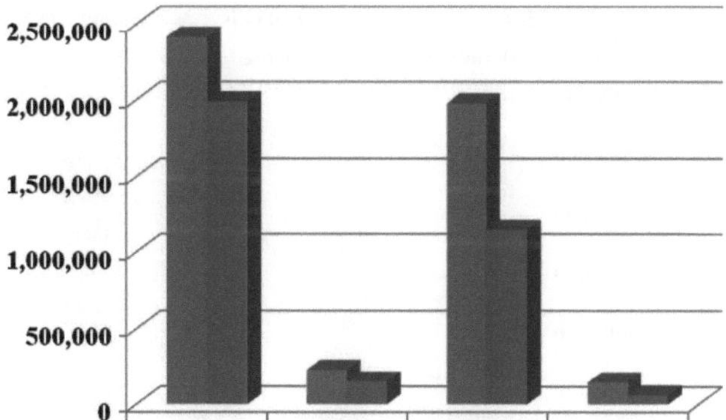

	Base Case	Heat Integration	Fuel Switching	Heat Integration + Fuel Switching
■ Total Operational Cost (RM/year)	2,415,135	229,790	1977761.6	147,907
■ Carbon Dioxide Emission (kg/year)	1,987,765	152,774	1,147,106	59,863

Fig. 13 Total operational cost and CO_2 emission for different scenarios

Figure 13 shows the total cost and CO_2 emission for four different combination cases, namely base case, heat integration, fuel switching and combination of heat integration and fuel switching.

Generally, the introduction of heat integration and fuel switching will reduce the total cost as well as CO_2 emission. From the figure, it can be seen that the implementation of heat integration will require less cost and release less CO_2 compared to base case scenario. This is due to the fact that energy is provided via the hot streams in the plant operation when process heat integration is implemented. Therefore, a reduction of energy consumption will significantly reduce the usage of diesel fuel, causing the cost of operation and CO_2 emission to be reduced greatly. Meanwhile, the implementation of fuel switching process reduces the operational cost and CO_2 emission concurrently compared to base case scenario. This is the fact that renewable fuel such as methane gas (POME) and low carbon intensive fuel such as natural gas used in fuel switching. Moreover, the switching of conventional fuels to renewable fuels will reduce the emission of CO_2, which will help to protect the environment.

Figure 14 illustrates the trend of CO_2 emission reduction after the implementation of heat integration and fuel switching process. From the figure, it can be seen that the total cost is increasing while the percentage of CO_2 emission is increased. In order to achieve significant reduction of CO_2 emission, biomass (EFB) is favored to be implemented to replace conventional fuel consumption (diesel and natural gas). The increment of the operational cost is due to the fact that the implementation of biomass boiler will increase the total cost for its high capital cost and lower heating value, which means that more fuels need to be consumed to generate same amount of energy compared to other fuels. Though zero CO_2 emission is highly

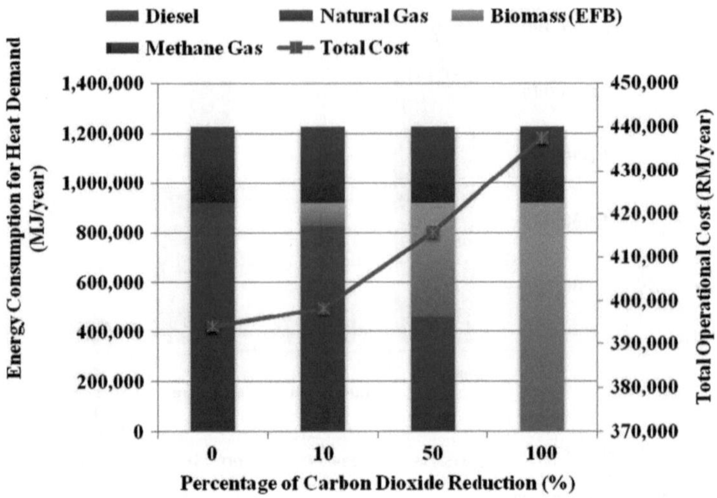

Fig. 14 Percentage of CO_2 reduction versus energy consumption and total

recommended to be introduced in the plant, it requires high amount of operational cost at the same time. Hence, a trade-off point in between operational cost reduction and CO_2 emission reduction is very important to achieve equilibrium while considering these two factors.

5 Conclusion

A new combined process integration and fuel switching technique are developed using graphical technique and mathematical approaches. This work is unique since it can be used to simultaneously design maximum energy recovery network and offer fuel selection options in order to satisfy CO_2 emission reduction target. The developed method has been effectively applied in two case studies of crude oil preheat train and palm oil refinery heat exchanger network. The method was prove to be an efficient integrated method for maximizing energy savings and targeting emissions reduction in chemical process plants.

References

1. Panjeshahi MH, Tahouni N (2008) Pressure drop optimisation in debottlenecking of heat exchanger networks. Energy 33(6):942–951
2. Delaby O, Smith R (1995) Minimization of flue gas emissions. Trans IChemE 73(B):21–32
3. Chaaban FB, Mezher T, Ouwayjan M (2004) Options for emissions reduction from power plants: an economic evaluation. Electr Power Energy Syst 26:57–63
4. Tjoe TN (1986) Retrofit of heat exchanger networks. Ph.D. thesis, UMIST

5. Van Reisen J, Grievink J, Polley G, Verheijen P (1995) The placement of two-stream and multi stream heat-exchangers in an existing network through path analysis. Comput Chem Eng 19:143–148
6. Varbanov PS, Klemes J (2000) Rules for paths construction for HENs debottlenecking. Appl Therm Eng 20:1409–1420
7. Polley GT, Panjehshahi MH, Jegede FO (1990) Pressure drop considerations in the retrofit of heat exchanger networks. Trans IChemE 68(A):211–220
8. Al-Riyami BA, Klemes J, Perry S (2001) Heat integration retrofit analysis of a heat exchanger network of a fluid catalytic cracking plant. Appl Thermal Eng 21(13–14):1449–1487
9. Van Reisen J (2008) Structured approach to heat exchanger network retrofit design. BOX Press, Oisterwijk
10. Mahmoud A, Shuhaimi M, Samed AM (2009) A combined process integration and fuel switching strategy for emissions reduction in chemical process plants. Energy 34:190–195
11. Manan ZA, Yusof FM (1994) Retrofit of palm oil refinery to improve energy consumption and reduce fouling costs, M.sc Thesis (UTM)
12. Chow MC, Wahid MB, Chan KW (2008) Availability and potential of biomass resources from the Malaysian palm oil industry for generating renewable energy. Oil Palm Bull 56:23–28

Index

DOI 10.1007/978-3-642-27431-2, © Springer-Verlag Berlin Heidelberg 2012

MIX
Papier aus verantwortungsvollen Quellen
Paper from responsible sources
FSC® C105338

If you have any concerns about our products,
you can contact us on
ProductSafety@springernature.com

In case Publisher is established outside the EU,
the EU authorized representative is:
Springer Nature Customer Service Center GmbH
Europaplatz 3, 69115 Heidelberg, Germany

Printed by Libri Plureos GmbH
in Hamburg, Germany